Oscillati

Oscillations and Waves

Suresh Garg
Professor of Physics
School of Sciences
Indira Gandhi National Open University (IGNOU)
New Delhi

C.K. Ghosh
Head, Student Services Centre
IGNOU
New Delhi

Sanjay Gupta
Senior Lecturer in Physics
School of Sciences
IGNOU
New Delhi

PHI Learning Private Limited
New Delhi-110001
2009

Rs. 195.00

OSCILLATIONS AND WAVES
Suresh Garg, C.K. Ghosh and Sanjay Gupta

ISBN-978-81-203-3921-7

The export rights of this book are vested solely with the publisher.

Published by Asoke K. Ghosh, PHI Learning Private Limited, M-97, Connaught Circus, New Delhi-110001 and Printed by Raj Press, New Delhi-110012.

To

Late Professor Krishnapada Ghosh

and

Professor Feroz Ahmed

for their inspiration and encouragement

Contents

Preface *xiii*

PART I — OSCILLATIONS

1. Mathematical Preliminaries **3–31**

Expected Learning Outcomes *3*

1.1 Introduction *3*

1.2 Trigonometric Formulae *4*

1.2.1 Features of Sinusoidal Functions *4*

1.2.2 General Solution of Equations $\sin \theta = \sin \alpha$ and $\cos \theta = \cos \alpha$ *5*

1.2.3 Transformation of Sums and Products *6*

1.3 Complex Numbers *7*

1.3.1 Purely Real and Purely Imaginary Numbers *8*

1.3.2 Geometrical Meaning of Number $j = \sqrt{-1}$ *8*

1.3.3 Geometrical Representation of Complex Numbers *9*

1.3.4 Fundamental Properties of Complex Numbers *10*

1.3.5 Demoivre's Theorem (DT) *12*

1.3.6 Using the Form $e^{j\theta}$ *14*

1.4 Differential Calculus *16*

1.4.1 Differential Coefficients of $\sin(\omega t + \delta)$ and $\cos(\omega t + \delta)$ *16*

1.4.2 Taylor's Theorem and Harmonic Approximation *17*

1.4.3 Partial Derivatives *19*

1.5 Integral Calculus *20*

1.5.1 Average Values of $\sin x$, $\cos x$, $\sin^2 x$, $\cos^2 x$ *20*

1.5.2 The Integrals $\int_0^T \sin^2(\omega t + \delta)\,dt$ and $\int_0^T \cos^2(\omega t + \delta)\,dt$ *21*

1.6 Differential Equations *22*

1.6.1 Differential Equations of the Form: $\ddot{x} + 2b\dot{x} + \omega^2 x = f(t)$ *23*

1.6.2 Partial Differential Equations of the Type $(\partial^2 y/\partial t^2) = v^2(\partial^2 y/\partial x^2)$ *23*

1.7 Fourier Analysis 25
1.7.1 Evaluation of a_n and b_n 27
1.7.2 Analysis of Periodic Waveforms 28
1.7.3 Complex Form of Fourier Series 29
Review Exercises 31

2. Simple Harmonic Motion **32–70**

Expected Learning Outcomes 32
2.1 Introduction 32
2.2 Description of SHM 33
2.2.1 Definition of SHM 34
2.2.2 Physical Systems Executing SHM Having One Degree of Freedom 35
2.3 Basic Characteristics of SHM 42
2.4 Phase of an Oscillator Executing SHM 43
2.5 Differential Equation of SHM and Its Solutions 44
2.6 Velocity and Acceleration in SHM 47
2.7 Energy of an Oscillating System 58
2.8 An Ideal LC-Circuit 61
2.9 Relation between Linear SHM and Uniform Circular Motion 64
2.10 Rotating Vector Representation of SHM 67
Review Exercises 68

3. Superposition of Harmonic Oscillations **71–86**

Expected Learning Outcomes 71
3.1 Introduction 71
3.2 Principle of Superposition and Linearity 72
3.3 Superposition of Two Collinear Simple Harmonic Oscillations of Same Frequency 73
3.4 Superposition of N Collinear Harmonic Oscillations of Same Frequency 75
3.5 Superposition of Two Collinear Harmonic Oscillations of Nearly Equal Frequencies 77
3.6 Superposition of Two Mutually Perpendicular Harmonic Oscillations 79
3.6.1 Oscillations of Equal Frequencies 79
3.6.2 Oscillations of Slightly Different Frequencies: Lissajous' Figures 82
3.7 Equivalence of Uniform Circular Motion and Mutually Perpendicular Harmonic Oscillations 83
Review Exercises 84

4. Damped Oscillations **87–110**

Expected Learning Outcomes 87
4.1 Introduction 87
4.2 Types of Damping Forces 88
4.3 Equation of Motion of a 1-D Damped Oscillator 89
4.4 Solutions of the Equation of Motion of a 1-D Damped Oscillator 91
4.4.1 Heavy Damping 93
4.4.2 Critical Damping 95
4.4.3 Weak Damping 96

4.5 Non-Mechanical Damped Systems *98*
4.5.1 A Series *LCR* Circuit *99*
4.5.2 A Ballistic Galvanometer *100*
4.6 Energy of a Weakly Damped System *101*
4.7 Characterising Weak Damping *102*
4.7.1 Logarithmic Decrement *103*
4.7.2 Quality Factor *105*
Review Exercises 109

5. Forced Oscillations **111–140**

Expected Learning Outcomes 111
5.1 Introduction *111*
5.2 Free and Forced Oscillations: Resonance *112*
5.2.1 Examples of Forced Vibrations and Resonance *113*
5.3 Forced Oscillations of a 1-D Weakly Damped Oscillator *114*
5.4 Steady State Behaviour of a 1-D Weakly Damped Forced Oscillator *115*
5.5 Amplitude and Velocity Resonance *122*
5.5.1 Low Driving Frequency *126*
5.5.2 High Driving Frequency *127*
5.6 Power Absorbed by a Weakly Damped Forced Oscillator *128*
5.7 Quality Factor: Sharpness of Resonance *129*
5.8 A Resonant *LCR* Circuit *133*
Review Exercises 138
Annexure 139

6. Coupled Oscillations **141–175**

Expected Learning Outcomes 141
6.1 Introduction *141*
6.2 Oscillations of Coupled Masses *142*
6.2.1 Longitudinal Oscillations *143*
6.2.2 Transverse Oscillations *155*
6.2.3 Small Oscillation Approximation *157*
6.3 Normal Mode Analysis of Other Oscillating Systems *159*
6.3.1 Coupled Simple Pendulums *160*
6.3.2 Inductively Coupled *LC* Circuits *162*
6.4 General Procedure for Calculating Normal Mode Frequencies *163*
6.5 Energy of Undamped Coupled Systems *166*
6.5.1 Coupled Masses *166*
6.5.2 Coupled Pendulums *167*
6.6 Normal Mode Analysis of a Forced Coupled Oscillator *169*
6.7 Longitudinal Oscillations of *N* Coupled Masses: Wave Equation *171*
Review Exercises 173

PART II WAVES

7. Wave Motion **179–222**

Expected Learning Outcomes *179*
7.1 Introduction *179*
7.2 Wave Formation and Propagation *181*
7.2.1 Transverse and Longitudinal Waves *185*
7.3 Describing Wave Motion *185*
7.3.1 Relation between Wave Velocity, Frequency and Wavelength *187*
7.3.2 Mathematical Description of Wave Motion *188*
7.4 Phase of a Wave *192*
7.4.1 Phase Velocity *196*
7.5 Energy Transported by Progressive Waves *199*
7.6 Intensity of a Wave *202*
7.7 Waves in Strings and Gases: 1-D Wave Equation *206*
7.7.1 Waves on Stretched Strings *206*
7.7.2 Longitudinal Waves in Solids *210*
7.7.3 Longitudinal Waves in Gases *212*
7.7.4 Sound Waves in Liquids *216*
7.8 Waves in Two and Three Dimensions *216*
7.9 The Doppler Effect *217*
7.9.1 Source Stationary and Listener in Motion *218*
7.9.2 Source in Motion and Observer Stationary *219*
7.9.3 Source and Listener both in Motion *220*
Review Exercises *221*

8. Reflection and Refraction of Waves **223–253**

Expected Learning Outcomes *223*
8.1 Introduction *223*
8.2 Wave Motion and Impedance *224*
8.2.1 Impedance Offered by Stretched Strings: Transverse Waves *225*
8.2.2 Impedance Offered by Gases: Sound Waves *227*
8.3 Reflection and Transmission Amplitude Coefficients *229*
8.3.1 Transverse Waves *229*
8.3.2 Longitudinal Waves *234*
8.4 Reflection and Transmission Power Coefficients *235*
8.5 Principle of Superposition of Waves *237*
8.5.1 Superposition of In-phase Waves of Different Amplitudes *239*
8.5.2 Superposition of Identical Out-of-phase Waves *239*
8.5.3 Superposition of identical Waves Moving in Opposite Directions *240*
8.5.4 Superposition of Waves of Slightly Different Frequencies *240*
8.6 Stationary Waves *240*
8.6.1 Velocity of a Particle in a Stationary Wave *242*
8.6.2 Harmonics in Stationary Waves *244*
8.6.3 Waves in Pipes *245*
8.6.4 Musical Sound and Noise *246*
8.7 Beats *248*
8.8 Wave Groups and Group Velocity *249*
Review Exercises *252*

9. Vibrations of Strings **254–279**

Expected Learning Outcomes *254*

9.1 Introduction *254*

9.2 Transverse Waves on a Stretched String *255*

9.2.1 Method of Separation of Variables *256*

9.2.2 Displacement of a Stretched String *258*

9.2.3 Eigenvalues and Eigenfunctions *261*

9.2.4 Mathematical Modelling of a String as a Stationary Wave *262*

9.2.5 Resonant Vibrations of Stretched Strings: Melde's Experiment *263*

9.2.6 Energy of a Vibrating String *265*

9.3 Vibrations of a Plucked String *269*

9.4 Vibrations of a Struck String *272*

9.5 Vibrations of a Bowed String *274*

Review Exercises *278*

10. Vibrations of Bars **280–294**

Expected Learning Outcomes *280*

10.1 Introduction *280*

10.2 Longitudinal Vibrations of a Bar *281*

10.3 Transverse Vibrations of a Bar *288*

10.4 Tuning Fork and Vibrations of a Bar *293*

Review Exercises *293*

11. Vibrations of Air Columns **295–307**

Expected Learning Outcomes *295*

11.1 Introduction *295*

11.2 Vibrations of Air Columns *296*

11.2.1 Vibrations in an Open Pipe *297*

11.2.2 Vibrations in a Closed Pipe *299*

11.2.3 End Correction *303*

11.3 Energy of a Vibrating Air Column *303*

Review Exercises *307*

12. Large Amplitude Oscillations **308–314**

Expected Learning Outcomes *308*

12.1 Introduction *308*

12.2 Free Oscillations of Large Amplitude *309*

12.3 Large Amplitude Oscillations Under Asymmetric Restoring Force *309*

12.4 Large Amplitude Oscillations Under Symmetric Restoring Force *311*

12.5 Combination Tones *311*

12.5.1 Production of Combination Tones *312*

12.5.2 Theories of Combination Tones *312*

Review Exercises *314*

Index ***315–318***

Preface

In every day life, we come across different types of motion—translational as well as rotational. While a satellite executes rotational motion, a body falling under the influence of gravity moves in a straight line. Oscillatory motion and wave motion are other important types of motion. These are responsible for our communication—seeing, speaking and hearing—with the outside world. When we speak, our vocal cords inside our throat vibrate. These vibrations cause the surrounding air molecules to vibrate and the effect—speech—manifests as sound. When this sound reaches the ears of another person, it sets her/his ear drums into vibrations making it heard. Sound is a form of energy and is carried by waves. The chirping of birds, humming of bees, whisper of streams, the fascinating music produced by orchestra or melody of Bharat Ratna Lata Mangeshkar is carried by sound waves. Music of any kind—vocal or instrumental—is made up of several simple harmonic oscillations. In general, harmonic oscillations can arise from isolated or coupled oscillators. Coupling leads to energy exchange and if the number of oscillators is large, it leads to wave propagation.

If you drop pebbles in a still pool of water, ripples propagate in the form of water waves. Our visual contact with the world around us depends on light waves. We are able to appreciate different hues and colours that the nature has created because light makes us to see. Though of entirely different nature, the principles, which help us to understand the physics of music, are equally applicable in spectroscopic techniques such as ECG, EEG, and MRI with important applications in medical science. In this book, we have made a sincere effort to present a unified approach. We have tried to introduce the subject as simply and succinctly as possible, with enough applications and examples from everyday life to establish the relevance of the results. The subject has been introduced from more familiar to less familiar situations. We have developed fundamental concepts gradually and discussed most general situations.

The book is organized in two parts: Part I (Chapters 1 through 6) deals with oscillations and Part II (Chapters 7 through 12) covers waves. In Chapter 1, we have presented Mathematical Preliminaries necessary for studying this course. Chapter 2 presents the concept of Simple Harmonic Motion executed by different physical systems. We have used analogies with mechanical spring-mass system to obtain expressions for frequency of oscillations as well as energy. The superposition of collinear and mutually orthogonal oscillations of same as well as nearly equal frequencies and/or amplitudes forms the subject matter of discussion of Chapter 3. In nature, most oscillations, left to them, die down gradually due to damping. It is, therefore, important to learn to characterise damping and how a system behaves under different conditions. Chapters 4 and 5 are respectively devoted to the discussion of damped free and forced oscillators. Here we will learn that when natural frequency of the forced oscillator equals the frequency of the driving force, a spectacular phenomenon of resonance occurs. Its applications in radio reception have been discussed in detail. Chapter 6 analyses coupled oscillators in terms of normal modes. And as the number of such oscillators becomes infinitely large, energy exchange between them leads to wave propagation in the system.

Part II begins with the discussion of wave formation, propagation and representation. The energy carried by waves is an important characteristic of wave motion. Chapter 7 discusses 1-D wave propagation in different media. Frequency is a fundamental characteristic of a wave but when the listener or source or both are in relative motion, it shows an apparent change. This is known as Doppler Effect and is also discussed here. Chapter 8 deals with reflection and transmission of waves at an interface. We have introduced the concept of impedance offered by the medium and obtained expressions for amplitude and power reflection and transmission coefficients. Chapters 9 to 12 cover vibrations of strings, air columns and bars under different conditions in small amplitude approximation. The analysis discussed here is useful in understanding formation of stationary waves, beats and vocal as well as instrumental music.

One of the acknowledged deficiencies in Indian higher science education is overemphasis on rote memorisation with little scope for innovations and creativity. We have structured the book in such a way that learners participate actively as they proceed; get ample opportunities to develop problem solving skills and discover answers to unknown situations. To ensure this, more than one hundred problems—numerical and reason based questions—with graded difficulty levels have been included as Practice Exercises and Review Exercises in each chapter. Moreover, Solved Examples have been interspersed in the text to facilitate clear understanding of the concepts involved in each section. Wherever possible, we have included different ways of solving a problem so that a reader has choice as per her/his preference and liking.

As such, great care has been taken at every stage of preparation of the manuscript to be rigorous and meticulous. Yet it is possible that some errors may have crept in the text. We will be grateful for constructive suggestions for the improvement in the book.

We wish to express sincere thanks to Ms Shivani Garg (Editor, PHI Learning) and her team for keen interest in this project. Dr. Suresh Garg, who is currently, Expert, Commonwealth Fund for Technical Cooperation, London and Director, Centre for Learning and Teaching, National University of Lesotho, Roma, Southern Africa, is grateful to Commonwealth Secretariat, London for the opportunity.

Suresh Garg
C.K. Ghosh
Sanjay Gupta

Part I

Oscillations

- Mathematical Preliminaries
- Simple Harmonic Motion
- Superposition of Harmonic Oscillations
- Damped Oscillations
- Forced Oscillations
- Coupled Oscillations

1

Mathematical Preliminaries

EXPECTED LEARNING OUTCOMES

In this Chapter, you will acquire capability to:

- state the specific areas of trigonometry, complex variables, differential and integral calculus which find applications in the study of oscillations and waves;
- master the basic techniques of solving ordinary and partial differential equations used in the study of oscillations and waves; and
- state Fourier's theorem and apply it to analyse harmonic functions.

1.1 INTRODUCTION

Mathematics is the language of science and its use is most pronounced in physics. You have applied mathematics to a reasonable extent while studying physics in your senior secondary classes. You must have appreciated how principles of physics get corroborated by mathematical derivations. This is true for all areas of physics. However, if you pause for a while and recall the applications of mathematics in the study of Oscillations and Waves, you would agree that trigonometry is one such subject. So, in this chapter, we start by identifying trigonometric formulae which find significant applications in the study of oscillations and waves. This is followed by specific applications of differential and integral calculus. For example, you will learn how average values of different physical parameters can be easily determined using integral calculus.

In classical physics, every equation of motion contains an acceleration term. (This follows from Newton's second law of motion.) You will note that acceleration is second order time-derivative of displacement. It signifies that a second order ordinary differential equation is of paramount importance in the study of oscillations and waves. We shall, in general, restrict ourselves to linear differential equations. For mathematical treatment of wave motion, you will need to know partial differential equations (PDEs) and how to solve them. For your convenience, we have discussed the basics of PDEs in Section 1.6.

In the last section, we have introduced Fourier analysis. It is an extremely fascinating mathematical tool for the analysis of all forms of harmonic oscillations. We would like you to master this tool because for mathematical treatment of propagation of waves through stretched strings as well as fluid media, it provides an ideal backup to partial differential equations.

As you proceed, you will realise that the mathematical tools discussed in this chapter give a prelude to the entire discussion in this book. We hope that a thorough study of this chapter will equip you with the basic mathematical methods and techniques required to learn this course well. In fact, you should feel confident of your mathematical capabilities required for proper understanding of the principles and theories of oscillations and waves. Therefore, you should study this chapter carefully before proceeding to the following chapters. In case you still have some difficulties, consult your mathematics books that you studied in your earlier classes. By solving the exercises given in this (and other) chapters yourself, you can develop the healthy practice of problem solving. This should give you an added advantage of enjoying the subject and be a deep learner.

1.2 TRIGONOMETRIC FORMULAE

You have studied trigonometry in your secondary and senior secondary mathematics course and learnt to apply trigonometric formulae in the study of physical phenomena. The most familiar applications are in the study of problems related to equilibrium of forces and oscillations and waves. You must have realised that the special characteristics of the sinusoidal functions (sine, cosine) are most vividly revealed in the study of oscillations and waves. In fact, you might have even be tempted at some stage to conclude that perhaps these functions were defined to facilitate the study of oscillations and waves. Let us now see some of the crucial links between trigonometry and physics.

1.2.1 Features of Sinusoidal Functions

As we know that the simplest form of time variation of displacement of a simple harmonic oscillator is given by

$$x = a \sin \omega t \tag{1.1a}$$

or

$$x = a \cos \omega t. \tag{1.1b}$$

Now, try to recall the different aspects of SHM that you had learnt earlier.

FEATURE 1: *The maximum and minimum values of* $\sin \theta$ *and* $\cos \theta$ *are* +1 *and* –1. It means that the maximum and minimum values of x are $+a$ and $-a$. This takes care of the definition of *amplitude*.

FEATURE 2: Refer to Fig. 1.1. It shows the plots of sin θ and cos θ versus θ. You will note that

$$\sin(2\pi + \theta) = \sin\theta \tag{1.2a}$$

and

$$\cos(2\pi + \theta) = \cos\theta. \tag{1.2b}$$

Note that Eqs. (1.2a) and (1.2b) take care of the *periodicity* exhibited by SHM. Similarly, the relation

$$\sin\left(\frac{\pi}{2} + \theta\right) = \cos\theta \tag{1.3}$$

takes care of the *phase difference* of 90° (π/2 radian) between displacement and velocity.

FEATURE 3: sin θ *and* cos θ undergo repeated changes in sign. When θ crosses the π-mark, sin θ changes sign from +ve to –ve and when θ crosses the π/2 and 3π/2 marks, cos θ changes sign respectively from +ve to –ve and –ve to +ve. The total range of sin θ and cos θ for which they are positive or negative are each equal to π. This signifies the fact that a simple harmonic oscillator spends equal time in the positive and negative cycles.

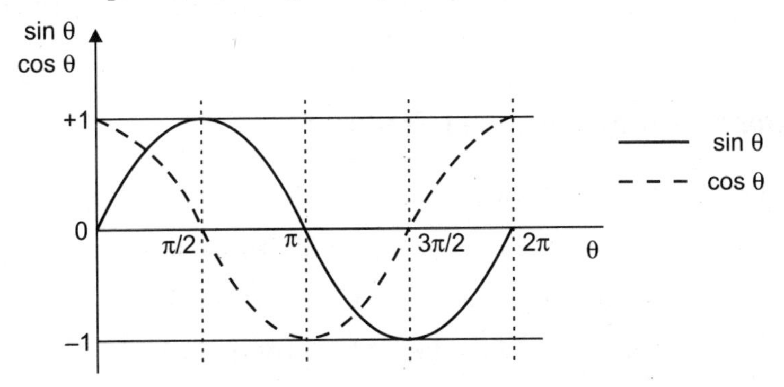

Fig. 1.1 Graph of sin θ and cos θ versus θ.

FEATURE 4: Sine and cosine functions are oscillatory. The sine and cosine functions oscillate periodically between the peak values of +1 and –1. The attainment of maximum and minimum values after a periodic interval signifies the turning points, which are so crucial in oscillatory motion.

You may now like to answer a Practice Exercise.

***Practice Exercise* 1.1** Establish Feature 4.

Now, we would like you to recall the general solutions of trigonometric equations. For simplicity, we shall confine ourselves to sinusoidal functions.

1.2.2 General Solution of Equations sin θ = sin α and cos θ = cos α

The general solution of the equation

$$\sin\theta = \sin\alpha \tag{1.4a}$$

is given by

$$\theta = n\pi + (-1)^n \alpha, \tag{1.4b}$$

whereas for

$$\cos \theta = \cos \alpha \tag{1.4c}$$

the general solution is

$$\theta = 2n\pi \pm \alpha, \tag{1.4d}$$

where n is an integer.
We can rewrite Eqs. (1.4b) and (1.4d) a little more elaborately as follows:

$$\theta = 2n\pi \pm \alpha$$

and

$$\theta = (2n + 1)\pi \pm \alpha.$$

The equation of motion of a mechanical system is bound to be of second order in displacement and time. And the solution of such an equation is obtained through integration in two stages; the first yields velocity and the second gives displacement.

FEATURE 5: For Eqs. (1.4a) and (1.4c) to be satisfied simultaneously, θ must be equal to $2n\pi + \alpha$.
This aspect has tremendous significance in mathematical description of phase, which is the mechanical condition of motion governed by the simultaneous consideration of displacement and velocity. We shall discuss it in detail in Section 2.5.
Now, we would like you to recall the relations pertaining to transformation of sums and products.

1.2.3 Transformation of Sums and Products

From your school mathematics, you may recall that

$$\sin (\alpha + \beta) + \sin (\alpha - \beta) = 2 \sin \alpha \cos \beta, \tag{1.5a}$$

$$\sin (\alpha + \beta) - \sin (\alpha - \beta) = 2 \cos \alpha \sin \beta, \tag{1.5b}$$

$$\cos (\alpha + \beta) + \cos (\alpha - \beta) = 2 \cos \alpha \cos \beta, \tag{1.5c}$$

and

$$\cos (\alpha - \beta) - \cos (\alpha + \beta) = 2 \sin \alpha \sin \beta. \tag{1.5d}$$

You must have worked out quite a few problems in trigonometry using Eqs. (1.5a–d). However, in this course, we shall deal more with relations given in Eqs. (1.5a) and (1.5c). On the left hand sides of these relations, α and β appear together (as sum or difference), whereas on the right hand sides they stand separated. We get such expressions to describe wave motion, where α and β represent spatial and temporal characteristics of a wave. If we consider a relation of the form $\sin (\alpha_1 + \beta_1) + \sin (\alpha_2 - \beta_2)$, the space and time variables do not get separated.

FEATURE 6: *For α and β to be separated, we must have $\alpha_1 = \alpha_2$ and $\beta_1 = \beta_2$.*
This feature has tremendous relevance in understanding of formation of stationary waves by superposition of two identical progressive waves moving in mutually opposite directions. You will learn about these in Section 8.11.

***Practice Exercise* 1.2** Prove that for any expression of the type $(p \cos \theta + q \sin \theta)$, it is always possible to find two constants r and γ, irrespective of the values of p and q, such that the expression is equal to $r \cos (\theta + \gamma)$.

Ans. $r = \sqrt{p^2 + q^2}$; $\gamma = \tan^{-1} \dfrac{q}{p}$

You will now learn about complex numbers. We are sure that you enjoyed working with these in your earlier classes. These facilitate mathematical treatments considerably and prove very useful mathematical tool.

1.3 COMPLEX NUMBERS

Refer to Fig. 1.2, where... $B_2B_1\ OA_1A_2$ ... represents a number line; O represents zero; A_1, A_2 represent +1 and +2 respectively and B_1, B_2 represent –1 and –2 respectively. As we move to the right of O, we get positive numbers and if we move to its left, we come across negative numbers.

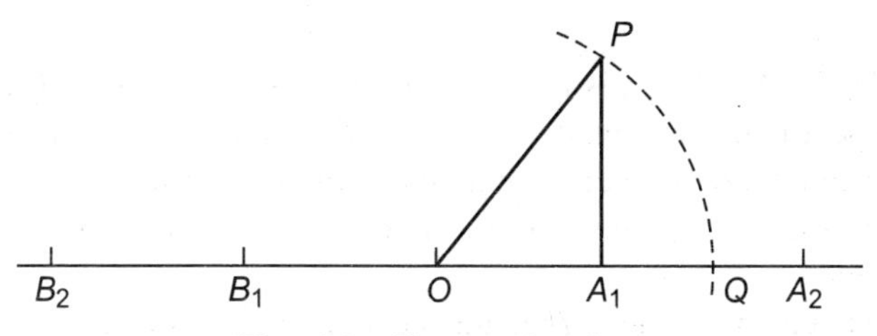

Fig. 1.2 The number line.

You know that all the rational numbers can be located on this line. Same can be said about irrational numbers. Now, study Example 1.1 carefully.

EXAMPLE 1.1

(a) Show that $\sqrt{2}$ lies on the number line shown in Fig. 1.2.

(b) Show that the solutions of the equation $x^2 + 4 = 0$ cannot be located on the number line.

Solution:

(a) Since A_1 represents +1, the length $OA_1 = 1$. At A_1, we draw A_1P (= OA_1) perpendicular to OA_1. Then for the right angled ΔOPA_1, we can write $OP^2 = OA_1^2 + A_1P^2 = 1^2 + 1^2 = 2$ so that $OP = \sqrt{2}$. Now, if we draw an arc of a circle with radius equal to OP, it will cut the number line at Q. So, $OQ = OP = \sqrt{2}$. Hence, 'Q' represents $\sqrt{2}$. That is, $\sqrt{2}$ can be located on the number line.

(b) We rewrite the given equation as $x^2 = -4$. You can easily convince yourself that its solutions are $x = \pm\sqrt{-4}$. In (a), we could construct a straight line OP such that $OP^2 = 2$. But we cannot draw a straight line, the square of whose length is –4. So we shall not be able to locate the solutions on the number line.

In Example 1.1(b), we could not draw a straight line, the square of whose length is –4. To cope with this inability, we draw another line perpendicular to the number line and passing through O (Fig. 1.3). One unit length of that line represents $\sqrt{-1}$, which is an imaginary quantity, as you cannot find any real number whose square would be negative. Now, for the line ... $D_2D_1\ O\ C_1C_2$ ..., if we go above the number line, we come across positive multiples of $\sqrt{-1}$ and if we go below the number line, we get negative multiples of $\sqrt{-1}$. Thus, we have made an arrangement for representing the imaginary quantities, the smallest unit of which is $\sqrt{-1}$.

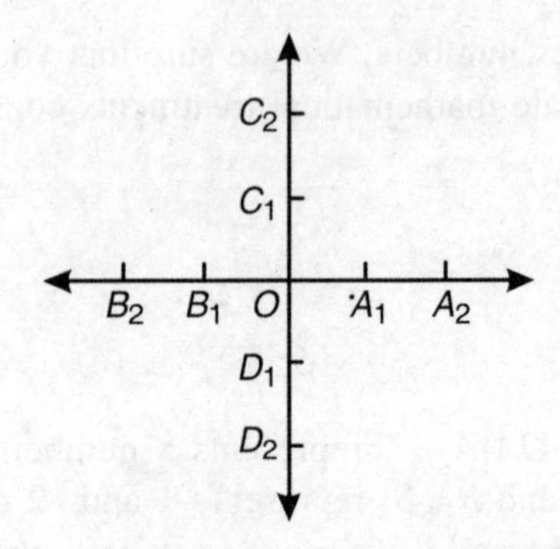

Fig. 1.3 Representation of the roots of the equation $x^2 + 4 = 0$ on number line.

Note that line … D_2D_1 O C_1C_2 … is independent of the line … B_2B_1 O A_1A_2 … The issue that these two lines are perpendicular is a matter of convention (and convenience as well). This is just like the case where we can express a vector quantity in two-dimensions as a linear combination of any two basis vectors in the same plane. But we choose to do so with two mutually perpendicular vectors of unit magnitude.

Now, that we have been able to depict imaginary numbers, we can define complex numbers. In general, a complex number is defined as

$$z = a + b\sqrt{-1}, \tag{1.6}$$

where a and b are real numbers.

1.3.1 Purely Real and Purely Imaginary Numbers

If we put $b = 0$ on the right hand side of Eq. (1.6), it reduces to $z = a$, which is a pure by real number. But if $a = 0$ in Eq.(1.6), then $z = b\sqrt{-1}$, which is a purely imaginary number. For this reason, 'a' is called the *real part of z*, while 'b' is called the *imaginary part of z*. We express these as

$$a = \text{Re}\,(z) \tag{1.7a}$$

and

$$b = \text{Im}\,(z). \tag{1.7b}$$

1.3.2 Geometrical Meaning of Number $j = \sqrt{-1}$

We use the symbol j to represent the square root of minus one, which is the basic unit of an imaginary quantity. Suppose that a purely real number 'N', which can be found on the line … B_2B_1 O A_1A_2 (Fig. 1.3), is multiplied by j. Then we obtain the number 'jN', which is located on the line … D_2D_1 O C_1C_2, as shown in Fig. 1.4. For this reason, the former is called the *real axis* and the latter, the *imaginary axis*. If the location of number N is Q on the real axis, the location R of is jN on the Imaginary Axis such that $OQ = OR$. In other words, multiplication by j amounts to rotating the line OQ by 90° in anticlockwise direction.

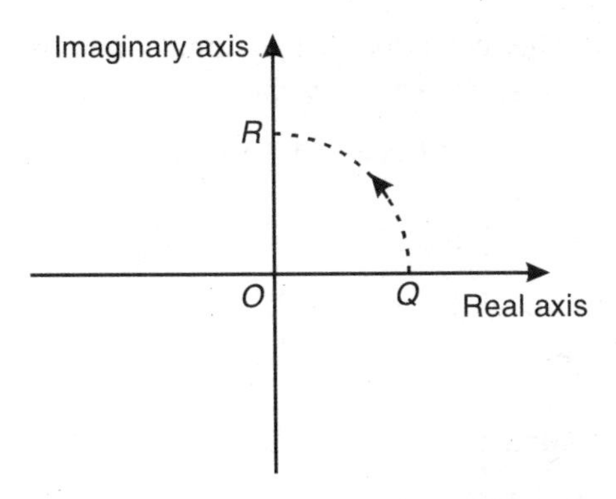

Fig. 1.4 Rotation under multiplication by j.

***Practice Exercise* 1.3** Locate the number $2j$ on a number line. What happens when this number is multiplied by j?

Ans. The –ve direction of real number line

1.3.3 Geometrical Representation of Complex Numbers

You now know that the real and the imaginary axes are perpendicular to each other. This fact is utilised for geometrical representation of a complex number. These axes are respectively made to coincide with the x- and y-axes of a rectangular Cartesian co-ordinate system. A complex number z (= $x + jy$) is represented by the point (x, y), say P, in Fig. 1.5. Using Pythagoras theorem, we can write

$$OP = r = \sqrt{x^2 + y^2}\,.$$

The length OP is called the *modulus* of the complex number, and angle $\angle POM = \theta = \tan^{-1}(y/x)$ is called the *argument* of the complex number. The geometrical representation (Fig. 1.5) is called the Argand diagram. Thus,

$$\text{Mod}\, z = |z| = r = \sqrt{x^2 + y^2} \tag{1.8a}$$

and

$$\text{Arg}\, z = \theta = \tan^{-1}\frac{y}{x}. \tag{1.8b}$$

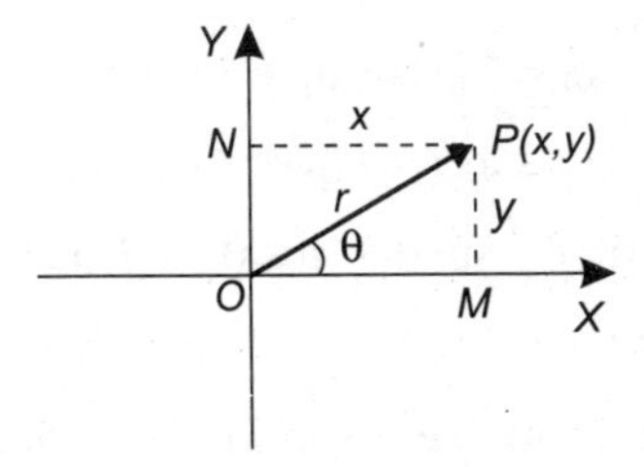

Fig. 1.5 Argand diagram: Geometrical representation of a complex number. (The xy-plane is called the Argand plane or simply the z-plane.)

From this discussion, we can easily conclude that a complex number can be expressed in terms of r and θ. As $x = r\cos\theta$ and $y = r\sin\theta$, we can write

$$z = r(\cos\theta + j\sin\theta). \tag{1.9a}$$

We can rewrite it as

$$z = |z|\{\cos(\text{Arg } z) + j\sin(\text{Arg } z)\}. \tag{1.9b}$$

You should now carefully go through Example 1.2.

EXAMPLE 1.2

Determine the modulus and argument of $z = -2 - 2j$.

Solution: From Eq. (1.8a), we have

$$|z| = \sqrt{(-2)^2 + (-2)^2} = \sqrt{8} = 2\sqrt{2} \quad (|z| \text{ is a length and hence always +ve})$$

$$\text{Arg } z = \tan^{-1}\left(\frac{-2}{-2}\right) = \tan{-1}\ 1$$

At this stage you have to use your judgement. Note that Arg z may not necessarily be the principal value of the inverse function. In the Argand diagram, the complex number under consideration lies in the third quadrant. So Arg $z = 5\pi/4$ and not $\pi/4$.

1.3.4 Fundamental Properties of Complex Numbers

Addition and subtraction

You are familiar with the addition and subtraction of vector quantities. The complex numbers follow analogous operations. You would recall that in vector addition and subtraction, we treat the x and y-components separately. And in case of complex numbers, we treat the real and imaginary parts separately. Therefore, if

$$z_1 = x_1 + jy_1$$

and

$$z_2 = x_2 + jy_2,$$

then

$$z_1 \pm z_2 = (x_1 \pm x_2) + j(y_1 \pm y_2). \tag{1.10}$$

Multiplication and division

To explain multiplication and division of complex numbers, we consider two complex numbers z_1 and z_2 and write

$$z_1 = r_1(\cos\theta_1 + j\sin\theta_1)$$

and

$$z_2 = r_2(\cos\theta_2 + j\sin\theta_2)$$

$\therefore$

$$z_m = z_1 z_2 = r_1 r_2 (\cos\theta_1 + j\sin\theta_1)(\cos\theta_2 + j\sin\theta_2)$$

$$= r_1 r_2 \{(\cos\theta_1\cos\theta_2 + j^2\sin\theta_1\sin\theta_2) + j(\sin\theta_1\cos\theta_2 + \cos\theta_1\sin\theta_2)\}.$$

Since $j^2 = -1$, using trigonometric formulae given in Eq. (1.5a, c), we can write

$$z_m = r_1 r_2\{\cos(\theta_1 + \theta_2) + j\sin(\theta_1 + \theta_2)\}. \tag{1.11}$$

$\therefore$

$$\text{Mod } z_m = r_1 r_2 = |z_1||z_2| \tag{1.11a}$$

and

$$\text{Arg } z_m = \theta_1 + \theta_2 = \text{Arg } z_1 + \text{Arg } z_2. \tag{1.11b}$$

For division of complex numbers, let us write

$$z_d = \frac{z_1}{z_2}.$$

You should convince yourself that

$$\text{Mod } z_d = \frac{|z_1|}{|z_2|} \tag{1.12a}$$

and

$$\text{Arg } z_d = \text{Arg } z_1 - \text{Arg } z_2. \tag{1.12b}$$

We would now like you to solve the following Practice Exercise and make sure that you have fixed your ideas about complex numbers.

***Practice Exercise* 1.4**

(a) Use Eq. (1.10) to represent $z_1 + z_2$ geometrically and prove that $|z_1| + |z_2| \geq |z_1 + z_2|$. When does the equality sign hold? Prove the same result using the form given in Eq. (1.9a) for representation of a complex number.

(b) Prove Eqs. (1.12a) and (1.12b).

(c) If $a_1 + jb_1 = a_2 + jb_2$, prove that $a_1 = a_2$, $b_1 = b_2$, provided a_1, a_2, b_1, b_2 are all real.

Complex conjugate

The conjugate of a complex number, $z = a + jb$ is defined as

$$z^* = a - jb. \tag{1.13}$$

Thus, if the original number is represented by the point (a, b), its conjugate is given by $(a, -b)$. This is shown in Fig. 1.6.

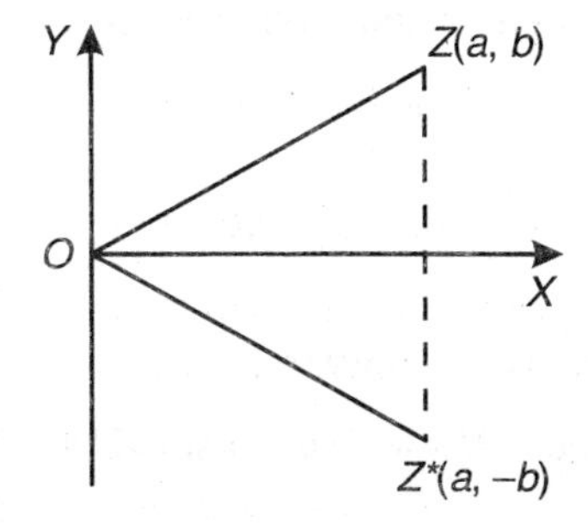

Fig. 1.6 Representation of a complex number and its conjugate.

Note that conjugate of a complex number is obtained by replacing j by $-j$ in the original number. It means that the conjugate of a conjugate will yield the original number:

$$(z^*)^* = a + jb = z \tag{1.14}$$

Also

$$zz^* = (a + jb)(a - jb) = a^2 + b^2 = |z|^2. \tag{1.15}$$

You may now like to answer the following Practice Exercise.

***Practice Exercise* 1.5** Show that the sum of a complex number with its conjugate is purely real, whereas the difference is purely imaginary.

Complex conjugation is used as a vital tool for expressing a complex number $(a_1 + jb_1)/(a_2 + jb_2)$ in the form $A + jB$. To prove this, let us put

$$\omega = \frac{a_1 + jb_1}{a_2 + jb_2}$$

By multiplying the numerator and the denominator by the complex conjugate of the denominator, we get

$$\omega = \frac{(a_1 + jb_1)(a_2 - jb_2)}{(a_2 + jb_2)(a_2 - jb_2)}$$

$$= \frac{(a_1a_2 + b_1b_2) + j(a_2b_1 - a_1b_2)}{a_2^2 + b_2^2}.$$

This can be expressed as

$$\omega = \left(\frac{a_1a_2 + b_1b_2}{a_2^2 + b_2^2}\right) + j\left(\frac{a_2b_1 - a_1b_2}{a_2^2 + b_2^2}\right).$$

1.3.5 Demoivre's Theorem (DT)

Demoivre's Theorem is a very useful theorem in respect of complex numbers. It states that $(\cos n\theta + j \sin n\theta)$ *is the value of* $(\cos \theta + j \sin \theta)^n$ *where* n *is a +ve or –ve integer and is one of the values of* $(\cos \theta + j \sin \theta)^n$, *when* n *is a fraction.*

Let us consider the case where n is a fraction, say $= p/q$, where p and q are integers prime to each other. Then according to DT, $[\cos (p\theta/q) + j \sin (p\theta/q)]$ is one of the values of $(\cos \theta + j \sin \theta)^{p/q}$. You may like to know about other values. To discover this, we write

$$z = (\cos \theta + j \sin \theta)^{p/q}$$

so that

$$z^q = (\cos \theta + j \sin \theta)^p = \cos p\theta + j \sin p\theta. \tag{1.16}$$

Note that Eq. (1.16) is a q-degree equation in z and will have q solutions. We have already obtained one solution. To obtain other solutions, we rewrite it as

$$z^q = \cos p\theta + j \sin p\theta = \cos (2k\pi + \theta)p + j \sin (2k\pi + \theta)p, \quad k = 0, 1, 2 \ldots$$

Hence

$$z = \cos (2k\pi + \theta)\frac{p}{q} + j \sin (2k\pi + \theta)\frac{p}{q}, \quad k = 1, 2, 3, \ldots, (q - 1) \tag{1.17}$$

Note that the $(q - 1)$ values of k will account for the remaining $(q - 1)$ solutions. $k = q$ will generate the solution identical to that of $k = 0$ and thereafter the cycle gets repeated like $k = q + 1, q + 2, q + 3, \ldots$ will correspond respectively to $k = 1, 2, 3, \ldots$ You should now study Example 1.3 carefully.

EXAMPLE 1.3

(a) Verify DT for (i) $n = 2$ and (ii) $n = -3$.
(b) Calculate the values of $(j)^{1/4}$.

Solution: For $n = 2$, we note that

(a) (i) $(\cos\theta + j\sin\theta)^2 = \cos^2\theta + j^2\sin^2\theta + 2j\cos\theta\sin\theta$

$$= (\cos^2\theta - \sin^2\theta) + j(2\cos\theta\sin\theta)$$

$$= \cos 2\theta + j\sin 2\theta$$

(ii) $(\cos\theta + j\sin\theta)^{-3} = \dfrac{1}{(\cos\theta + j\sin\theta)^3}$

Now,

$$(\cos\theta + j\sin\theta)^3 = (\cos\theta + j\sin\theta)^2(\cos\theta + j\sin\theta)$$

$$= (\cos 2\theta + j\sin 2\theta)(\cos\theta + j\sin\theta)$$

$$= (\cos 2\theta\cos\theta - \sin 2\theta\sin\theta) + j(\sin 2\theta\cos\theta + \cos 2\theta\sin\theta)$$

$$= \cos(2\theta + \theta) + j\sin(2\theta + \theta)$$

$$= \cos 3\theta + j\sin 3\theta$$

$$(\cos\theta + j\sin\theta)^{-3} = \frac{1}{(\cos\theta + j\sin\theta)^3}$$

On multiplying the numerator and denominator by $\cos 3\theta - j\sin 3\theta$, we get

$$(\cos\theta + j\sin\theta)^{-3} = \frac{(\cos 3\theta - j\sin 3\theta)}{(\cos 3\theta + j\sin 3\theta)(\cos 3\theta - j\sin 3\theta)}$$

$$= \frac{\cos(-3\theta) + j\sin(-3\theta)}{\cos^2 3\theta + \sin^2 3\theta}$$

$$= \cos(-3)\theta + j\sin(-3)\theta \qquad (\because \cos^2 3\theta + \sin^2 3\theta = 1)$$

(b) We can write the given number as

$$j = 0 + j.1. = \cos\frac{\pi}{2} + j\sin\frac{\pi}{2}$$

$$= \cos\left(2k\pi + \frac{\pi}{2}\right) + j\sin\left(2k\pi + \frac{\pi}{2}\right)$$

$$\therefore\ j^{1/4} = \cos\left\{\frac{1}{4}\left(2k\pi + \frac{\pi}{2}\right)\right\} + j\sin\left\{\frac{1}{4}\left(2k\pi + \frac{\pi}{2}\right)\right\}, \quad \text{for } k = 0, 1, 2, 3,$$

So, the required values are: $\left(\cos\dfrac{\pi}{8} + j\sin\dfrac{\pi}{8}\right), \left(\cos\dfrac{5\pi}{8} + j\sin\dfrac{5\pi}{8}\right),$

$$\left(\cos\frac{9\pi}{8} + j\sin\frac{9\pi}{8}\right) \text{ and } \left(\cos\frac{13\pi}{8} + j\sin\frac{13\pi}{8}\right).$$

1.3.6 Using the Form $e^{j\theta}$

For complex numbers, Euler's Theorem states that

$$e^{j\theta} = \cos\theta + j\sin\theta \tag{1.18}$$

$$\therefore \quad e^{-j\theta} = \cos\theta - j\sin\theta \tag{1.19}$$

From Eqs. (1.18) and (1.19), we can write

$$\cos\theta = \frac{e^{j\theta} + e^{-j\theta}}{2} \tag{1.20a}$$

and

$$\sin\theta = \frac{e^{j\theta} - e^{-j\theta}}{2j}. \tag{1.20b}$$

As you proceed, you will realise that working with complex numbers becomes very elegant by using the exponential form. We now illustrate it by considering multiplication and division of two complex numbers.

You may recall that for product of two complex numbers, we can write the modulus as

$$|z_m| = |z_1||z_2|$$

and argument as

$$\text{Arg } z_m = \text{Arg } z_1 + \text{Arg } z_2.$$

On the other hand, for division of two complex numbers, the modulus is given by

$$|z_d| = \frac{|z_1|}{|z_2|}.$$

And the argument is given by

$$\text{Arg } z_d = \text{Arg } z_1 - \text{Arg } z_2.$$

The results derived for multiplication and division for complex numbers can be obtained quite elegantly by using the exponential form. To show this, we write

$$z_1 = r_1 e^{j\theta_1} \quad \text{and} \quad z_2 = r_2 e^{j\theta_2}.$$

From these substitutions, it readily follows that

$$|z_1| = r_1, |z_2| = r_2, \text{Arg } z_1 = \theta_1, \text{Arg } z_2 = \theta_2.$$

Hence the product of given complex numbers can be expressed as

$$z_m = z_1 z_2 = r_1 r_2 e^{j(\theta_1+\theta_2)}$$

$$\therefore \quad |z_m| = r_1 r_2 = |z_1||z_2|$$

and $$\text{Arg } z_m = \theta_1 + \theta_2 = \text{Arg } z_1 + \text{Arg } z_2.$$

Similarly, division of these numbers gives

$$z_d = \frac{z_1}{z_2} = \frac{r_1}{r_2} e^{j(\theta_1-\theta_2)}$$

$$\therefore \quad |z_d| = \frac{r_1}{r_2} = \frac{|z_1|}{|z_2|}$$

and

$$\text{Arg } z_d = \theta_1 - \theta_2 = \text{Arg } z_1 - \text{Arg } z_2.$$

Let us now consider differentiation and integration of complex numbers.

Differentiation and integration

Let

$$u = re^{j\omega t},$$

where ω is a constant having dimension of reciprocal of time. The first time-derivative of u can be easily written as

$$\frac{du}{dt} = j\omega re^{j\omega t} = j\omega u.$$

Similarly, integration of u with time gives

$$\int u dt = r \int e^{j\omega t} dt = \frac{r}{j\omega} e^{j\omega t}$$

$$= -\frac{jr}{\omega} e^{j\omega t}.$$

Note that we have ignored the constant of integration. Physically, we can say that differentiation of a complex number with respect to time introduces rotation by $\pi/2$ in the anticlockwise sense, whereas integration with respect to time introduces the same effect in the clockwise sense. Thus, $re^{j\omega t}$ can be considered a rotating vector of constant magnitude r and constant angular velocity ω. Moreover, $e^{j\omega t}$ is a periodic function in time. Since $e^{\pm j2\pi} = 1$, its time period is $2\pi/\omega$ and frequency is $\omega/2\pi$.

You should now carefully read the following Example.

EXAMPLE 1.4

(a) Express $3 + j4$ in the form $re^{j\theta}$.

(b) The supply voltage in an alternating current circuit is expressed as $V = V_0 e^{j\omega t}$. It is fed to a capacitor of capacitance C. Show that the magnitude of current through the capacitor at any time is 'ωC' times the voltage and it is a vector oriented at an angle 90° in the anticlockwise sense with reference to the voltage.

Solution:

(a) $z = 3 + j4 = r(\cos\theta + j\sin\theta)$, where $r = \sqrt{3^2 + 4^2} = 5$

and $$\theta = \tan^{-1}\frac{4}{3}.$$

$\therefore$ $$z = 5e^{j\tan^{-1}(4/3)}.$$

(b) The charge on the capacitor, $q = CV_0 e^{j\omega t}$

$\therefore$ $$\text{Current, } i = \frac{dq}{dt} = j\omega CV_0 e^{j\omega t}.$$

It is a vector having magnitude ωC times V_0. Since it is multiplied by j, it is oriented at 90° (anticlockwise) with reference to the voltage.

1.4 DIFFERENTIAL CALCULUS

In physics, use of calculus facilitates mathematical formulation. This is particularly true of oscillations and waves. In this chapter, we have briefly discussed those aspects of differential and integral calculus and differential equations that you will use more frequently in this course.

1.4.1 Differential Coefficients of sin $(\omega t + \delta)$ and cos $(\omega t + \delta)$

The time variation of displacement of a simple harmonic oscillator can be expressed by either of the following two equations:

$$x = a \sin(\omega t + \delta) \tag{1.21a}$$

and

$$x = a \cos(\omega t + \delta), \tag{1.21b}$$

where a, ω and δ are constants.

Let us first consider Eq. (1.21a). The first time-derivative of displacement gives particle velocity:

$$v = \frac{dx}{dt} = a\omega \cos(\omega t + \delta) \tag{1.22}$$

and second time-derivative of displacement gives particle acceleration:

$$\text{acc} = \frac{dv}{dt} = -a\omega^2 \sin(\omega t + \delta) \tag{1.23}$$

By combining Eqs. (1.21a) and (1.23), we get

$$\text{acc} = -\omega^2 x. \tag{1.24}$$

This result shows that acceleration is directly proportional to displacement. The negative sign indicates that it is directed against displacement, i.e. it points towards a fixed point, which is a basic characteristic of SHM.

Further, Eq. (1.22) can be rewritten as

$$v = a\omega \sin\left(\omega t + \delta + \frac{\pi}{2}\right). \tag{1.25}$$

From the above discussion we can say that displacement, velocity and acceleration are sinusoidal. However, velocity leads the displacement by $\pi/2$. Since, the maximum value of the sinusoidal function is always unity, we can easily determine maximum values of these physical quantities.

You may now like to work out a Practice Exercise.

***Practice Exercise* 1.6** Derive Eqs. (1.22), (1.24) and (1.25) using Eq. (1.21b).

Now, you will learn about another interesting mathematical tool, known as Taylor's Theorem.

1.4.2 Taylor's Theorem and Harmonic Approximation

Taylor's Theorem states that if a function $f(x)$ is continuous and its derivatives of all orders exist at $x = x_0$, then the value of the function for $x = x_0 + h$, where h is an infinitesimal increment over x_0, is given by

$$f(x_0 + h) = f(x_0) + hf'(x_0) + \frac{h^2}{2!} f''(x_0) + \frac{h^3}{3!} f'''(x_0) + \cdots \tag{1.26}$$

EXAMPLE 1.5

Let $f(\theta) = \sin \theta$, find $f(\theta + \alpha)$ and hence show that

$$\sin \alpha = \alpha - \frac{\alpha^3}{3!} + \frac{\alpha^5}{5!} - \cdots$$

Solution:

$$f(\theta + \alpha) = f(\theta) + \alpha f'(\theta) + \frac{\alpha^2}{2!} f''(\theta) + \frac{\alpha^3}{3!} f'''(\theta) + \frac{\alpha^4}{4!} f^{iv}(\theta) + \frac{\alpha^5}{5!} f^{v}(\theta) + \cdots$$

$$f(\alpha) = f(0 + \alpha) = f(0) + \alpha f'(0) + \frac{\alpha^2}{2!} f''(0) + \frac{\alpha^3}{3!} f'''(0) + \frac{\alpha^4}{4!} f^{iv}(0) + \frac{\alpha^5}{5!} f^{v}(0) + \cdots$$

$$f(\theta) = \sin \theta$$

$$\therefore \quad f'(\theta) = \cos\theta,\ f''(\theta) = -\sin\theta,\ f'''(\theta) = -\cos\theta,\ f^{iv}(\theta) = \sin\theta,\ f^{v}(\theta) = \cos\theta$$

$$\therefore \quad f(0) = 0,\ f'(0) = 1,\ f''(0) = 0,\ f'''(0) = -1,\ f^{iv}(0) = 0,\ f^{v}(0) = 1$$

$$\therefore \quad \sin \alpha = f(\alpha) = 0 + \alpha \cdot 1 + \frac{\alpha^2}{2!} \cdot 0 + \frac{\alpha^3}{3!}(-1) + \frac{\alpha^4}{4!} \cdot 0 + \frac{\alpha^5}{5!}(1) - \cdots$$

$$\therefore \quad \sin \alpha = \alpha - \frac{\alpha^3}{3!} + \frac{\alpha^5}{5!} - \cdots \tag{1.27}$$

***Practice Exercise* 1.7** Following the steps used in deriving Eq. (1.27), show that

$$\cos \alpha = 1 - \frac{\alpha^2}{2!} + \frac{\alpha^4}{4!} - \cdots \tag{1.28}$$

In all problems on oscillations, we come across potential energy (PE) function and the mean position of oscillation coincides with its minimum. For example, in case of a simple pendulum it is the gravitational PE function. You may recall that PE is minimum at the position where its first derivative vanishes (i.e., net force acting on the oscillator is zero and hence it is the position of equilibrium) and the second derivative is positive. Now, let us work out an expansion of a PE function $V(X)$ about its equilibrium position, x_0:

$$V(X) = V(x_0 + x) = V(x_0) + xV'(x_0) + \frac{x^2}{2!} V''(x_0) + \frac{x^3}{3!} V'''(x_0) + \frac{x^4}{4!} V^{iv}(x_0) + \cdots$$

$$V'(x_0) = 0$$

The effective PE at $X = x_0 + x$, determined with respect to the equilibrium position $x = x_0$, is given by

$$V_{eff}(X) = V(x) - V(x_0) = \frac{x^2}{2!}V''(x_0) + \frac{x^3}{3!}V'''(x_0) + \frac{x^4}{4!}V^{iv}(x_0) + \cdots$$

If x is very small such that higher order powers are negligible, then

$$V_{eff}(X) = \frac{1}{2}V''(x_0)\,x^2. \tag{1.29}$$

In such a situation, the oscillator executes simple harmonic oscillations, as

$$\text{Force} = -\frac{dV}{dx} = -kx, \tag{1.30}$$

where

$$k = V''(x_0).$$

Thus, if an oscillator is displaced by a very small extent, so that the terms involving power of displacement higher than two can be neglected then the PE function takes the form 1/2 (a constant) $(\text{displacement})^2$, [as is shown in Eq. (1.29)] and the oscillator executes simple harmonic motion. This mathematical intervention is called *harmonic approximation*.

If the PE function has terms in number of powers of order higher than two, then the oscillations becomes *anharmonic*.

EXAMPLE 1.6

Show that for a simple pendulum of length l (Fig. 1.7) displaced by an extremely small angle θ from its equilibrium position, the PE function is given by (1/2) (a constant) θ^2.

Solution: By referring to Fig. 1.7, we can write

$$\text{PE} = \text{mg}l\,(1 - \cos l). \tag{1.31}$$

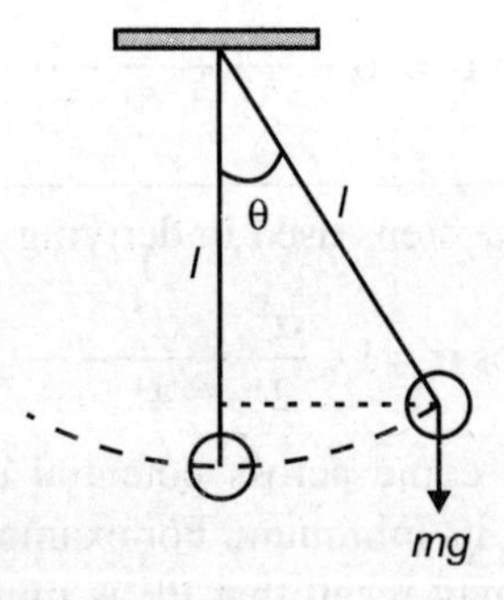

Fig. 1.7 A simple pendulum.

Using Eq. (1.28), for small θ, we get

$$1 - \cos\theta = \frac{\theta^2}{2}.$$

$$\therefore \qquad \text{PE} = \frac{1}{2}\text{mg}l\,\theta^2. \tag{1.32}$$

***Practice Exercise* 1.8** Show that the PE function can be expressed as in Eq. (1.29) for small oscillations of (a) torsional pendulum, (b) oscillation magnetometer.

1.4.3 Partial Derivatives

We come across many situations where a variable depends on more than one variable. In an examination, marks secured by you are a function of factors such as the number of hours per day you put in for your study, your cognitive level, your retention capability, etc. The internal energy of a gas depends on its temperature, pressure and volume. The potential of an electrostatic field depends on the co-ordinates (x, y, z). In such situations, we study variation of the dependent variable with respect to one of the independent variables at a time, keeping other variable(s) constant. Mathematically, we say that we calculate what is known as the partial derivative.

In wave motion, displacement/phase varies with position as well as time. You will be required to study the nature of variation of a wave parameter with respect to position and time separately using partial derivatives. For example, the standard equation of a plane progressive harmonic wave travelling along the +ve direction of x-axis is

$$y(x, t) = a \sin \frac{2\pi}{\lambda}(vt - x), \tag{1.33}$$

where y is displacement of a particle from its mean position, a, v and λ are respectively the amplitude, velocity and wavelength of the wave, x and t are respectively the co-ordinates of the mean position of the particle and time. The particle velocity and acceleration are $(\partial y/\partial t)_x$ and $(\partial^2 y/\partial t^2)_x$, i.e., the first and second order partial derivatives of y with respect to t—not the total derivatives—at constant x. Similarly, the partial derivate $(\partial y/\partial x)_t$ becomes crucial in the study of stresses in waves. The bulk stress in a medium through which a wave propagates is given by

$$\Delta p = -K\frac{\partial y}{\partial x},$$

where K is the bulk modulus.

So we obtain the following expression from Eq. (1.33):

$$\text{Particle velocity} = \frac{\partial y}{\partial t} = \frac{2\pi a v}{\lambda}\cos\frac{2\pi}{\lambda}(vt - x), \tag{1.34}$$

$$\text{Particle acceleration} = \frac{\partial^2 y}{\partial t^2} = \frac{4\pi^2 a v^2}{\lambda^2}\sin\frac{2\pi}{\lambda}(vt - x), \tag{1.35}$$

$$\text{Bulk stress } \Delta p = -K\frac{\partial y}{\partial x} = \frac{2\pi a K}{\lambda}\cos\frac{2\pi}{\lambda}(vt - x). \tag{1.36}$$

Likewise, you will come across several expressions using the knowledge of partial derivatives. You may now like to solve a Practice Exercise.

***Practice Exercise* 1.9** If $y = f(vt \pm x)$, prove that $\dfrac{\partial^2 y}{\partial t^2} = v^2 \dfrac{\partial^2 y}{\partial x^2}$.

We shall now discuss mathematical tools pertaining to integral calculus.

1.5 INTEGRAL CALCULUS

In Section 1.2, we discussed the features of sinusoidal functions. From Fig. 1.1, we recall that both sine and cosine functions are bounded in the range (–1, 1). They are also periodic with a period 2π. It would be worthwhile to examine their average values over the cycle. Along with $\sin x$ and $\cos x$, we shall examine the average values of $\sin^2 x$ and $\cos^2 x$.

1.5.1 Average Values of sin *x*, cos *x*, sin²*x*, cos²*x*

The average value of a function $f(x)$ in the range $[a, b]$ is given by

$$\overline{f(x)} = \frac{1}{b-a} \int_a^b f(x)\, dx. \tag{1.37}$$

It means that for the range $0 \le x \le 2\pi$, we get

$$\overline{\sin x} = \frac{1}{2\pi} \int_0^{2\pi} \sin x\, dx = -\frac{1}{2\pi} \left| \cos x \right|_0^{2\pi} = -\frac{1}{2\pi}(1-1) = 0, \tag{1.38}$$

$$\overline{\cos x} = \frac{1}{2\pi} \int_0^{2\pi} \cos x\, dx = \frac{1}{2\pi} \left| \sin x \right|_0^{2\pi} = \frac{1}{2\pi}(0-0) = 0, \tag{1.39}$$

and

$$\overline{\sin^2 x} = \frac{1}{2\pi} \int_0^{2\pi} \sin^2 x\, dx = \frac{1}{4\pi} \int_0^{2\pi} (1-\cos 2x)\, dx = \frac{1}{4\pi} \left[\int_0^{2\pi} dx - \int_0^{2\pi} \cos 2x\, dx \right]$$

$$= \frac{1}{4\pi} \left[2\pi - \left| \frac{\sin 2x}{2} \right|_0^{2\pi} \right] = \frac{2\pi}{4\pi} = \frac{1}{2}. \tag{1.40}$$

Similarly, you can convince yourself that

$$\overline{\cos^2 x} = \frac{1}{2}. \tag{1.41}$$

The results contained in Eqs. (1.38), (1.39), (1.40) and (1.41) can also be understood from the graphs shown in Fig. 1.8.

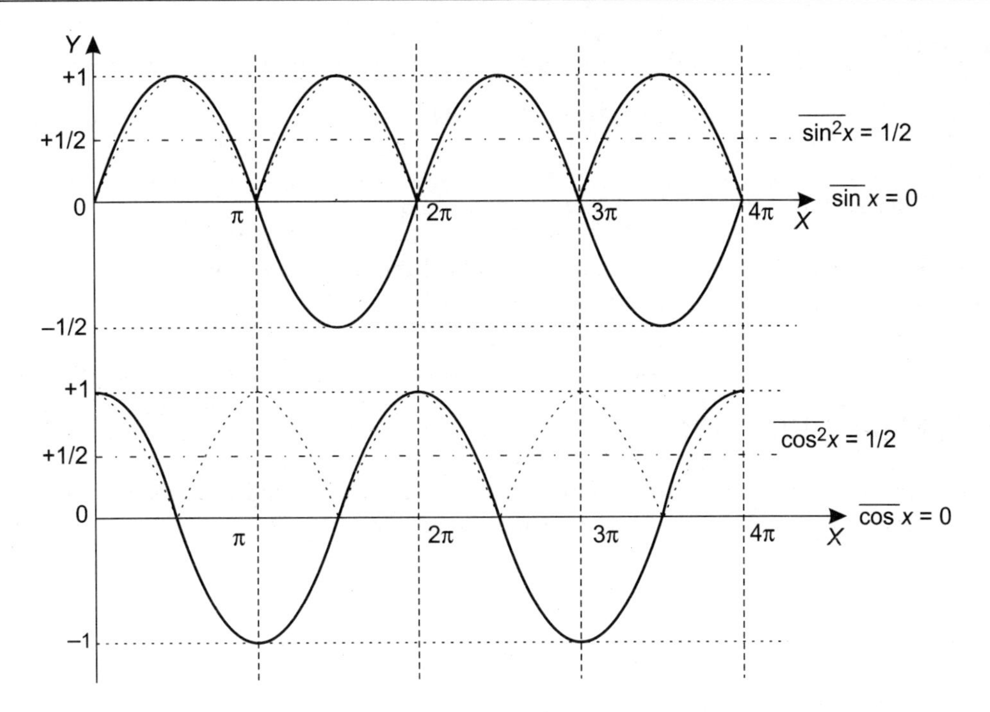

Fig. 1.8 Plots of sin x, cos x, $\sin^2 x$ and $\cos^2 x$ versus x. It clearly shows that average value of sin x and cos x over a cycle is zero, while average value of $\cos^2 x$ and $\sin^2 x$ is one-half.

1.5.2 The Integrals $\int_0^T \sin^2(\omega t + \delta)\, dt$ and $\int_0^T \cos^2(\omega t + \delta)\, dt$

Consider the integrals $\int_0^T \sin^2(\omega t + \delta)\, dt$ and $\int_0^T \cos^2(\omega t + \delta)\, dt$. You will come across these integrals quite often in this course. Here T is the time period given by $T = (2\pi/\omega)$. These integrals can be expressed as

$$I = \frac{1}{2}\int_0^T \{1 \pm \cos 2(\omega t + \delta)\}\, dt$$

$$= \frac{1}{2}\left[\int_0^T dt \pm \int_0^T \cos 2(\omega t + \delta)\}\, dt\right].$$

Proceeding further, we note that

$$\int_0^T \cos 2(\omega t + \delta)\, dt = \frac{1}{2\omega}\left|\sin 2(\omega t + \delta)\right|_0^T = \frac{1}{2\omega}\left|\sin 2(\omega T + \delta) - \sin 2\delta\right|$$

$$= \frac{1}{2\omega}\left|\sin(4\pi + 2\delta) - \sin 2\delta\right| = 0.$$

It means that

$$\int_0^T \sin^2(\omega t + \delta)\, dt = \int_0^T \cos^2(\omega t + \delta)\, dt = \frac{T}{2}. \tag{1.42}$$

***Practice Exercise* 1.10** Prove the result contained in Eq. (1.41).

We shall now familiarise you with some of the typical differential equations that you will come across in the study of oscillations and waves.

1.6 DIFFERENTIAL EQUATIONS

Any equation involving differential coefficients, the independent and dependent variables is a differential equation. For example, let us consider the relation

$$y = \sin^{-1} x. \tag{1.43}$$

$$\therefore \qquad \frac{dy}{dx} = \frac{1}{\sqrt{1-x^2}}$$

or

$$\sqrt{1-x^2}\,\dot{y} = 1,$$

where dot over y denotes its first order derivative with respect to x. On squaring both sides, we get

$$(1-x^2)\,\dot{y}^2 = 1.$$

On differentiating both sides of this equation with respect to x, we get

$$(1-x^2)\,2\dot{y}\ddot{y} + (-2x)\,\dot{y}^2 = 0$$

or

$$(1-x^2)\,\ddot{y} - x\dot{y} = 0. \tag{1.44a}$$

Equation (1.44a) is a differential equation. It can also be expressed as

$$(1-x^2)\,y_2 - xy_1 = 0, \tag{1.44b}$$

where suffixes indicate the order of derivative of y with respect to x. The highest order derivative appearing in an equation is called its *order*. Thus, Eq. (1.44b) is a second order differential equation.

A differential equation is said to be *linear* when the dependent variable and its derivatives of all orders occur only in the first degree. In other words, their squares, cubes, mixed products, etc. do not appear in the equation. Thus, a linear differential equation of the order n is of the form

$$a_n(x)\frac{d^n y}{dx^n} + a_{n-1}(x)\frac{d^{n-1}}{dx^{n-1}} + \cdots + a_1(x)\frac{dy}{dx} + a_0(x)y = f(x), \tag{1.45}$$

where the coefficients a_0, a_1, ..., a_n may be function of the independent variable x and $a_n \neq 0$.

1.6.1 Differential Equations of the Form: $\ddot{x} + 2b\dot{x} + \omega^2 x = f(t)$

In the study of oscillations and waves, we most often come across second order linear differential equations, i.e., $n = 2$ in Eq. (1.45) and the differential equation is of the form

$$a_2(x)\frac{d^2y}{dx^2} + a_1(x)\frac{dy}{dx} + a_0(x)y = f(x). \tag{1.45a}$$

If the coefficients a_0, a_1, a_2 are constants, with $f(x) = 0$, the equation is said to be homogeneous, otherwise it is inhomogeneous. We shall come across differential equations describing time-variation of displacement with time in the following three forms:

$$\ddot{x} + \omega^2 x = 0, \tag{1.46a}$$

$$\ddot{x} + 2b\dot{x} + \omega^2 x = 0, \tag{1.46b}$$

and

$$\ddot{x} + 2b\dot{x} + \omega^2 x = f(t). \tag{1.46c}$$

Note that Eq. (1.46c) has the most generalised form; Eq. (1.46b) is obtained with $f(t) = 0$ and for Eq. (1.46a), we must also have $b = 0$. Equations (1.46a), (1.46b) and (1.46c) respectively represent the motion of a harmonic oscillator, a damped oscillator and a forced oscillator. We can also treat Eq. (1.46a) as a special case of Eq. (1.46b). You will learn to solve these in Chapter 4. However, you should try to solve Review Exercise 1.4.

In the study of wave motion, we come across second order partial differential equations in x and t. Let us consider these now.

1.6.2 Partial Differential Equations of the Type $(\partial^2 y/\partial t^2) = v^2(\partial^2 y/\partial x^2)$

You now know that when the dependent variable is a function of more than one independent variable, we get partial derivatives. And equations involving partial derivatives are called *partial differential equations*. We come across such equations in the study of wave motion. Wave motion is basically propagation of phase. In other words, wave parameters like displacement and velocity of medium particle supporting wave motion is a function of space as well as time. For a wave propagating along the x-axis with a velocity v, we get

> We see in Eq. (1.48) that $y = f(x, t)$. However, the argument of the function must be unidimensional. So it would either have the dimension of space or that of time. In other words, it would have the form $(x \pm vt)$ or $[t \pm (x/v)$. The translational factor between space and time is the velocity of the wave.

$$\frac{\partial^2 y}{\partial t^2} = v^2 \frac{\partial^2 y}{\partial x^2}. \tag{1.47}$$

Let us pause for a while and ask: How is Eq. (1.47) different from Eq. (1.45)? While Eq. (1.45) is an ordinary second order linear differential equation, Eq. (1.47) is a linear second order partial differential equation. Its most general solution is of the form

$$y = f_1(vt + x) + f_2(vt - x). \tag{1.48}$$

An important feature of this solution is that x and t are inseparable. It has turned out as expected because every wave parameter is a time varying quantity. However, x and t can be separated under special circumstances (see Feature 6 of section 1.2.3). Then, we can write solution of Eq. (1.47) in the form

$$y = X(x)\ T(t). \tag{1.49}$$

The procedure adopted for solving a partial differential equation is called *method of separation of variables.* Let us see how we go about that. On substituting Eq. (1.49) in Eq. (1.47), we get

$$X\frac{d^2T}{dt^2} = v^2 T\frac{d^2X}{dx^2}.$$

Dividing both sides by XT, we have

$$\frac{1}{T}\frac{d^2T}{dt^2} = v^2\frac{1}{X}\frac{d^2X}{dx^2}. \tag{1.50}$$

Note that in Eq. (1.50), the expression on the left hand side is a function only of t, whereas that on the right hand side is a function of x only. So either side can be equated to a constant. Now, what would be the sign of the constant—positive or negative?

If the constant is taken as positive, $T(t)$ turns out to be exponential in time, whereas if it is negative, $T(t)$ will be harmonic in time. In wave motion, $T(t)$ is expected to be harmonic in time. So, we rewrite Eq. (1.50) as

$$\frac{1}{T}\frac{d^2T}{dt^2} = v^2\frac{1}{X}\frac{d^2X}{dx^2} = -\omega^2, \tag{1.51}$$

where ω is real and has the dimension of reciprocal time. Then, we get

$$\frac{d^2T}{dt^2} + \omega^2 T = 0 \tag{1.52a}$$

and

$$\frac{d^2X}{dx^2} + \left(\frac{\omega}{v}\right)^2 X = 0. \tag{1.52b}$$

The solutions of the Eqs. (1.52a) and (1.52b) are given by

$$T = A_1 \cos \omega t + A_2 \sin \omega t \tag{1.53a}$$

and

$$X = B_1 \cos\frac{\omega x}{v} + B_2 \sin\frac{vx}{v}. \tag{1.53b}$$

So, the general solution is of the form

$$y = XT = (A_1 \cos\omega t + A_2 \sin\omega t)\left(B_1\cos\frac{\omega x}{v} + B_2\sin\frac{\omega x}{v}\right). \tag{1.54}$$

The solution given in Eq. (1.54) will be extremely useful in the analysis of vibration of strings, bars and air columns which we have discussed in Chapters 9, 10 and 11.

In 3-D, we get the Laplacian on the right hand side of Eq. (1.47). It takes the form

$$\frac{\partial^2 U}{\partial t^2} = v^2\nabla^2 U. \tag{1.55}$$

Depending on the nature of the problem/geometry, we expand $\nabla^2 U$ in rectangular Cartesian, spherical polar or cylindrical polar co-ordinates.

Now, we discuss the method of Fourier analysis.

1.7 FOURIER ANALYSIS

When we analyse waves in physical systems, we have to solve partial differential equations of the type given in Eq. (1.47) with different initial and boundary conditions. In such situations, knowledge of Fourier series and Fourier analysis comes in very handy. For example, if we consider the vibrations of a string fixed at both ends, it can vibrate at a number of different frequencies. You have studied in your senior secondary physics course that the lowest frequency is called the *fundamental*. If we denote it as v, other frequencies $2v$, $3v$, are called higher harmonics. These frequencies can be excited simultaneously and the resulting vibration has a complex form. However, it will still be periodic and the time period T would be $1/v$. It means that the aggregate of a number of frequencies that are related harmonically gives rise to a complex *but periodic* waveform. Looked from another angle, you may say that a complex periodic waveform can be treated as a sum of a number of sinusoidal vibrations. This issue is central to the concept of Fourier analysis. Let us learn about it now.

The general form of Fourier's Theorem is given by the statement: Any arbitrary function $f(\theta)$, where θ is a real variable, defined in an interval, $c < \theta < d$ and satisfying the conditions (i) it is single-valued, (ii) continuous, but may have a number of finite discontinuities, and (iii) have a finite number of maxima and minima, can be expanded as

$$f(\theta) = \frac{1}{2}a_0 + a_1 \cos \omega\theta + a_2 \cos 2\omega\theta + \cdots + b_1 \sin \omega\theta + b_2 \sin 2\omega\theta + \cdots$$

where

$$\omega = \frac{2\pi}{d - c}$$

Consider a periodic function $f(t)$, such that

$$f(T + t) = f(t). \tag{1.56}$$

Then $f(t)$ can be represented as a series.

$$f(t) = \frac{1}{2}a_0 + a_1 \cos \omega t + a_2 \cos 2\omega t + a_3 \cos 3\omega t + \cdots + b_1 \sin \omega t + b_2 \sin 2\omega t + b_3 \sin 3\omega t + \cdots, \tag{1.57}$$

where

$$\omega = \frac{2\pi}{T}. \tag{1.58}$$

(The factor 1/2 in the first term is a matter of choice for mathematical convenience). Equation (1.57) is the analytical form of the Fourier's Theorem, which can be stated as under:

Any single valued, periodic function which is continuous or has a number of finite discontinuities may be expressed as a sum of simple harmonic terms, having frequencies which are multiples of the frequency of the given function.

Equation (1.57) can be rewritten as

$$f(t) = \frac{1}{2}a_0 + \sum_{n=1}^{\infty} (a_n \cos n\omega t + b_n \sin n\omega t). \tag{1.59a}$$

Using Eq. (1.58), we can rewrite Eq. (1.59a) as

$$f(t) = \frac{1}{2}a_0 + \sum_{n=1}^{\infty} \left(a_n \cos \frac{2\pi}{T} nt + b_n \sin \frac{2\pi}{T} nt \right). \tag{1.59b}$$

Equation (1.57) signifies that the functions (1, cos ωt, cos $2\omega t$, ...) and (sin ωt, sin $2\omega t$, ...) form a set of linearly independent infinite number of functions and $f(t)$ is expressed as their linear combination. In other words, $f(t)$ is an infinite dimensional vector where base vectors are 1, cos ωt, cos $2\omega t$, ... and sin ωt, sin $2\omega t$, ...

Using the relations involving sine and cosine functions, which you learnt in Section 1.5, you can easily work out the following integrals, with m and n being non-zero integers and $\omega = (2\pi/T)$:

$$\int_0^T \cos m\omega t \, dt = 0 \tag{1.60a}$$

$$\int_0^T \sin m\omega t \, dt = 0 \tag{1.60b}$$

$$\int_0^T \cos^2 m\omega t \, dt = \frac{T}{2} \tag{1.60c}$$

$$\int_0^T \sin^2 m\omega t \, dt = \frac{T}{2} \tag{1.60d}$$

$$\int_0^T \cos m\omega t \cos n\omega t \, dt = 0 \quad (m \neq n) \tag{1.60e}$$

$$\int_0^T \sin m\omega t \sin n\omega t \, dt = 0 \quad (m \neq n) \tag{1.60f}$$

$$\int_0^T \cos m\omega t \sin n\omega t \, dt = 0 \tag{1.60g}$$

Equations (1.60c and e) and (1.60d and f) may be combined to write

$$\int_0^T \cos m\omega t \cos n\omega t \, dt = \frac{T}{2} \delta_{mn}, \tag{1.60h}$$

$$\int_0^T \sin m\omega t \sin n\omega t \, dt = \frac{T}{2} \delta_{mn}, \tag{1.60i}$$

where δ_{mn} is known as Kronecker's delta. It is given by

$$\delta_{mn} = 0, \quad \text{for } m \neq n \tag{1.61a}$$

$$= 1, \quad \text{for } m = n \tag{1.61b}$$

Equations (1.60e, f and g) indicate that the functions cos $m\omega t$ and sin $n\omega t$ (with m and n being integers) form an orthogonal set of functions. Thus, they form an infinite dimensional orthogonal vector space (like $\hat{\mathbf{i}}, \hat{\mathbf{j}}, \hat{\mathbf{k}}$ form a finite (three) dimensional orthogonal vector space).

We shall now discuss how to evaluate the coefficients a_n and b_n.

1.7.1 Evaluation of a_n and b_n

By combining Eqs. (1.59a) and (1.60a, b), we can write

$$\int_0^T f(t)dt = \frac{1}{2}a_0 \int_0^T dt + \sum_{n=1}^{\infty}\left(a_n \int_0^T \cos n\omega t\, dt + b_n \int_0^T \sin n\omega t\, dt\right)$$

$$= \frac{1}{2}a_0 \cdot T + 0 = \frac{a_0}{2}T.$$

$$\therefore \qquad a_0 = \frac{2}{T}\int_0^T f(t)dt. \tag{1.62}$$

Again, from Eqs. (1.59a), (1.60a), (1.60h) and (1.60g), we get

$$\int_0^T f(t)\cos m\omega t\, dt = \frac{1}{2}a_0 \int_0^T \cos m\omega t\, dt + \sum_{n=1}^{\infty}\left(a_n \int_0^T \cos m\omega t \cos n\omega t\, dt\right)$$

$$+ \sum_{n=1}^{\infty}\left(b_n \int_0^T \sin n\omega t \cos m\omega t\, dt\right)$$

$$= \frac{1}{2}a_0 \cdot 0 + \sum_{n=1}^{\infty} a_n \frac{T}{2}\delta_{mn} + 0 = \frac{T}{2}a_m.$$

$$\therefore \qquad a_m = \frac{2}{T}\int_0^T f(t)\cos m\omega t\, dt.$$

i.e.

$$a_n = \frac{2}{T}\int_0^T f(t)\cos n\omega t\, dt. \tag{1.63}$$

Similarly, using Eqs. (1.59a), (1.60b), (1.60i) and (1.60g), we get

$$b_n = \frac{2}{T}\int_0^T f(t)\sin n\omega t\, dt. \tag{1.64}$$

For evaluation of a_0, a_n and b_n, we have considered the range $t = 0$ to $t = T$. By shifting the origin, we may change the range to $t = -T/2$ to $t = T/2$ and put $\omega = 2\pi/T$. Then you will readily arrive at the following results:

$$a_0 = \frac{2}{T}\int_{-T/2}^{T/2} f(t)\, dt, \tag{1.65a}$$

$$a_n = \frac{2}{T}\int_{-T/2}^{T/2} f(t)\cos \frac{2\pi}{T}nt\, dt, \tag{1.65b}$$

and

$$b_n = \frac{2}{T}\int_{-T/2}^{T/2} f(t)\sin \frac{2\pi}{T}nt\, dt. \tag{1.65c}$$

Note that,

$$\int_{-a}^{+a} f(x) = 0,$$

if $f(x)$ is an odd function, that is $f(-x) = -f(x)$.

Note that cos $(2\pi/t)\, nt$ is an even function and sin $(2\pi/T)\, nt$ is an odd function. Hence, the integrands of Eqs. (1.63) and (1.64) are odd functions of t, if $f(t)$ is an odd and even function, respectively. Thus, if $f(t)$ is odd, $a_n = 0$ and we have a sine series only. And, if $f(t)$ is even, $b_n = 0$. We only have a cosine series. We shall see these features in Section 1.7.2.

1.7.2 Analysis of Periodic Waveforms

We shall now discuss how Fourier analysis can be used to analyse a typical periodic square waveform. Refer to Fig. 1.9. It shows a square waveform whose amplitude is unity and period is T. Due to the particular choice of origin, the waveform is asymmetric.

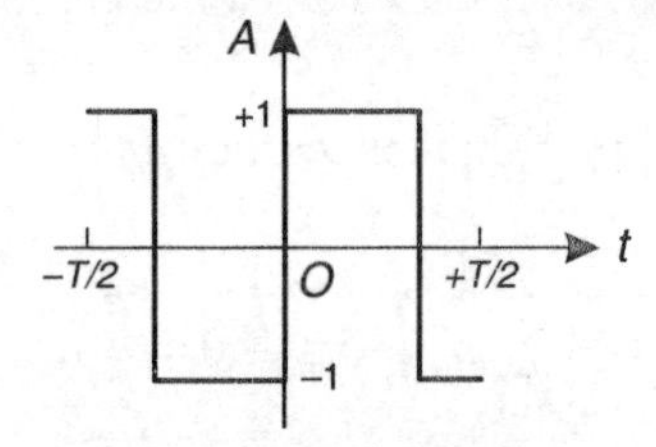

Fig. 1.9 A square wave.

Mathematically, we can represent the function as

$$f(t) = -1, \quad -\frac{T}{2} < t < 0$$

$$= 1, \quad 0 < t < \frac{T}{2}$$

and

$$f(t + T) = f(t).$$

Since, the function is odd, $a_n = 0$. Then, from Eq. (1.65c), we can write

$$b_n = \frac{2}{T} \int_{-T/2}^{+T/2} f(t) \sin \frac{2\pi}{T} nt\, dt$$

$$= \frac{4}{T} \int_{0}^{+T/2} f(t) \sin \frac{2\pi}{T} nt\, dt$$

$$= \frac{4}{T} \cdot \frac{T}{2\pi n} \left| -\cos \frac{2\pi}{T} nt \right|_0^{T/2}$$

$$= \frac{2}{n\pi}(1 - \cos n\pi) = \frac{2}{n\pi}\{1 - (-1)^n\}.$$

Hence, from Eq. (1.59b), we can write

$$f(t) = \frac{4}{\pi}\left(\sin\frac{2\pi t}{T} + \frac{1}{3}\cdot\sin\frac{6\pi t}{T} + \frac{1}{5}\sin\frac{10\pi t}{T} + \cdots\right). \tag{1.66}$$

We have included another familiar waveform in the Review Exercises.

1.7.3 Complex Form of Fourier Series

Refer to Eq. (1.59a):

$$f(t) = \frac{1}{2}a_0 + \sum_{n=1}^{\infty}(a_n \cos n\omega t + b_n \sin n\omega t)$$

In Practice Exercise 1.2, you learnt to express $a_n \cos n\omega t + b_n \sin n\omega t$ as

$$(a_n \cos n\omega t + b_n \sin n\omega t) = \sqrt{a_n^2 + b_n^2}\left\{\frac{a_n}{\sqrt{a_n^2 + b_n^2}}\cos n\omega t + \frac{b_n}{\sqrt{a_n^2 + b_n^2}}\sin n\omega t\right\}$$

$$= \sqrt{a_n^2 + b_n^2}\cos(n\omega t - \theta_n), \tag{1.67}$$

where

$$\tan\theta_n = \frac{b_n}{a_n}.$$

Using Eq. (1.20a), we get

$$a_n \cos n\omega t + b_n \sin n\omega t = \frac{(a_n^2 + b_n^2)^{1/2}}{2}\{e^{j(n\omega t - \theta_n)} + e^{-j(n\omega t - \theta_n)}\}.$$

So far we have considered n to have only positive integral values. If we also take into account the negative integral values of n, then we shall have (keeping in mind that cosine and sine are even and odd functions respectively)

$$a_{-n} = a_n, \tag{1.68a}$$

$$b_{-n} = -b_n. \tag{1.68b}$$

And

$$\tan\theta_{-n} = \frac{b_{-n}}{a_{-n}} = -\frac{b_n}{a_n} = -\tan\theta_n.$$

$$\therefore \qquad \theta_{-n} = -\theta_n$$

Hence, from Eq. (1.67), we get

$$a_n \cos n\omega t + b_n \sin n\omega t = \frac{(a_n^2 + b_n^2)^{1/2}}{2}[e^{j(n\omega t - \theta_n)} + e^{j\{(-n)\omega t - \theta_{-n}\}}]$$

$$\therefore \quad f(t) = \frac{1}{2}a_0 + \sum_{n=1}^{\infty} \frac{(a_n^2 + b_n^2)^{1/2}}{2} [e^{j(n\omega t - \theta_n)} + e^{j\{(-n)\omega t - \theta_{-n}\}}]$$

$$= \frac{1}{2}a_0 + \sum_{n=1}^{\infty} r_n e^{j(n\omega t - \theta_n)} + \sum_{n=1}^{\infty} r_n e^{j\{(-n)\omega t - \theta_{-n}\}},$$

where

$$r_n = \frac{1}{2}(a_n^2 + b_n^2)^{1/2} = r_{-n}.$$

$$\therefore \quad f(t) = \frac{1}{2}a_0 + \sum_{n=1}^{\infty} r_n e^{j(n\omega t - \theta_n)} + \sum_{n=-1}^{-\infty} r_n e^{j(n\omega t - \theta_n)}.$$

Now, noting that $b_0 = 0$, $\theta_0 = \tan^{-1}(b_0/a_0) = \tan^{-1} 0 = 0$, we have

for $n = 0$, $$\frac{1}{2}(a_n^2 + b_n^2)^{1/2} e^{j(n\omega t - \theta_n)} = \frac{1}{2}a_0$$

$$\therefore \quad f(t) = \sum_{n=-\infty}^{\infty} r_n e^{j(n\omega t - \theta_n)}$$

$$= \sum_{n=-\infty}^{\infty} r_n e^{-j\theta_n} e^{jn\omega t}$$

$$= \sum_{n=-\infty}^{+\infty} R_n e^{jn\omega t}, \tag{1.69}$$

where

$$R_n = r_n e^{-j\theta_n} = \frac{1}{2}(a_n^2 + b_n^2)^{1/2} e^{-j\theta_n}. \tag{1.70}$$

$$\therefore \quad R_n = r_n (\cos \theta_n - j \sin \theta_n)$$

$$= \frac{1}{2}(a_n^2 + b_n^2)^{1/2} \left\{ \frac{a_n}{(a_n^2 + b_n^2)^{1/2}} - j\frac{b_n}{(a_n^2 + b_n^2)^{1/2}} \right\}$$

$$= \frac{1}{2}(a_n - jb_n). \tag{1.71}$$

From Eqs. (1.65b) and (1.65c), we get

$$R_n = \frac{1}{2} \cdot \frac{2}{T} \int_{-T/2}^{T/2} f(t) \left\{ \cos \frac{2\pi}{T} nt - j \sin \frac{2\pi}{T} nt \right\} dt$$

or

$$R_n = \frac{1}{T}\int_{-T/2}^{T/2} f(t)\, e^{-j\cdot(2\pi/T)nt}\, dt. \tag{1.72}$$

Equations (1.69) and (1.72) represent the complex form of the Fourier series. We expect you to find the physical significance of the negative value of n.

REVIEW EXERCISES

1.1 Obtain the cube roots of unity using Demoivre's Theorem.

$$\left[\textbf{Ans.}\quad 1, \frac{1}{2}\left(-1 + j\sqrt{3}\right),\ -\frac{1}{2}\left(1 + j\sqrt{3}\right)\right]$$

1.2 Use Taylor's theorem to write the expansion of $\log_e (1 + x)$.

$$\left[\textbf{Ans.}\quad x - \frac{x^2}{2} + \frac{x^3}{3} - \frac{x^4}{4} + \cdots\right]$$

1.3 (a) If $y = \cos 3x$, show that $\dfrac{d^2y}{dt^2} + 9y = 0$.

(b) If $y = \tan^{-1} x$, show that $(1 + x^2)\dfrac{d^2y}{dx^2} + 2x\dfrac{dy}{dx} = 0$.

1.4 (a) Show that $x = a \cos \omega t + b \sin \omega t$ is a solution of Eq. (1.46a).

(b) Show that $x = e^{-bt}(A_1 e^{\alpha t} + A_2 e^{-\alpha t})$ is a solution of Eq. (1.46b).

1.5 A saw-tooth wave is defined as follows:

$$f(t) = \frac{2}{T}t;\quad -\frac{T}{2} < t < \frac{T}{2}$$

and

$$f(t + T) = f(t).$$

Obtain the Fourier series.

$$\left[\textbf{Ans.}\quad f(t) = \frac{2}{\pi}\left(\sin\frac{2\pi t}{T} - \frac{1}{2}\sin\frac{4\pi t}{T} + \frac{1}{3}\sin\frac{6\pi t}{T} - \frac{1}{4}\sin\frac{8\pi t}{T} + \cdots\right)\right]$$

2

Simple Harmonic Motion

EXPECTED LEARNING OUTCOMES

In this Chapter, you will acquire capability to:

- define simple harmonic motion in three different ways and establish consistency between them;
- state the basic characteristics of linear and angular simple harmonic motions and realise them through examples of various physical systems;
- establish the differential equations for mechanical and electromagnetic systems executing simple harmonic motion, solve them and identify the commonality between them;
- write down the general equation of simple harmonic motion and solve it;
- derive expressions for kinetic, potential and total energies of a system executing simple harmonic motion; and
- explain the link between uniform circular motion and SHM and apply it to represent an SHM using a rotating vector.

2.1 INTRODUCTION

You must have enjoyed listening to chirping of birds, humming of bees and musical performance using flute, tabla or orchestra in a musical concert many a time since you were a child. But did you ever bother to learn or even know the physics underlying musical performances? Probably it was not considered important. But we cannot help appreciating the fact that nature loves symmetry and unity. You will be fascinated to know that the principles which help us to understand the physics of music are equally applicable in spectroscopic techniques such as ECG, EEG, MRI, etc. with important applications in medical science. This could be the reason why several eminent physicists were attracted to music. During their leisure, Albert Einstein used to play violin, S.N. Bose used to play sitar while Richard Feynman had a fancy for playing percussion instruments. Music of any kind—vocal or instrumental—is made up of several simple harmonic oscillations. Usually the word 'oscillation' is preceded by two adjectives 'harmonic' and 'simple'. You will learn the meaning of these terms in this chapter. While doing so, you will need to recall the knowledge acquired by you about Simple Harmonic Motion (SHM) in your senior secondary physics course.

Simple pendulum and spring-mass system are two illustrative examples of SHM. You have studied about the motion of these systems in your school. Yet for simplicity and ease in understanding, we will begin with an elaborate discussion on these. In Section 2.3, you will learn to analyse the motion of compound pendulum, torsional pendulum, oscillation magnetometer and an electric circuit consisting of a pure inductor and a pure capacitor by analogy. Using the discussion of an *LC*–circuit, we shall also draw analogies between mechanical and electromagnetic oscillations.

We shall discuss the variation of energy of an oscillator with position as well as time in Section 2.4. It presents an example of a conservative system and provides useful insights into the interplay of different forms of energy possessed by the systems undergoing oscillation.

Note that this chapter forms the basis for understanding physics of oscillations and waves in subsequent chapters. So we have included many solved examples, practice exercises and quite a few challenging problems. You must try to solve these after reading this chapter. We hope you will be able to do so with ease but if you fail to do so, discuss with your teacher.

2.2 DESCRIPTION OF SHM

You are quite familiar with the motion of a simple pendulum. In your senior secondary physics, you used it to perform an experiment on the variation of time period with length. That is why we have used this example to introduce SHM. An ideal simple pendulum consists of a heavy point mass suspended from a rigid suspension by a weightless inextensible string (Fig. 2.1).

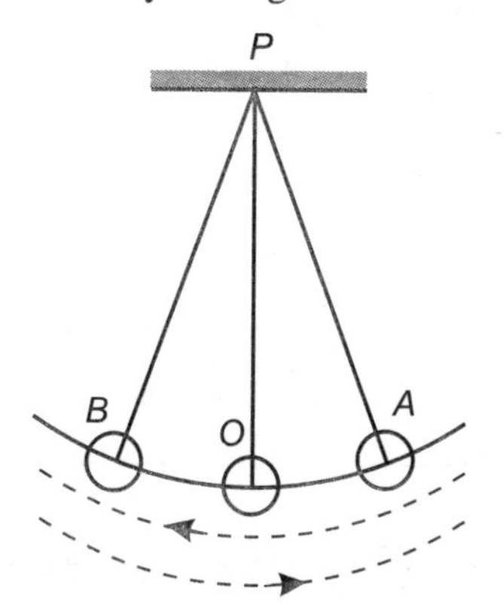

Fig. 2.1 A simple pendulum.

You know that a simple pendulum can be realised in practice by hanging a metal bob by a thread from a stand. Left to itself, the bob occupies the position *PO*, which is known as the mean equilibrium position. From this position, the pendulum is drawn towards an extremity *A*. On doing so, the bob gains some finite potential energy (PE). When it is released from *A*, its PE begins to decrease; it becomes minimum as the bob approaches *O*. At *O* its kinetic energy (KE) is maximum and due to inertia of motion, it overshoots this position gaining PE. At the other extremity B, its KE is minimum and PE is maximum. As a natural tendency, the PE again gets minimised and thus the bob retraces its path back from B to O to A. That is, the energy of the bob keeps alternating between potential and kinetic forms as it executes to and fro motion.

Under ideal conditions, that is in the absence of air resistance, losses due to dissipative forces such as friction do not affect the to and fro motion and it is supposed to continue eternally. In practice, however, it does not happen like that. You may like to know as to what is responsible for this deviation. You will learn about it in detail in Chapter 4. The point which we want to highlight is that under ideal conditions, the bob will repeatedly keep tracing and retracing the paths AOB and BOA indefinitely. That is, the bob will execute a periodic motion; the *time period* being equal to the time taken by the bob to move from O to A to B and again back from B to O. Such to and fro motion, where the periodicity is maintained, is a simple harmonic motion (SHM). Several systems other than a simple pendulum also execute SHM. These include the spring-mass system, the compound pendulum, the torsional pendulum, the LC-circuit and so on. You will learn to draw analogies between these.

2.2.1 Definition of SHM

The SHM can be defined in a number of ways:

(i) If the force acting on a particle is proportional to its displacement with reference to a fixed point and is always directed towards it, the particle executes SHM about the fixed point.

(ii) If the PE of a particle is proportional to the square of its displacement with reference to a fixed point, the particle executes SHM about that fixed point.

(iii) If the displacement versus time curve of a particle is sinusoidal, it executes SHM.

The correlation between definitions given in (i) and (ii) is rather simple and if you have required analytical capabilities/mathematical orientation, you must have mentally worked it out. We can mathematically express (i) as

$$\mathbf{F} = -k\mathbf{x}, \tag{2.1}$$

where $\mathbf{x}$ is displacement from a fixed point, O (Fig. 2.2), $\mathbf{F}$ is the force acting on the particle and k is a constant, known as *force constant.*

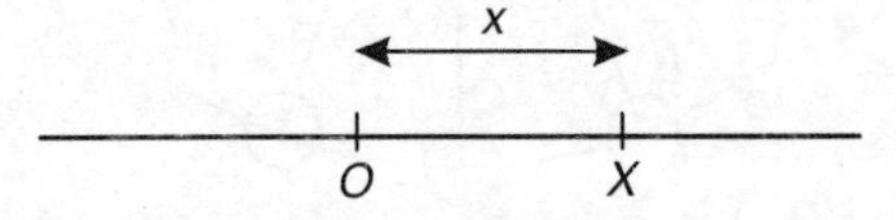

Fig. 2.2 Force and displacement are related as $F = -kx$.

If the unit vector along OX is $\mathbf{x}$, we can rewrite Eq. (2.1) as

$$F\mathbf{x} = -\, kx\,\mathbf{x}$$

or

$$F = -kx. \tag{2.2}$$

If V is the PE, then we can write

$$F = -\frac{dV}{dx}. \tag{2.3}$$

On combining Eqs. (2.2) and (2.3), we get

$$\frac{dV}{dx} = kx$$

or

$$dV = kx\,dx.$$

By integrating with respect to x, we get

$$V = \frac{1}{2}kx^2 + A,$$

where A is constant of integration. The mean position ($x = 0$) corresponds to minimum value of PE and we standardise it as $V = 0$ at $x = 0$. This gives $A = 0$. Then the expression for potential energy takes the form

$$V = \frac{1}{2}kx^2. \tag{2.4}$$

This corroborates definition (ii).

We shall establish the equivalence of the definition (iii) with (i) and (ii) in Section 2.4. Let us now study some physical systems which execute SHM.

> The degrees of freedom of a physical system are the total number of independent co-ordinates required to specify the position of the system. The motion of a simple pendulum takes place on a vertical plane represented by *XPY*. If the Cartesian co-ordinates of the point *C* are (x, y), then $x^2 + y^2 = l^2$ is the equation of constraint. That is, x and y are not independent. It means that the number of independent co-ordinates is one. Now, to describe the motion of simple pendulum, we neither consider x nor y; we switch over to polar co-ordinates and observe that for every point, the radial coordinate is 'l'. In other words, it remains invariant. The polar angle θ changes from point to point and so we describe the motion using θ and not x and y. This corroborates the fact that the simple pendulum executes angular motion and it is in the fitness of things that the motion is described using the angular co-ordinate.

2.2.2 Physical Systems Executing SHM Having One Degree of Freedom

Simple pendulum

Refer to Fig. 2.3 and consider the situation when the string makes an angle θ (= ∠*OPC*) with the mean position *PX*. The effective length of the pendulum is l and the mass of the bob is m. The weight of the bob, mg, has two components mg cos θ along the string and mg sin θ perpendicular to the string. The tension in the string adjusts itself by getting equated to mg cos θ plus the centrifugal force. The mg sin θ component provides the restoring torque, which makes the bob to return to its mean equilibrium position.

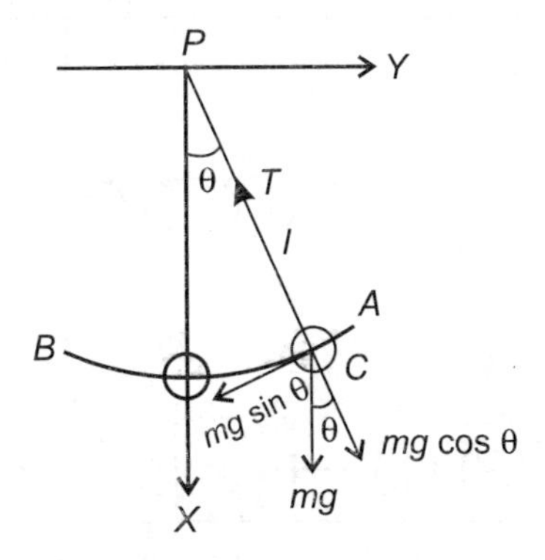

Fig. 2.3 Resolution of forces acting on the bob of a simple pendulum into components.

$$\sin\theta = \theta - \frac{\theta^3}{3!} + \frac{\theta^5}{5!} - \cdots$$

If θ is very small, $\sin\theta = \theta$.

Some authors mention that θ should be within 4°. We do not prescribe any such upper limit. It turns out that SHM is an approximation. Closer is θ to zero, more valid is the approximation. It may be worthwhile taking a look at the value of sin18°, which is $\frac{\sqrt{5}-1}{4} = \frac{2.236-1}{4} = \frac{1.236}{4} = 0.309$. That is, sin 18° = 0.31 (correct up to two decimal places). And $18° = \frac{180°}{10} = \frac{\pi}{10} = \frac{3.142}{10} = 0.3142 = 0.31$ (correct up to two decimal places). So, how do you react to the approximation, $\sin\theta = \theta$?

You must have noted that we have used the term torque rather than force. This is because of the fact that the motion is angular. It suggests that definition (i) would have to be modified by considering rotational analogues of force and displacement. The modified form would read as: the torque is proportional to angular displacement and directed towards the mean position. The component $mg\sin\theta$ is indeed directed towards the mean position and so is the torque. That is why the adjective 'restoring' is used with torque.

Note that angular motion takes place about an axis perpendicular to the plane of this paper and passing through P. The torque $mg\sin\theta$ about that axis is given by $mg\sin\theta \times l = mgl\sin\theta$.

When θ is small, $\sin\theta \approx \theta$. So the magnitude of the torque in such a situation is equal to $mgl\theta$ and the equation relating torque to angular displacement θ is given by

$$\tau = -mgl\theta. \tag{2.5}$$

From Eq. (2.5), we note that τ is proportional to θ. The negative sign indicates that torque is directed towards the mean position. So the motion is simple harmonic.

Now, refer to Eq. (1.31). You will note that PE is proportional to the square of θ, which agrees with definition (ii).

Spring-mass system

You must have studied about spring-mass system in your senior secondary curriculum. It consists of a spring connected with a mass m which is free to move on a frictionless horizontal surface, as shown in Fig. 2.4(a). The stretched and compressed configurations of the spring are shown in Fig, 2.4(b) and 2.4(c), respectively.

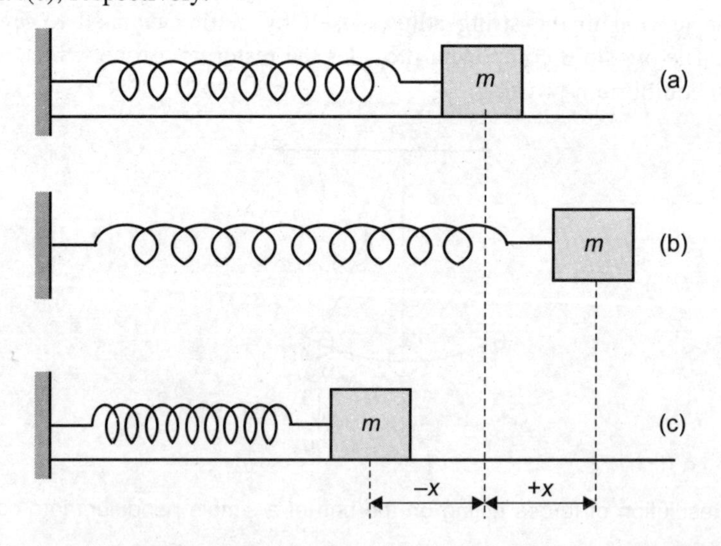

Fig. 2.4 (a) Normal (b) stretched and (c) compressed configurations of a spring-mass system.

You may obtain such a system by hanging a mass from a vertical spring. On being stretched or compressed it behaves like an elastic system under stress. You may recall that if a wire of length L is stretched through a distance x by a force of magnitude F, the Young's Modulus Y of the material of the wire is given by

$$Y = \frac{F/\alpha}{x/L}, \tag{2.5a}$$

where α is the cross-sectional area of the wire. You can easily rearrange terms and write

$$Y = \frac{FL}{x\alpha},$$

so that

$$F = \frac{Y\alpha}{L}x. \tag{2.5b}$$

We know that elasticity is that property by virtue of which a body offers resistance to any change in its size or shape or both and makes the body regain its original condition when the deforming force, applied within a certain maximum limit, is removed. In other words, in the deformed condition, the body develops a restoring force and according to Newton's third law of motion, this force is equal in magnitude but opposite in direction to the deforming force. Equation (2.5b) implies that the restoring force is proportional to elongation, and is directed towards the mean position. Similarly, for the spring-mass system, the restoring force is proportional to the displacement of the spring in case of stretched as well as compressed configurations. Hence, we can write

$$F = -kx, \tag{2.6}$$

where k is referred to as the spring constant. Using Newton's second law of motion, we can write the differential equation of motion of the spring-mass system as

$$m\ddot{x} = -kx. \tag{2.7}$$

This example is extremely illustrative because Eq. (2.7) is analogous to the general differential equation of simple harmonic motion. You must have also observed the similarity between Eqs. (2.6) and (2.1). You will learn to solve this equation in the next section.

If the PE of the system is V, then by expressing F as $-dV/dx$, Eq. (2.6) takes the form

$$dV = kx\,dx.$$

On integrating it over x and using the condition, $V = 0$ at $x = 0$, we get

$$V = \frac{1}{2}kx^2. \tag{2.8}$$

It shows that PE is proportional to the square of displacement.

A compound pendulum (or physical pendulum)

Any rigid body mounted on a suspension in such a manner that it can oscillate freely in a vertical plane (Fig. 2.5) about a horizontal axis passing through the suspension is called a *compound* or a *physical pendulum*. As a matter of fact, all real pendulums are compound in nature.

> In an ideal simple pendulum, we assume that the mass suspended from a rigid suspension by a weightless, inextensible string is point mass and heavy. We make a similar idealisation in the definition of a compound pendulum in that it is a rigid body.

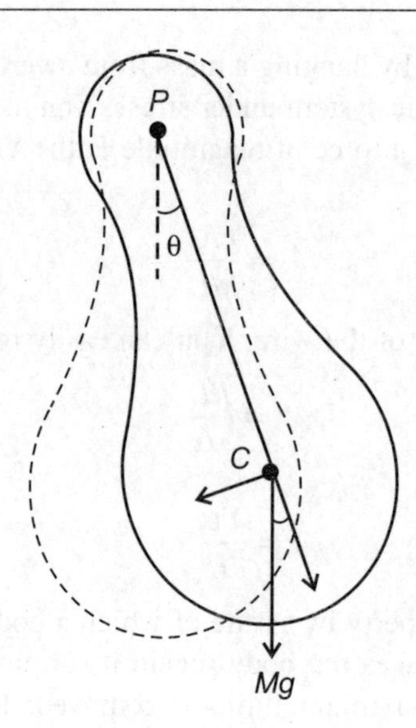

Fig. 2.5 A compound pendulum oscillating about an axis passing through *P* normal to the plane of the paper. The dotted line indicates the equilibrium position and the continuous line shows the instantaneous position where the line *PC* makes an angle θ with the vertical.

If the mass of the pendulum is *M*, the component of its weight along the line *PC* is $Mg \cos \theta$ and the perpendicular component is $Mg \sin \theta$. If $PC = l_0$, the system experiences a restoring torque $Mg \sin \theta l_0$ about the axis passing through *P*. When θ is small, the torque on the system can be expressed as

$$\tau = -Mgl_0\theta. \tag{2.9}$$

From Eq. (2.9) we can say that in this case also, torque is proportional to angular displacement and directed towards the equilibrium position. Hence we can conclude that a compound pendulum executes simple harmonic motion.

Oscillation magnetometer

Refer to Fig. 2.6(a), which shows a bar magnet suspended freely from the point of suspension *P* with the help of a very light non-magnetic cradle. This arrangement constitutes an oscillation magnetometer. By free suspension, we mean that the suspension wire *PQ*, when produced, passes through the centre *C* of the bar magnet. Since the suspension is free, the magnet will orient itself along the magnetic meridian. And when a small piece of magnetic substance is brought close to one end of the suspended magnet, the magnet will attract it. In the process, it gets deflected from the equilibrium position, i.e., from the magnetic meridian. This situation is depicted in Fig. 2.6b.

In the deviated position, the N-pole of the magnet experiences a force whose magnitude is mH along the direction of **H**. The S-pole experiences the same force but in the direction opposite to **H**. You may recall that two equal and unlike parallel forces constitute a couple. In the instant case, it is in the anticlockwise sense and makes the magnet realign in the direction of **H**. Hence, it is a restoring torque. If the magnet is of length $2l$, the magnitude of the torque is $mH \cdot 2l \sin \theta = 2lm H \sin \theta$. We know that magnetic moment of the magnet is given by $2lm = M$. Hence the torque experienced by the coil can be expressed as

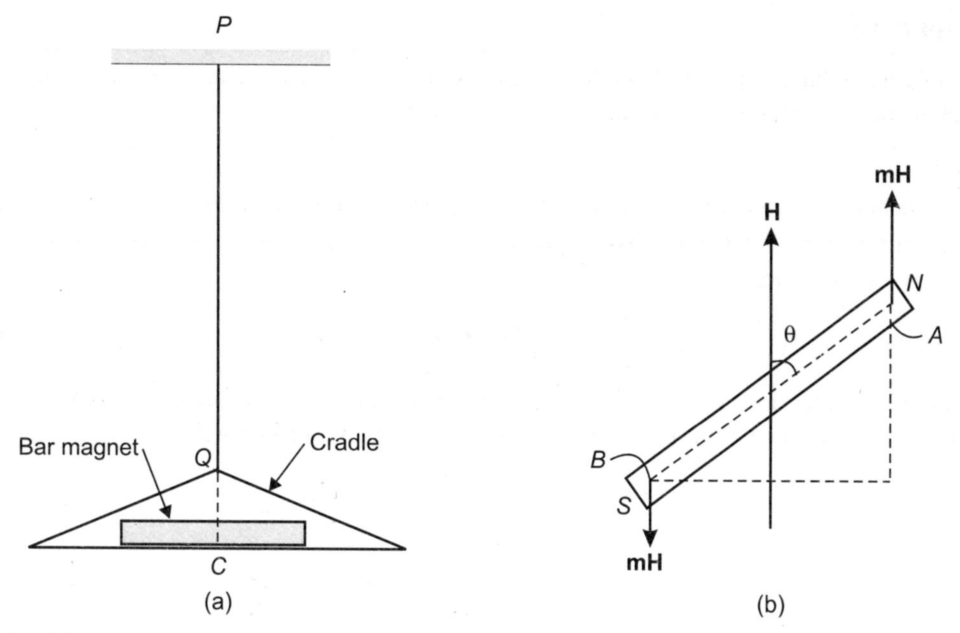

Fig. 2.6 (a) An oscillation magnetometer; and (b) the torque experienced by the magnet. H is magnitude of the horizontal component of earth's magnetic intensity and m is the pole strength of magnet.

$$\tau = -MH \sin \theta. \tag{2.10a}$$

For small θ,

$$\tau = -MH\,\theta. \tag{2.10b}$$

Note that Eq. (2.10b) satisfies the condition of SHM. Hence the system will execute simple harmonic oscillations. What is the equilibrium position in this case? If you think that it is magnetic meridian, you are thinking correctly. The PE of the system in the deviated position is given by

$$V = -\int_0^\theta \tau\, d\theta$$

$$= \int_0^\theta MH \sin\theta\, d\theta$$

$$= MH\,(1 - \cos\theta). \tag{2.11a}$$

For small θ, $\cos\theta = 1 - (\theta^2/2)$ and the expression for PE simplifies to

$$V = \frac{1}{2} MH\theta^2. \tag{2.11b}$$

This indicates that PE is proportional to the square of displacement.

You should now read the Example 2.1 carefully.

EXAMPLE 2.1

The simple pendulum, compound pendulum and oscillation magnetometer execute angular SHM. Now, let us derive differential equation of motion of an oscillation magnetometer.

Solution:

Let the moment of inertia of the system of the suspended magnet about the axis of rotation be I. In the deviated position, its KE = (1/2) $I\dot{\theta}^2$. Hence, from Eq. (2.11b), we can write the expression for total energy of the system as

$$E = \frac{1}{2} I\dot{\theta}^2 + \frac{1}{2} MH\theta^2 = \text{Constant.} \tag{2.12}$$

In writing this expression, we have assumed that there is no air resistance or any other frictional force. It means that there is no loss of energy with time. Then, we can write

$$\frac{dE}{dt} = 0,$$

i.e.

$$I\dot{\theta}\ddot{\theta} + MH\theta\dot{\theta} = 0,$$

or

$$\dot{\theta}\left(\ddot{\theta} + \frac{MH}{I}\theta\right) = 0.$$

Since $\dot{\theta} \neq 0$, we find that

$$\ddot{\theta} + \frac{MH}{I}\theta = 0. \tag{2.13}$$

Equation (2.13) is the differential equation of motion of a magnetometer. MH/I has dimensions of $1/T^2$, therefore, Eq. (2.13) can be expressed as

$$\ddot{\theta} + \omega^2\theta = 0, \tag{2.14}$$

where the dimension of ω is that of reciprocal of time.

You can also arrive at Eq. (2.13) by using the rotational analogue of Newton's second law of motion. The expression for torque at the deflected position is given by

$$(\text{Moment of inertia}) \times (\text{Angular acceleration}) = I\ddot{\theta}.$$

∴ From Eq. (2.10b), we can write

$$I\ddot{\theta} = -MH\theta,$$

which leads to Eq. (2.13).

You must have observed that in arriving at Eq. (2.14), we have taken θ to be small and have hence used the expressions given in Eqs.(2.10b) and (2.11b) respectively for the torque and PE. If we had not done so, Eq. (2.12) would have taken the form

$$E = \frac{1}{2} I\dot{\theta}^2 + MH(1 - \cos\theta) = \text{Constant.} \tag{2.15}$$

Using the condition $(dE/dt) = 0$, we obtain

$$I\dot{\theta}\ddot{\theta} + MH\sin\theta\dot{\theta} = 0.$$

For $\dot{\theta} \neq 0$, it takes the form

$$\ddot{\theta} + \frac{MH}{I}\sin\theta = 0.$$

If θ is very small, $\sin\theta \approx \theta$ and we get

$$\ddot{\theta} + \frac{MH}{I}\theta = 0,$$

which is identical to Eq. (2.13).

You should now answer Practice Exercise.

***Practice Exercise* 2.1** (a) Starting from Eq. (2.10a), derive Eq. (2.13) by using rotational analogue of Newton's second law of motion.

(b) Obtain the differential equation of motion of a simple pendulum by starting from the expression of the restoring torque and assuming it to be a conservative system.

We shall now consider yet another example of angular SHM.

Torsional pendulum

A torsional pendulum is schematically depicted in Fig. 2.7. It is suspended from a rigid support at P. The suspension passes through the centre O of the disc. A point A is taken as reference on the circumference of the disc. When set to oscillate, it reaches extreme positions OA_1 and OA_2 and $\angle AOA_1 = \angle AOA_2 = \theta_0$.

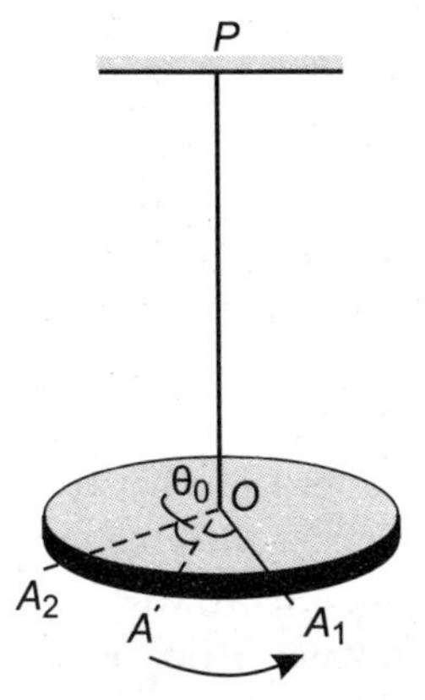

Fig. 2.7 A torsional pendulum.

If the system is rotated about OP in the sense shown by the arrow, the wire OP gets twisted. If left to itself, it would tend to return to its original configuration. But due to inertia, as in case of a simple pendulum, it will overshoot the equilibrium position and in the process the wire again gets twisted. It would tend to unwind itself and get back to the equilibrium position. Thus, the system will keep oscillating and under ideal conditions, the oscillations will continue indefinitely. These oscillations, occurring due to alternate twisting and untwisting of the wire are called torsional oscillations and the system is known as *torsional pendulum*.

If the instantaneous angular twist is θ, the pendulum experiences restoring torque $C\theta$, where C is given by

$$C = \frac{\pi \eta r^4}{2L}. \tag{2.15a}$$

Here η is modulus of rigidity of the material of the wire, r is radius of the wire and L is its length. Hence, the torque experienced by the system for angular twist θ is given by

$$\tau = C\theta. \tag{2.15b}$$

Note that this relation is similar to that given in Eq. (2.9). Hence we can say that a torsional oscillator executes simple harmonic motion.

You may now ask: Whether or not smallness of θ needs to be considered in the instant case, as in the case of simple or compound pendulum and the oscillation magnetometer. On the basis of mathematical analysis you may conclude that it is not required. But here also, the assumption that θ is small appears by way of default. Equation (2.15b) is valid for pure *shear*, i.e., when the twist of the wire is ideal. In other words, the change in shape should occur without any accompanying change in size and the approximation becomes progressively more valid as $\theta \to 0$.

You may now like to answer a Practice Exercise.

***Practice Exercise* 2.2** Obtain the differential equation of motion of a torsional pendulum.

Let us now recapitulate the basic characteristics of SHM.

2.3 BASIC CHARACTERISTICS OF SHM

You have studied the basic characteristics of SHM in your senior secondary classes: Amplitude, frequency, time period and phase. Let us recall the definition and significance of these characteristics.

Amplitude is the maximum value of displacement. While describing SHM, we mentioned that energy of a system executing SHM alternates between kinetic and potential forms. At the extremities of the oscillation, which are also called the turning points, the KE is zero and the PE is maximum. In other words, in case of linear SHM, when $x = \pm a$, where a is amplitude, the KE, and hence the velocity, is zero. In case of angular SHM, as in case of torsional pendulum, the angular velocity of the system becomes zero at the positions OA_1 and OA_2 (Fig. 2.7).

Frequency is the number of complete oscillations executed per second. It is expressed in cycles per second or simply s^{-1} or hertz (Hz). If the frequency is ν, the system executes ν oscillations per second. We also define a term called *angular frequency*, denoted by ω_0, which is given by

$$\omega_0 = 2\pi\nu. \tag{2.16}$$

Time period is the time taken by a system to execute one complete oscillation. Thus,

$$T = \frac{1}{\nu}, \tag{2.17a}$$

or

$$\nu = \frac{1}{T}. \tag{2.17b}$$

And

$$\omega_0 = 2\pi\nu = \frac{2\pi}{T}. \tag{2.17c}$$

We have been using the expression 'one complete oscillation' time and again. With reference to the simple pendulum (Fig. 2.1), it refers to the distance covered in completing trajectories *OAOBO* and the system is in the same state of oscillation. To understand it further, we have to study the *phase of* an oscillation of a system.

2.4 PHASE OF AN OSCILLATOR EXECUTING SHM

The word 'phase' is synonymous to what we call 'state'. In thermodynamics it refers to the solid, liquid and vapour states of a substance. In mechanics and oscillations, it refers to the *mechanical condition of motion governed by simultaneous consideration of position and velocity of a system.* You may recall that the equation of motion of a system is of second order. To obtain the displacement, we integrate the differential equation twice with respect to time; first integration provides description of velocity and the second integration gives us displacement. If you read the italicised statement again, you will realise its significance. To help you fix your concept of phase, we take the following example.

Refer to Fig. 2.8, where we have again taken the example of a simple pendulum. In its trajectory from O to A, A to B, and B to A, the pendulum goes through the point C thrice; at instants $t = t_1$, $t = t_2$ and $t = t_3$. At all these instants, the angular displacement is same. But the velocities are same, both in magnitude and direction, at $t = t_1$ and $t = t_3$. Had displacement been the only consideration, the phase of the oscillator would have been same at all times. But this is not so; the phase is same only at $t = t_1$ and $t = t_3$. In fact, the phases at $t = t_1$, t_3 and $t = t_2$ are opposite.

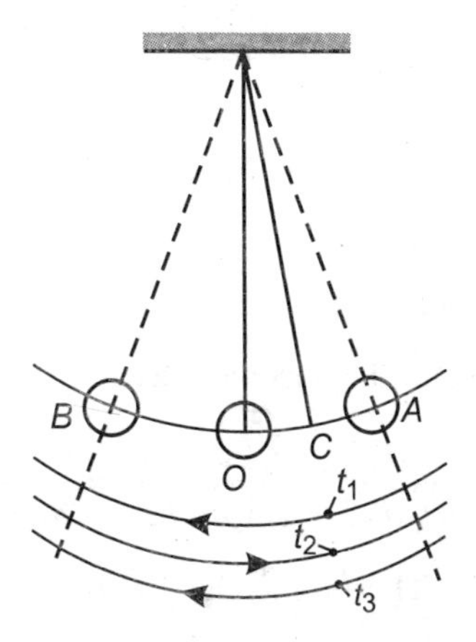

Fig. 2.8 Same phase at $t = t_1$ and $t = t_3$; opposite phases at $t = t_1$, t_3 and $t = t_2$.

The time-interval between two consecutive instants at which the oscillator is in the same phase of motion defines the time period. In other words, if we can express both the displacement (x or θ) and the velocity v as functions of t, then the mathematical definition of time period is given by simultaneous consideration of the following equations:

$$x(t + T) = x(t) \tag{2.18a}$$

and

$$v(t + T) = v(t). \tag{2.18b}$$

For an ideal system, the oscillations continue indefinitely. So more generalised forms of these equations are

$$x(t + nT) = x(t) \tag{2.18c}$$

and

$$v(t + nT) = v(t), \tag{2.18d}$$

where $n \in I$.

We shall now establish the differential equation of the SHM.

2.5 DIFFERENTIAL EQUATION OF SHM AND ITS SOLUTIONS

Let us reconsider Eq. (2.2) and express it in differential form using Newton's second law of motion: $F = m\ddot{x}$. So we can rewrite it as

$$m\ddot{x} = -kx$$

or

$$m\ddot{x} + kx = 0$$

or

$$\ddot{x} + \frac{k}{m}x = 0. \tag{2.19a}$$

Equation (2.19a) is the differential equation of SHM. k is the force constant (also called spring constant with reference to the spring-mass system) and has the dimensions $(MLT^{-2}/L) = MT^{-2}$. So the dimension of k/m is T^{-2}, i.e., square of reciprocal of time. We can replace k/m by ω^2, where ω has dimension of reciprocal of time. Then Eq. (2.19a) takes the form

$$\ddot{x} + \omega^2 x = 0 \tag{2.19b}$$

The general solution of this equation is given in Section 1.6. We now solve it in a different manner and also find out the physical meaning of ω.

Solution of the differential equation

We write the second time derivative of displacement as

$$\ddot{x} = \frac{d^2 x}{dt^2} = \frac{d}{dt}\left(\frac{dx}{dt}\right) = \frac{d}{dx}\left(\frac{dx}{dt}\right)\frac{dx}{dt} = \frac{dx}{dt}\frac{d}{dx}\left(\frac{dx}{dt}\right).$$

But dx/dt defines the velocity v. Therefore, the expression for second time derivative of displacement takes the form

$$\ddot{x} = v\frac{dv}{dx} = \frac{d}{dx}\left(\frac{v^2}{2}\right) = \frac{1}{2}\frac{d}{dx}(v^2). \tag{2.20}$$

Using this result in Eq. (2.19b), we get

$$\frac{d}{dx}(v^2) + 2\omega^2 x dx = 0$$

or

$$\frac{d}{dx}(v^2 + \omega^2 x^2) = 0.$$

On integrating it with respect to x, we get

$$v^2 + \omega^2 x^2 = A_0 = \text{constant}. \tag{2.21}$$

Using the condition $v = 0$ for $x = \pm a$ in this expression (Section 2.3), we get

$$A_0 = \omega^2 a^2.$$

Using this result in Eq. (2.21) and rearranging terms, we get

$$v^2 = \omega^2 (a^2 - x^2), \tag{2.22a}$$

so that

$$v = \pm\omega\sqrt{a^2 - x^2}. \tag{2.22b}$$

This is the expression for velocity of a particle executing SHM. We now proceed to calculate the displacement of a particle executing SHM. To this end, we note that

$$v = \frac{dx}{dt}$$

$$\therefore \qquad \frac{dx}{dt} = \pm\omega\sqrt{a^2 - x^2}.$$

We rearrange terms to write

$$\pm\frac{dx}{\sqrt{a^2 - x^2}} = \omega\, dt \tag{2.23}$$

and integrate with respect to t. This gives

$$\sin^{-1}\frac{x}{a} = \omega t + \delta \tag{2.24a}$$

and

$$\cos^{-1}\frac{x}{a} = \omega t + \delta, \tag{2.24b}$$

where δ is a dimensionless constant.

Note that solution in Eq. (2.24a) corresponds to the '+' sign in Eq. (2.23) and that in Eq. (2.24b) to the '–' sign. In other words, Eqs. (2.24a) and (2.24b) respectively correspond to +ve and –ve values of dx/dt. Hence, time variation of displacement corresponding to increasing and decreasing x with t are respectively given by

$$x(t) = a \sin(\omega t + \delta) \tag{2.24a$'$}$$

and

$$x(t) = a \cos(\omega t + \delta). \tag{2.24b$'$}$$

Equations (2.24a′) and (2.24b′) are the solutions of the differential equation of SHM. These involve three constants, a, ω and δ. You know the physical significance of a. Let us discover the significances of ω and δ.

Let us reconsider Eq. (2.24a′) and differentiate it with time. This gives

$$v(t) = \frac{d}{dt}\{x(t)\} = a\omega \cos(\omega t + \delta). \tag{2.25}$$

Using Eq. (2.18c), we can write

$$a \sin(\omega t + \delta) = a \sin\{\omega(t + nT) + \delta\}$$

so that

$$a\omega \cos(\omega t + \delta) = a\omega \cos\{\omega(t + nT) + \delta\}.$$

From these results, we can write

$$\sin \alpha = \sin(\alpha + \omega nT), \tag{2.26a}$$

and

$$\cos \alpha = \cos(\alpha + \omega nT) \tag{2.26b}$$

where

$$\alpha = \omega t + \delta.$$

From Feature (5) of section 1.2.2, we recall that for Eqs. (2.26a) and (2.26b) to be valid simultaneously, we must have

$$\alpha + \omega nT = \alpha + 2n\pi$$

or

$$\omega = \frac{2\pi}{T} = 2\pi v.$$

This result shows that ω provides a measure of the frequency and on comparison with Eq. (2.16), we call it *angular frequency*. (It is also called *circular frequency*. We shall explain this nomenclature in Section 2.9.).

Recall that when we defined ω and wrote Eq. (2.19b), our knowledge about the constant was restricted to the fact that it had dimension of reciprocal of time. Now, we find that it signifies angular frequency of the SHM. In the process, we have mastered the mathematical tool. We summarise the steps for you.

We first write the differential equation of motion and satisfy ourselves that it is of the form

$$\ddot{x} + C_1 x = 0 \qquad \text{(for linear SHM)} \tag{2.27a}$$

and

$$\ddot{\theta} + C_2 \theta = 0 \qquad \text{(for angular SHM)} \tag{2.27b}$$

Then the motion is simple harmonic and angular frequency is given by

$$\omega = \sqrt{C}, \tag{2.27c}$$

where C (= C_1 or C_2)

$$\therefore \qquad v = \frac{\sqrt{C}}{2\pi}, \tag{2.27d}$$

and

$$T = \frac{2\pi}{\sqrt{C}}. \tag{2.27e}$$

Hence, from Eqs. (2.7) and (2.13), we can say that the time periods of oscillation of the spring-mass system and the oscillation magnetometer are respectively given by

$$T = 2\pi\sqrt{\frac{m}{k}}, \tag{2.28}$$

and

$$T = 2\pi\sqrt{\frac{I}{MH}}. \tag{2.29}$$

You may now like to answer a Practice Exercise.

***Practice Exercise* 2.3** Apply the results of Practice Exercise 2.1 to obtain expressions for time periods of a simple pendulum, physical pendulum and torsional pendulum.

Ans. $2\pi\sqrt{\frac{l}{g}}, 2\pi\sqrt{\frac{k^2 + l^2}{g\lambda}}, 2\pi\sqrt{\frac{I}{C}}$

Now, we know how to determine amplitude, frequency and time period of SHM. But you may like to know: How do we measure phase? We know that it is the mechanical condition of motion governed by simultaneous consideration of displacement and velocity. By examining the expressions for displacement and velocity [Eqs. (2.25a) and (2.25b)], we find that the quantity $(\omega t + \delta)$ appears in the argument of both the sine and cosine functions. So phase is measured by $(\omega t + \delta)$. At $t = 0$, the phase is equal to δ, which is called the *epoch* of the SHM.

***Practice Exercise* 2.4** Starting from Eq. (2.24b′), obtain the physical significance of ω and δ.

On solving the differential equation of SHM, we find that displacement versus time equations [Eqs. (2.24a′) and (2.24b′)] are sinusoidal, which corroborates definition (iii) given in Section 2.2. We shall now study how velocity and acceleration of a particle executing SHM are related.

2.6 VELOCITY AND ACCELERATION IN SHM

Recall the expression for velocity given in Eq. (2.25):

$$v = a\omega \cos(\omega t + \delta).$$

From this we note that velocity amplitude is given by

$$v_{max} = a\omega. \tag{2.30}$$

Hence, acceleration is given by

$$f = \frac{dv}{dt} = -a\omega^2 \sin(\omega t + \delta). \tag{2.31a}$$

We can rewrite it as

$$f = -\omega^2 x. \tag{2.31b}$$

Did you not expect this result? The expression for force acting on a particle executing SHM is given by

$$F = mf = -m\omega^2 x. \tag{2.31c}$$

This agrees with the expression given in Eq. (2.2).

The maximum value of acceleration is given by

$$f_{\max} = \omega^2 a. \tag{2.31d}$$

You may now like to answer a Practice Exercise.

***Practice Exercise* 2.5** Derive Eqs. (2.30) and (2.31d) starting from Eq. (2.24b′).

The equations representing time variation of displacement, velocity and acceleration with time, i.e., Eqs. (2.24a′), (2.24b′) and (2.31a), can be expressed as

$$MP = (AMP) \text{ sine or cosine } (\theta_P), \tag{2.32}$$

where MP is the mechanical parameter (displacement, velocity or acceleration), AMP is amplitude of mechanical parameter and θ_p is the phase angle, $\omega t + \delta$.
Equations. (2.24a′), (2.24b′) and (2.31a) can also be written as

$$x = a \sin \theta_p, \tag{2.32a}$$

$$v = a\omega \sin\left(\theta_p + \frac{\pi}{2}\right), \tag{2.32b}$$

and

$$f = a \sin (\theta_p + \pi). \tag{2.32c}$$

Equations (2.32a, b, c) exhibit the phase relationship between displacement, velocity and acceleration of a particle executing SHM. Note that velocity is ahead of displacement of $\pi/2$ and acceleration is ahead of displacement by π. In other words, the acceleration is opposite in phase to displacement. Figure 2.9 shows the time variation of these variables based on the following relations:

$$x = a \sin \omega t. \tag{2.33a}$$

$\therefore$

$$v = \frac{dx}{dt} = a\omega \cos \omega t = a\omega \sin\left(\omega t + \frac{\pi}{2}\right), \tag{2.33b}$$

and

$$f = \frac{dv}{dt} = -a\omega^2 \sin \omega t = a\omega^2 \sin (\omega t + \pi). \tag{2.33c}$$

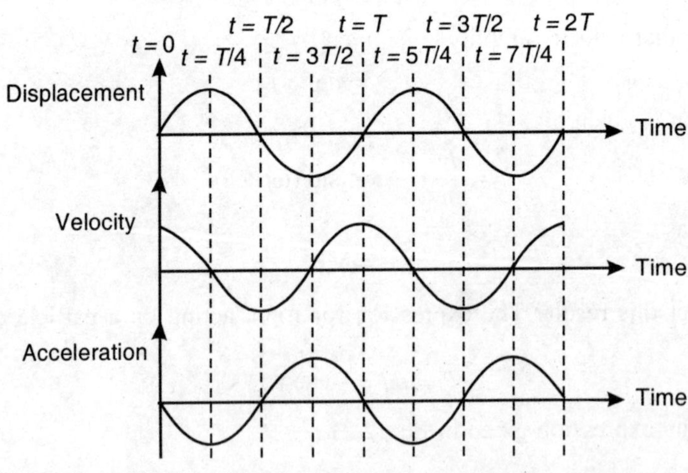

Fig. 2.9 Time variation of displacement, velocity and acceleration of a particle executing SHM.

EXAMPLE 2.2

The speed v of a particle moving along x-axis is given by $v^2 = \mu^2(6bx - x^2 - 5b^2)$. Show that it executes SHM with its centre at $x = 3b$ and amplitude $2b$. Also calculate the time period and the time taken in going from $x = 4b$ to $x = 5b$.

Solution: We know that

$$v^2 = \mu^2(6bx - x^2 - 5b^2) = -\mu^2(x^2 - 6bx + 5b^2)$$

$$= -\mu^2(x - b)(x - 5b). \tag{2.34}$$

And $$\frac{d}{dx}(v^2) = -\mu^2(2x - 6b) = -2\mu^2(x - 3b)$$

or $$2v\frac{dv}{dx} = -2\mu^2(x - 3b)$$

$$v\frac{dv}{dx} = -\mu^2(x - 3b)$$

$\therefore$ $$f = -\mu^2(x - 3b) \tag{2.35a}$$

or $$\ddot{x} = -\mu^2(x - 3b). \tag{2.35b}$$

By putting $X = x - 3b$ in Eq. (2.35b), we get

$$\ddot{X} = -\mu^2 X \tag{2.35c}$$

Equation (2.35c) represents SHM of period

$$T = \frac{2\pi}{\mu}. \tag{2.36}$$

The SHM takes place about $X = 0$, i.e., $x - 3b = 0$ or $x = 3b$.

We know that extremities of an oscillation correspond to $v = 0$. So from Eq. (2.34), we find that these are located at $x = b$ and $x = 5b$. Therefore, the amplitude = $(5b - 3b)$ or $(3b - b) = 2b$. This is depicted in Fig. 2.10.

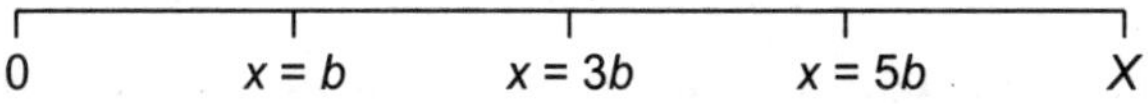

0 $x = b$ $x = 3b$ $x = 5b$ X

Fig. 2.10 The centre of oscillation is at $x = 3b$ and the extremities are at $x = b$ and $x = 5b$.

Thus, the displacement versus time equation will be

$$X = 2b \cos \mu t. \tag{2.37}$$

(The cosine function takes care of the fact that the extremities correspond to $X = -2b$ and $2b$, i.e., $x = b$ and $5b$). The time taken from $x = 4b$ to $x = 5b$ is equal to the time taken from $x = 5b$ to $x = 4b$.

When $x = 4b$, $X = b$; $\therefore$ $$t_{X=b} = \frac{1}{\mu}\cos^{-1}\frac{1}{2} = \frac{1}{\mu}\cdot\frac{\pi}{3} = \frac{\pi}{3\mu}, \text{ and}$$

when $x = 5b$, $X = 2b$; $\therefore$ $$t_{X=2b} = \frac{1}{\mu}\cos^{-1}1 = \frac{1}{\mu}\cdot 0 = 0.$$

$\therefore$ The required time interval $= \dfrac{\pi}{3\mu}$.

EXAMPLE 2.3

A horizontal shelf moves vertically in a simple harmonic manner with a period one second. Calculate the greatest amplitude (in metre) that it can have so that the books resting on it may always be in contact with it. (Take $g = 10\ \text{ms}^{-2}$).

Solution: We know that acceleration amplitude is given by

$$f_{max} = \omega^2 a.$$

Now, in order that the books do not lose contact with the shelf, we must have

$$\omega^2 a \ngtr g$$

or

$$a \ngtr \frac{g}{\omega^2}$$

or

$$a \ngtr \frac{gT^2}{4\pi^2}$$

or

$$a_{max} = \frac{gT^2}{4\pi^2} = \frac{10.1^2}{4\pi^2}$$

Taking $\pi^2 = 10$ (apx.), we have $a_{max} = 0.25$ m.

EXAMPLE 2.4

A particle is in equilibrium under the action of two repulsive forces whose magnitudes are inversely proportional to third power of distance. The force per unit mass at unit distance are μ and μ', respectively. If the particle is slightly displaced towards one of them, show that the motion is approximately simple harmonic. Also calculate the time period of oscillations.

Solution: Refer to Fig. 2.11. Suppose that O corresponds to the equilibrium position under the action of forces at Q and R. Suppose that $OQ = q$, $OR = r$ and let P be displaced from O through OP $(=x)$. According to the given condition, we can write

$$\frac{\mu}{q^3} = \frac{\mu'}{r^3}.$$

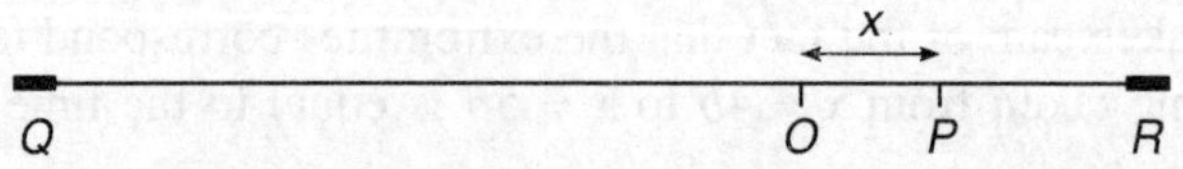

Fig. 2.11 *O* is the position of equilibrium corresponding to the centres of forces at *Q* and *R*.

If the force of repulsion due to R is greater than that due to Q, the net force experienced per unit mass at the position P will be directed along PO:

$$\frac{\mu'}{(r-x)^3} - \frac{\mu}{(q+x)^3},$$

Hence, the equation of motion can be written as

$$\ddot{x} = -\frac{\mu'}{(r-x)^3} + \frac{\mu}{(q+x)^3} \tag{2.38}$$

The RHS of Eq. (2.38) can be written as

$$\text{RHS} = -\frac{\mu'}{r^3}\left(1-\frac{x}{r}\right)^{-3} + \frac{\mu}{q^3}\left(1+\frac{x}{q}\right)^{-3}$$

For small x, we retain terms only up to first power of x to obtain

$$\text{RHS} = -\frac{\mu'}{r^3}\left(1+\frac{3x}{r}\right) + \frac{\mu}{q^3}\left(1-\frac{3x}{q}\right)$$

$$= C_0 - 3\left(\frac{\mu}{q^4}+\frac{\mu'}{r^4}\right)x$$

$$= -3\left(\frac{\mu}{q^4}+\frac{\mu'}{r^4}\right)(x-C),$$

where C_0 and C are constants.

To proceed further, we put

$$u = x - C.$$

This gives

$$\ddot{u} = \ddot{x}.$$

Hence, we can rewrite Eq. (2.38) as

$$\ddot{u} = -ku,$$

where

$$k = 3\left(\frac{\mu}{q^4}+\frac{\mu'}{r^4}\right).$$

This can be rewritten as

$$\ddot{u} + ku = 0,$$

which is the differential equation of SHM whose time period is given by

$$T = \frac{2\pi}{\sqrt{k}} = 2\pi\Big/\sqrt{3\left(\frac{\mu}{q^4}+\frac{\mu'}{r^4}\right)}.$$

EXAMPLE 2.5

A weightless elastic spring of unstretched length l and modulus of elasticity E is fixed at one end and a mass m hangs at the lower end. If the system is displaced vertically from its equilibrium position, calculate the period of its oscillation. If the mass m is pulled vertically downwards to a position where the extension of the string is e, discuss its motion and determine the maximum value of e so that during its subsequent motion, the string may never be slack.

Solution: Refer to Fig. 2.12. It shows a spring of natural length *OP*. Suppose that *PQ* signifies the extended length when the mass *m* is in equilibrium under the influence of weight *mg*.

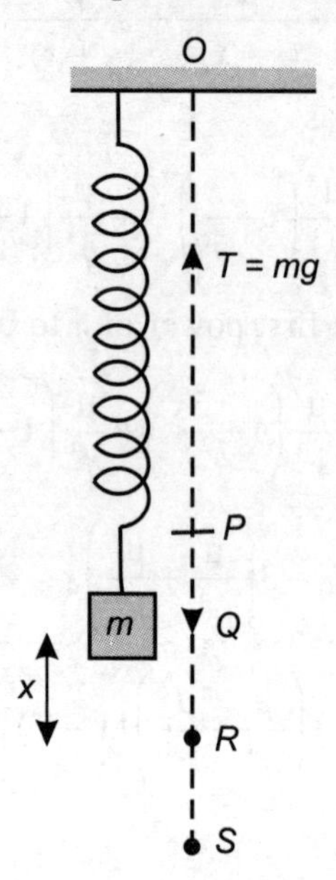

Fig. 2.12 A spring-mass system executing vertical oscillations.

Suppose that tension in the spring is *T*. Obviously, it will be equal to *mg*. Since the string is extended to the position *PQ*, we can also express the tension in the spring as $(E \times PQ)/l = mg$ so that $PQ = (mgl/E)$. Let us denote the instantaneous displacement (extension) from the equilibrium position by *x*. This corresponds to position *R*. At this stage, the tension in the spring will be Ex/a. Hence, the equation of motion of the particle is

$$m\ddot{x} = mg - \frac{Ex}{l}$$

or

$$\ddot{x} = -\frac{E}{ml}\left(x - \frac{mgl}{E}\right).$$

Let us introduce a change of variable by defining $x - (mgl/E) = u$ so that $\dot{x} = \dot{u}$ and $\ddot{x} = \ddot{u}$. Then the equation of motion of the spring-mass system simplifies to

$$\ddot{u} + \frac{E}{ml}u = 0. \tag{2.39}$$

This equation shows that spring-mass system executes SHM about the centre of oscillation at *Q*. The time period is given by

$$T = \frac{2\pi}{\sqrt{E/ma}} = 2\pi\sqrt{\frac{ma}{E}}$$

Multiplying both sides of Eq. (2.39) by $2\dot{u}$, we get

$$\frac{d}{dt}(\dot{u}^2) + \frac{E}{ml}\frac{d}{dt}(u^2) = 0$$

or $$\frac{d}{dt}\left(\dot{u}^2 + \frac{E}{ml}u^2\right) = 0$$

On integrating over time, we get

$$\dot{u}^2 + \frac{E}{ml}u^2 = \text{a constant} = A, \text{ say}$$

where A is constant of integration.

In terms of x, we can rewrite this result as

$$\dot{x}^2 + \frac{E}{ml}\left(x - \frac{mgl}{E}\right)^2 = A.$$

We know that $\dot{x} = 0$ when $x = e$. Using this condition in above expression, we get the value of constant of integration:

$$A = \frac{E}{ml}\left(e - \frac{mgl}{E}\right)^2$$

so that $$\dot{x}^2 = \frac{E}{ml}\left\{\left(e - \frac{mgl}{E}\right)^2 - \left(x - \frac{mgl}{E}\right)^2\right\} \tag{2.40a}$$

$$= \frac{E}{ml}(e - x)(e + x - f), \tag{2.40b}$$

where $$f = \frac{2mgl}{E}.$$

At the beginning $x = e$, and thereafter it begins to move up.

***Case* (i)** $e < f$. From Eq. (2.40b), we find that $\dot{x}$ again becomes zero when $x = f - e$ and the mass m oscillates between $x = e$ and $x = f - e$:

$$x = e \quad \text{and} \quad x = \frac{2mgl}{E} - e,$$

which correspond to

$$u = e - \frac{mgl}{E} \quad \text{and} \quad u = \frac{mgl}{E} - e.$$

That is, the spring-mass system executes vertical oscillations about Q.

***Case* (ii)** $e > f$. From Eq. (2.40b), we note that when $x = 0$ (i.e., at P),

$$\dot{x}^2 = \frac{Ee}{ml}(e - f).$$

It means that upward velocity is still not zero. Beyond this, x is –ve, i.e., the spring becomes slack and Eq. (2.39) fails to describe the motion of the system. In fact, the mass then moves up like a free body under the influence of gravity. It rises up to the highest point and then comes down reaching P with the same velocity $\sqrt{\frac{Ee}{ml}\left(e - \frac{2mgl}{E}\right)}$ as before.

After this, Eqs. (2.39), (2.40a and b) again become valid. You must have noted that $\dot{x}$ is again zero for $x = e$, that is when the particle reaches S. The motion is then repeated. Thus, the maximum value of e in order that the string may never be slack is $e = (2mgl/E)$ and then the particle will oscillate exactly between P and S.

EXAMPLE 2.6

A mass m is at rest on a smooth horizontal surface. It is connected to three vertices of an equilateral triangle of side $2\sqrt{3}l$ through elastic strings, each of modulus of elasticity E and natural length l. The mass is displaced slightly along one of the strings and then released. Discuss the nature of its motion and calculate its time period.

Solution: Refer to Fig. 2.13, which shows the original as well as instantaneous configuration of the system. ABC is an equilateral triangle. AD is perpendicular drawn from A on BC. O is centroid of ΔABC. The particle is at rest at O and the original positions of the strings are OA, OB, and OC. It is displaced slightly $(=x)$ along OA. BXC is an isosceles triangle, where XB and XC are the displaced positions of the strings. $\angle BXD = \angle CXD = \theta$, say. Note that at the displaced position, tensions in the strings will not be equal. However, on account of symmetry, let us suppose that tensions along XB and XC are equal to T, say and that along XA is T_0. It means that the resultant force influencing displacement is given by

$$F = T_0 - 2T\cos\theta. \tag{2.41}$$

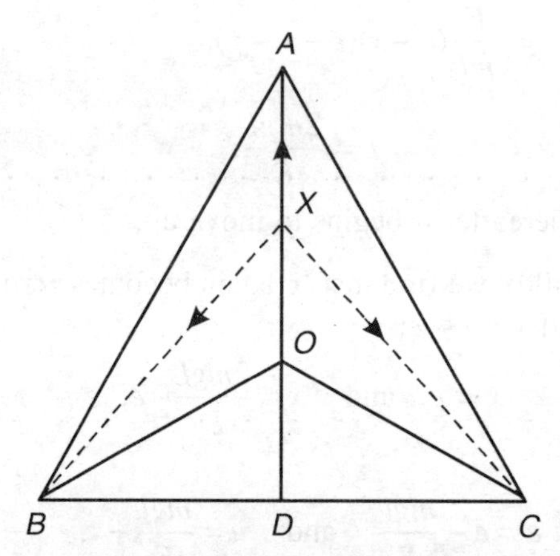

Fig. 2.13 The initial and instantaneous configurations of a mass m connected to three springs.

Now, $OA = OB = OC = R$, where R is the circumradius of the triangle. From properties of triangles, we can write

$$\frac{BC}{\sin\angle BAC} = 2R.$$

$$\therefore \qquad R = \frac{2\sqrt{3}l}{2\sin 60^\circ} = 2l.$$

$$\therefore \qquad OA = OB = OC = 2l.$$

It means that $\frac{OD}{OA} = \frac{1}{2}$ so that $OD = l$.

Since $OX = x$, and $XA = 2l - x$, we can write

$$\begin{aligned} XC^2 &= XA^2 + AC^2 - 2XA \cdot AC \cos \angle XAC \\ &= (2l - x)^2 + (2\sqrt{3}l)^2 - 2 \cdot (2l - x) \cdot 2\sqrt{3}l \cos 30^\circ \\ &= 4l^2 + x^2 - 4lx + 12l^2 - (2l - x)\, 6l \\ &= 4l^2 + x^2 + 2lx = 4l^2\left(1 + \frac{x}{2l} + \frac{x^2}{4l^2}\right). \end{aligned}$$

$$\therefore \qquad XC = 2l\left(1 + \frac{x}{2l} + \frac{x^2}{4l^2}\right)^{1/2} = 2l\left(1 + \frac{x}{4l}\right),$$

where we have neglected second and higher order terms in x. Hence

$$\begin{aligned} \cos\theta &= \frac{XD}{XC} = \frac{OX + OD}{XC} = \frac{l + x}{2l[1 + (x/4l)]} \\ &= \frac{1}{2}\left(1 + \frac{x}{l}\right)\left(1 + \frac{x}{4l}\right)^{-1} = \left(1 + \frac{x}{l}\right)\left(1 - \frac{x}{4l}\right) = \frac{1}{2}\left(1 + \frac{3x}{4l}\right). \end{aligned}$$

Note that as before, we have retained only linear terms in x.

Now,

$$E = \frac{T_0}{[(2l - x) - l]/l} \quad \text{or} \quad T_0 = \frac{lE - x}{l} = E\left(1 - \frac{x}{l}\right).$$

Similarly,

$$T = E\frac{XC - l}{l} = \frac{E}{l}\left(l + \frac{x}{2}\right) = E\left(1 + \frac{x}{2l}\right).$$

On inserting these results in Eq. (2.41), we get

$$F = E\left\{\left(1 - \frac{x}{l}\right) - 2\left(1 + \frac{x}{2l}\right)\frac{1}{2}\left(1 + \frac{3x}{4l}\right)\right\}.$$

If we retain only first order terms in x, we can write

$$\begin{aligned} F &= E\left\{\left(1 - \frac{x}{l}\right) - \left(1 + \frac{5x}{4l}\right)\right\} \\ &= E\left(-\frac{9x}{4l}\right) = -\frac{9E}{4l}x. \end{aligned}$$

$\therefore$ The differential equation of motion is

$$m\ddot{x} = -\frac{9E}{4l}x$$

or

$$\ddot{x} + \frac{9E}{4lm}x = 0, \tag{2.42}$$

which represents SHM whose time period is given by

$$T = \frac{2\pi}{\sqrt{9E/4lm}} = \frac{4\pi}{3}\sqrt{\frac{lm}{E}}.$$

EXAMPLE 2.7

An elastic string, whose modulus of elasticity is E and natural length l, takes the shape of a circle when placed on a smooth horizontal plane. It is then acted upon by a force from the centre of the circle whose magnitude is μ times the distance per unit mass of the string. Prove that for an extremely small displacement, the radius will vary harmonically about a mean value r_0, such that $2\pi E\ (2\pi r_0 - l) = m\mu l r_0$, where m is the mass of the string and $4\pi^2 E > m\mu l$. Also calculate the frequency of the simple harmonic motion.

Solution: Refer to Fig. 2.14. Suppose that original radius of the circle is a. Then $l = 2\pi a$. Let $OB = OC = r$ denote the radius of the circle at time t. Then, arc $BC = r\delta\theta$. Since the displacement is small, we can assume that $\delta\theta$ is also very small.

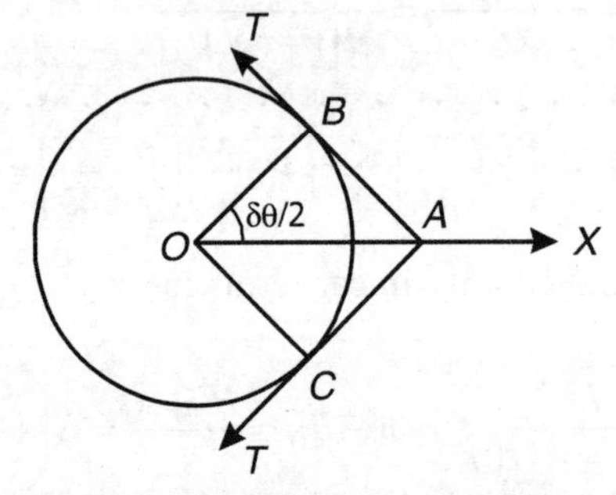

Fig. 2.14 The circle with centre at O represents the original shape of the string. BC is an arc of the circle which subtends an angle $\delta\theta$ at the centre. OX bisects $\angle BOC$.

Mass of the arc $BC = \dfrac{m}{2\pi r}\cdot r\delta\theta = \dfrac{m\delta\theta}{2\pi}$

Instantaneous tension in the string, $T = E\cdot\dfrac{2\pi r - 2\pi a}{2\pi a} = E\left(\dfrac{r-a}{a}\right)$

The resultant of the tensions T in the string as shown in the figure will provide the necessary restoring force which would oppose the applied force, $\mu r\,(m/2\pi)\,d\theta$ along OX. Therefore,

Final resultant force along $OX = \frac{m\mu r}{2\pi}\delta\theta - 2T\cos\left(\frac{\pi}{2} - \frac{\delta\theta}{2}\right)$

$$= \frac{m\mu r}{2\pi}\delta\theta - 2T\sin\frac{\delta\theta}{2}$$

$$= \frac{m\mu r}{2\pi}\delta\theta - 2T\cdot\frac{\delta\theta}{2} \quad (\because\ \delta\theta \text{ is very small})$$

$$= \left(\frac{m\mu r}{2\pi} - E\frac{r-a}{a}\right)\delta\theta$$

The equation of motion of PQ along OX can be written as

$$\frac{m}{2\pi}\delta\theta\cdot\frac{d^2r}{dt^2} = \left(\frac{m\mu r}{2\pi} - E\frac{r-a}{a}\right)\delta\theta$$

or
$$\frac{d^2r}{dt^2} = \mu r - \frac{2\pi E}{ma}(r-a) = \mu r - \frac{2\pi E}{ma}r + \frac{2\pi E}{m}$$

or
$$\ddot{r} = \left(\mu - \frac{2\pi E}{ma}\right)r + \frac{2\pi E}{m}$$

$$= \left(\mu - \frac{2\pi E}{ma}\right)\left\{r + \frac{2\pi E/m}{(\mu ma - 2\pi E)/ma}\right\}$$

or
$$\ddot{r} = -\left(\frac{2\pi E}{ma} - \mu\right)\left(r - \frac{2\pi aE}{2\pi E - \mu ma}\right)$$

or
$$\ddot{r} = -\left(\frac{2\pi E - ma\mu}{ma}\right)(r - r_0), \tag{2.42a}$$

where
$$r_0 = \frac{2\pi aE}{2\pi E - \mu ma}.$$

Let $r - r_0 = r_1$, and we get

$$\ddot{r}_1 + Cr_1 = 0, \tag{2.42b}$$

where
$$C = \frac{2\pi E - ma\mu}{ma}.$$

Equation (2.42b) represents simple harmonic oscillations about $r_1 = 0$, i.e., $r = r_0$, and

$$r_0 = \frac{2\pi aE}{2\pi E - \mu ma} = \frac{El}{2\pi E - (\mu ma/2\pi)}$$

or
$$r_0 = \frac{2\pi El}{4\pi^2 E - \mu ml}$$

or
$$4\pi^2 Er_0 - \mu mlr_0 = 2\pi El$$

or
$$2\pi E(2\pi r_0 - l) = m\mu lr_0.$$

The –ve sign in Eq. (2.42a) is of crucial importance. Equation (2.42b) becomes a valid differential equation of SHM, provided the sign is genuinely negative, meaning thereby that

$$2\pi E - ma\mu > 0$$

or

$$2\pi E - m\mu \frac{l}{2\pi} > 0$$

i.e.

$$4\pi^2 E > m\mu l.$$

From Eq. (2.42b), we note that the frequency of oscillation is given by

$$\nu = \frac{\sqrt{C}}{2\pi} = \frac{1}{2\pi}\sqrt{\frac{2\pi E - ma\mu}{ma}} = \frac{1}{2\pi}\sqrt{\frac{4\pi^2 E - lm\mu}{ml}}.$$

We have worked out quite a few examples and arranged these in increasing order of degree of difficulty. The basic objective is to provide you practice in framing equations of motion and appreciating that those represent SHM under some approximation, the commonality being small oscillations.

2.7 ENERGY OF AN OSCILLATING SYSTEM

While discussing the motion of typical physical systems, we discovered that the energy of oscillation alternates between potential and kinetic forms; the PE being minimum at the mean position and maximum at the extremities. On the other hand, KE is maximum at the mean equilibrium position and minimum at the extremities. It would, therefore, be worthwhile to derive an expression for total energy in SHM.

Let us start with Eq. (2.25):

$$x = a \sin(\omega t + \delta)$$

$\therefore$

$$v = \frac{dx}{dt} = a\omega \cos(\omega t + \delta) = a\omega\sqrt{1 - \left(\frac{x}{a}\right)^2},$$

or

$$v = \omega\sqrt{a^2 - x^2}.$$

This is same as Eq. (2.22b).

Hence,

$$\text{KE} = \frac{1}{2}mv^2 = \frac{1}{2}m\omega^2(a^2 - x^2). \tag{2.43}$$

We can also express the kinetic energy as

$$\text{KE} = \frac{1}{2}mv^2 = \frac{1}{2}ma^2\omega^2\cos^2(\omega t + \delta). \tag{2.44}$$

Therefore, average KE in a complete cycle is obtained by integrating it over t from 0 to T:

$$<KE> = \frac{1}{2}ma^2\omega^2\overline{\cos^2(\omega t + \delta)}$$

$$= \frac{1}{2}ma^2\omega^2\left[\frac{1}{T}\int_0^T \cos^2(\omega t + \delta)\, dt\right]$$

$$= \frac{1}{2}ma^2\omega^2 \cdot \frac{1}{2} = \frac{1}{4}ma^2\omega^2. \tag{2.45}$$

Let us now estimate the PE. We go back to Eq. (2.6) and write

$$F = -kx = -m\omega^2 x. \tag{2.46}$$

The PE of the oscillator at a particular position is equal to the work done against the force in taking the oscillation from the mean position to the instantaneous position. In other words,

$$PE = -\int_0^x F dx = \int_0^x m\omega^2 x dx = \frac{1}{2}m\omega^2 x^2 \tag{2.47}$$

$$= \frac{1}{2}m\omega^2 a^2 \sin^2(\omega t + \delta). \tag{2.48}$$

As before, average potential energy in a complete cycle is obtained by integrating it over time from 0 to T. This gives

$$<PE> = \frac{1}{2}ma^2\omega^2\overline{\sin^2(\omega t + \delta)}$$

$$= \frac{1}{2}m\omega^2 a^2\left[\frac{1}{T}\int_0^T \sin^2(\omega t + \delta)\, dt\right]$$

$$= \frac{1}{2}m\omega^2 a^2 \cdot \frac{1}{2} = \frac{1}{4}m\omega^2 a^2. \tag{2.49}$$

From Eqs. (2.45) and (2.49), we note that average potential energy is equal to average kinetic energy:

$$<KE> = <PE> = \frac{1}{4}m\omega^2 a^2. \tag{2.50}$$

On combining Eqs. (2.44) and (2.48), we get the required expression for total energy of a particle executing SHM:

$$\text{Total energy} = E = PE + KE = \frac{1}{2}m\omega^2 a^2. \tag{2.51}$$

Note that total energy is constant, equal to $\frac{1}{2}m\omega^2 a^2 = \frac{1}{2}ka^2$. Moreover, it is directly proportional to the square of the angular frequency as well as the amplitude. In Fig. 2.15a and b, we have depicted the variation of KE and PE respectively with displacement based on Eqs. (2.43) and (2.47) and with time based on Eqs. (2.44) and (2.48).

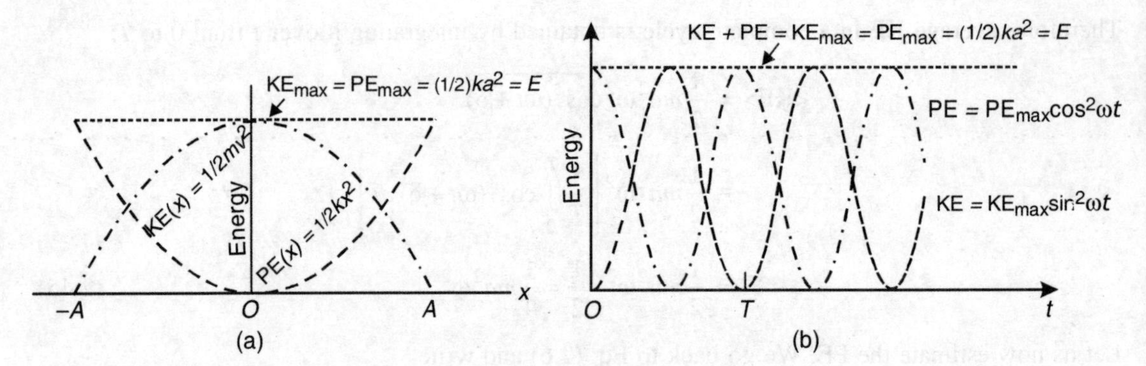

Fig. 2.15 Variation of total energy of a simple harmonic oscillation (a) with displacement (b) with time. Note that when displacement is maximum, PE is maximum but KE is zero. At the mean position, PE is zero, but KE is maximum.

The variation of KE and PE during a cycle can be studied more analytically by observing the motion of a spring-mass system at intervals of $T/8$ (Fig. 2.16).

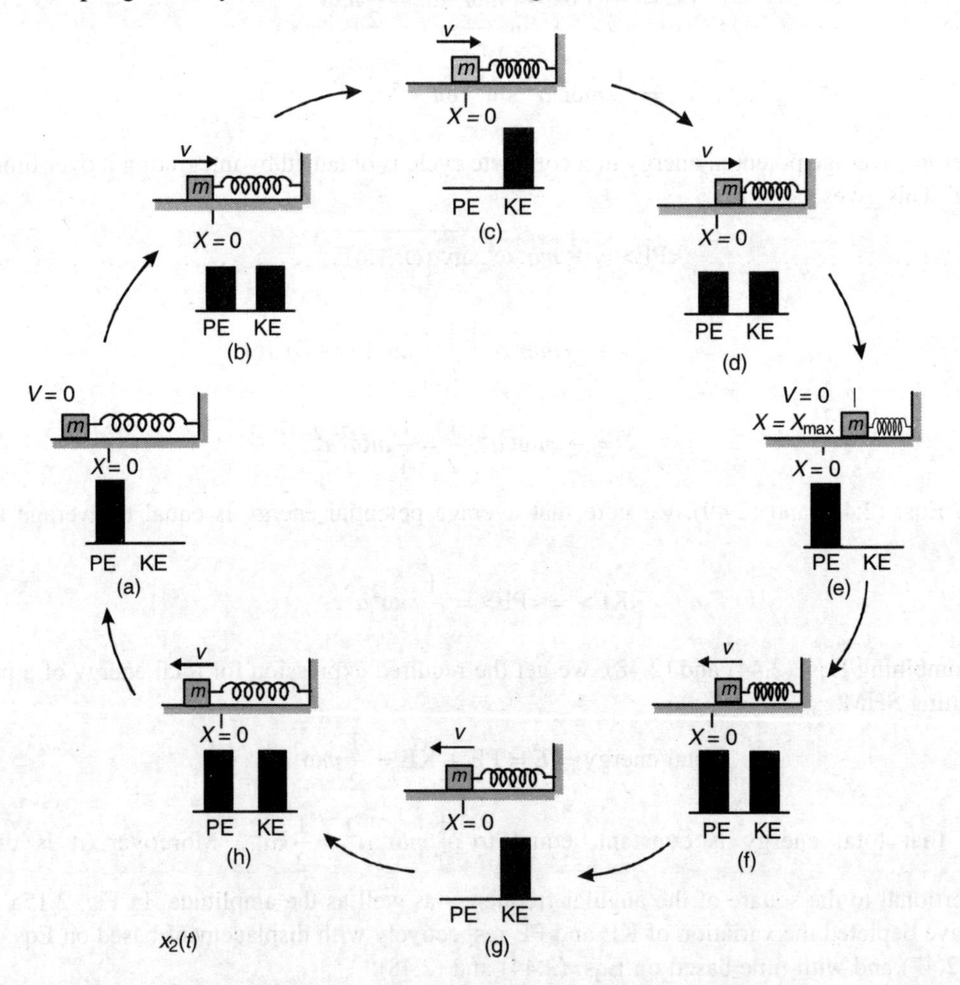

Fig. 2.16 Variation of kinetic energy and potential energy with the configuration of a spring-mass system at intervals of $T/8$.

Let us assume that at the beginning of the cycle, the spring is elongated the most so that KE = 0 and PE is maximum (Fig. 2.16a). Then KE begins to grow at the expense of PE. After one-eighth of the time period, the values of KE and PE are identical (Fig 2.16b). At $t = T/4$, PE becomes zero and KE takes maximum value, as shown in Fig. 2.16c. Another $T/8$ seconds later, the PE and KE again become equal (Fig. 2.16d). However, KE continues to decrease and the PE keeps growing till at $t = T/2$, the former becomes zero and the latter attains the maximum value because the spring is compressed the most (Fig. 2.16e). For $t > T/8$, the mass starts moving in the opposite direction and KE begins to grow at the expense of the PE. These become equal at $t = 5T/8$ (Fig. 2.16f). At $t = 3T/4$, KE becomes maximum again and PE becomes zero (Fig. 2.16g). Another $t = T/8$ seconds later, PE and KE become equal (Fig. 2.16h) and at the end of the cycle, $t = T$, the original condition is obtained.

We can corroborate the time variations of KE and PE depicted in Fig. 2.15 with the illustration shown in Fig. 2.16. From Eq. (2.50), we recall that the average values of KE and PE taken over a complete cycle are each equal to (1/4) ka^2.

We shall now analyse energy equation of the *LC*-circuit; electromagnetic analogue of the spring-mass system. You must have studied it in your senior secondary physics classes.

2.8 AN IDEAL LC-CIRCUIT

In Section 2.1, we pointed out that oscillations are not restricted to mechanical systems only. These are observed in electromagnetic systems also. The most instructive example of an electromagnetic system is a circuit having a combination of a pure inductor and a pure capacitor (Fig. 2.17). We shall study the phenomenon using energy considerations.

> A 'pure' inductor has zero resistance and a 'pure' capacitor has infinite resistance.

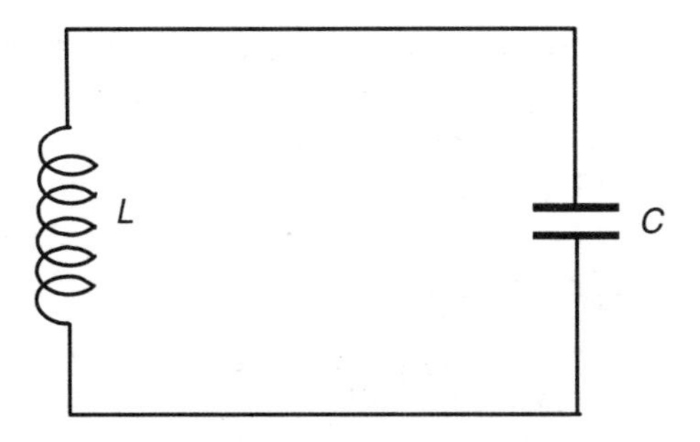

Fig. 2.17 An ideal *LC*-circuit.

From Eq. (2.8), we recall that for a spring-mass system, total energy is given by

$$E = \frac{1}{2}m\dot{x}^2 + \frac{1}{2}kx^2. \tag{2.52}$$

Note that the first term defines KE and the second term defines PE. When the spring is compressed, energy is stored in the spring in the form of PE. We know that every system has a natural tendency to move towards the state of minimum PE. And the spring is no exception. The PE stored in the spring is converted into KE of the mass and becomes maximum at its mean position. This forces the mass to overshoot the mean position and is responsible for to and fro motion. As discussed earlier, the energy alternates between potential and kinetic forms obeying Eq. (2.52) and the system oscillates about its mean equilibrium position. Now, let us see what happens in an *LC*-circuit.

The energy of a capacitor having charge q at an instant t is $q^2/2C$, where C is its capacitance. The instantaneous current i in the *LC*-circuit due to the discharge of the capacitor is given by $i = dq/dt$. The corresponding energy of the inductor is $(1/2)\, Li^2$. So in an ideal situation, when there is no resistance in the circuit, the total energy can be expressed as

$$E = \frac{1}{2} Li^2 + \frac{q^2}{2C} = \frac{1}{2} L \left(\frac{dq}{dt}\right)^2 + \frac{q^2}{2C}. \tag{2.53}$$

Note that Eqs. (2.52) and (2.53) are analogous. In fact, if we replace x by q, m by L and k by $1/C$ in Eq. (2.52), we arrive at Eq. (2.53). It means that inductor plays the same role in an electric circuit as does mass in a mechanical system. And we expect the charge in an electrical circuit to oscillate in time.

To visualise this, we consider that a fully charged capacitor is connected with an inductor, as shown in Fig. 2.17. The capacitor will discharge through the inductor. The charge stored in it will reduce gradually and current will begin to flow through the inductor. Due to growth of current in the inductor, an e.m.f. will be induced. According to Lenz's law, it will oppose the very cause which produces it. That is, it will oppose the growth of current. As a result, the current will take some time to reach its maximum value. (This is analogous to the situation of spring-mass system, where mass attains maximum value of KE.) After attaining the maximum value, the current through the inductor begins to decrease and in due course of time, its energy becomes zero. At that instant, the charge on the capacitor becomes maximum again. This cycle is repeated and energy of the system fluctuates alternately between electric and magnetic (equal to $q^2/2C$ and $(1/2)\, Li^2$, respectively).

Let us pause for a while and discover the similarities and dissimilarities between a mechanical and an electrical system. Let us first compare energy transformations in a spring-mass system. In a mechanical system, the displacement varies with time in a sinusoidal manner. In an electrical system, we expect the charge to exhibit same behaviour. Let us go back to Eq. (2.53) and differentiate its both sides with respect to time with the knowledge that $dE/dt = 0$, since total energy is constant. Then, Eq. (2.53) takes the form

$$\frac{L}{2} \cdot 2\dot{q}\ddot{q} + \frac{1}{2C} 2q\dot{q} = 0$$

We can rewrite it as

$$\dot{q}\left(L\ddot{q} + \frac{q}{C}\right) = 0,$$

where number of dots over q indicate the number of times it has been differentiated with respect to time.

Since $\dot{q}$ denotes current in the circuit and it is finite in general, this expression leads us to the result

$$\ddot{q} + \frac{1}{LC} q = 0. \tag{2.54}$$

Equation (2.54) shows that in an LC circuit, charge varies with time in a sinusoidal manner and obeys the differential equation of SHM. The frequency of oscillation is given by

$$\nu_q = \frac{1}{2\pi}\sqrt{\frac{1}{LC}}. \tag{2.55}$$

We are now in a position to summarize the analogies between mechanical and electromagnetic oscillations. It is given in Table 2.1.

Table 2.1 Analogies between Mechanical and Electromagnetic Oscillations

Mechanical	**Electromagnetic**
Displacement (x)	Charge (q)
Velocity ($\dot{x}$)	Current ($\dot{q}$)
Mass (m)	Inductance (L)
Stiffness $\left(\frac{1}{k}\right)$	Capacitance (C)
Potential energy $\left(\frac{1}{2}kx^2\right)$	Electrical energy $\left(\frac{q^2}{2C}\right)$
Kinetic energy $\left(\frac{1}{2}m\dot{x}^2\right)$	Magnetic energy $\left(\frac{1}{2}L\dot{q}^2\right)$

We arrived at the differential equation (2.54) related to the phenomenon of electromagnetic oscillation using energy consideration. You can derive these results from the first principle. This is left as a Practice Exercise for you.

***Practice Exercise* 2.6** In an LC circuit, the e.m.f. induced in an inductor is given by $e = -L(di/dt)$ when current i flows through it and the potential difference across the plates of a capacitor holding charge q is q/C. Derive Eq. (2.54) by establishing a differential equation on the basis of the situation described earlier.

We now work out an example to calculate the averages values of KE and PE with respect to position over one cycle.

EXAMPLE 2.8

Calculate the average values of KE and PE of an SHM taken with respect to position, over a cycle.

Solution:

$$<\text{KE}> = \frac{1}{a}\int_0^a \frac{1}{2}m\omega^2(a^2 - x^2)\,dx$$

$$= \frac{m\omega^2}{2a}\left(a^3 - \frac{a^3}{3}\right) = \frac{1}{3}m\omega^2 a^2 = \frac{1}{3}ka^2$$

$$<\text{PE}> = \frac{1}{a}\int_0^a \frac{1}{2}m\omega^2 x^2 dx$$

$$= \frac{m\omega^2}{2a}\int_0^a x^2 dx = \frac{m\omega^2}{2a}\cdot\frac{a^3}{3} = \frac{1}{6}m\omega^2 a^2 = \frac{1}{6}ka^2.$$

Note that <KE> + <PE> = (1/2) ka^2 as expected. But <KE> is not equal to <PE>. We expect you to discuss the factor responsible for this inequality of <KE> and <PE>.

In your senior secondary classes, you may have studied the connection between uniform circular motion and SHM. We discuss it now briefly.

2.9 RELATION BETWEEN LINEAR SHM AND UNIFORM CIRCULAR MOTION

Consider a particle executing uniform circular motion with angular speed ω along the circumference of a circle defined by $x^2 + y^2 = a^2$ (Fig. 2.18). If we choose $t = 0$ at the instant the particle is at A and if the particle moves to the point P in time t, $\angle AOP = \omega t$. From P, we drop a perpendicular PN on the x-axis. Thus, $x = ON$. Similarly, if PM is perpendicular from P on the y-axis, we can write $y = NP = OM$.

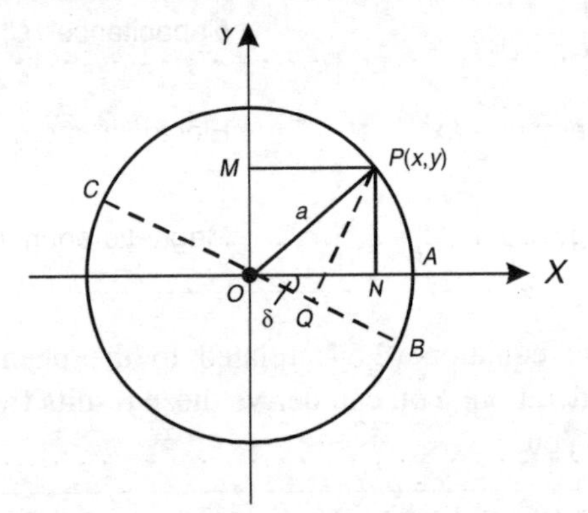

Fig. 2.18 A particle executing uniform circular motion with angular speed ω along the circumference of the circle, $x^2 + y^2 = a^2$.

By referring to Fig. 2.18, we can write

$$x = OP \cos \angle PON = a \cos \omega t \tag{2.55}$$

and

$$y = OP \sin \angle PON = a \sin \omega t. \tag{2.56}$$

Note that these equations represent time variation of displacement for SHM of amplitude a and angular frequency ω. If P moves on the circumference of the circle, the points N and M move along the x and y-axes, respectively and their motion is simple harmonic.

Proceeding further, we note that with reference to the circle, the x and y-axes are along two diameters and so is BC. Let $\angle AOB = \delta$. Then

$$\angle POQ = \angle AOP + \angle AOB = \omega t + \delta.$$

$$\therefore \qquad OQ = OP \cos \angle POQ = a \cos(\omega t + \delta).$$

If we denote OQ by x', we get

$$x' = a \cos(\omega t + \delta). \tag{2.57}$$

Equation (2.57) shows that x' varies sinusoidally with time. Also recall that it has the same form as Eq. (2.24b′). So we can say that with the motion of P along the circumference of the circle and the motion of Q along BC is simple harmonic. That is, if a particle executes uniform circular motion, the foot of the perpendicular from its position on a diameter executes SHM. The magnitude of the angular speed of the uniform circular motion is equal to that of the angular frequency of the SHM.

Does this suggest a link between the centripetal force associated with uniform circular motion and the force causing SHM? Read Example 2.9 and draw your own conclusion.

EXAMPLE 2.9

Starting from the expression of the centripetal acceleration of a particle P executing uniform circular motion, obtain the expression for force causing SHM of Q and show that it is proportional to the displacement and directed towards the centre.

Solution: The centripetal force experienced by P is equal to $m\omega^2 a$ and it is directed along PO. Here m is mass of the particle and ω is its angular frequency. The component of centripetal force along the negative direction of x-axis is given by

$$m\omega^2 a \cos \omega t = m\omega^2 x$$

so that

$$F = -m\omega^2 x.$$

This result shows that F is proportional to x and is directed towards O.
We work out another example to help you fix your ideas.

EXAMPLE 2.10

(a) A satellite revolves in a circular orbit around the earth at a certain height above it. (The height is significantly less than the radius (R) of the earth.) Calculate its time period of revolution.

(b) A tunnel is bored through the earth, say from the North Pole to the South Pole. A small mass is dropped in the tunnel. Show that the particle executes S.H.M. Calculate its time period.

Solution:

(a) Refer to Fig. 2.19. Let the angular speed of the satellite be ω. The centripetal force required for its rotation is provided by the gravitational attraction between the satellite and the earth, whose masses are m_s and m_e, respectively. Hence we can write

$$\frac{Gm_s m_e}{(R+h)^2} = m_s \omega^2 (R+h)$$

or

$$\omega^2 = \frac{Gm_e}{(R+h)^3}.$$

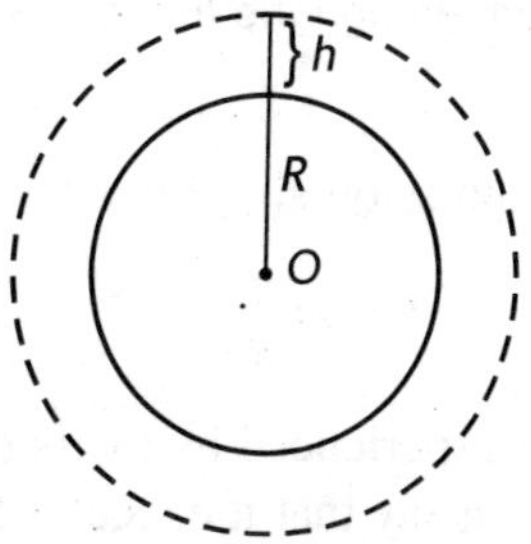

h
R
O

Fig. 2.19 A satellite of mass m_s is revolving in an orbit at a height h above the earth ($h << R$).

Since $h << R$, we can take $R + h \approx R$. Hence

$$\omega^2 = \frac{Gm_e}{R^3} = \frac{gR^2}{R^3} = \frac{g}{R}.$$

In writing the last equality, we have used the relation $Gm_e = gR^2$. Therefore, the time period of revolution of the satellite is given by

$$T_a = \frac{2\pi}{\omega} = 2\pi\sqrt{\frac{R}{g}}. \tag{2.58}$$

(b) Refer to Fig. 2.20. At point P, the particle experiences gravitational attraction due to the sphere of radius x. So the force experienced by the particle directed towards PO is given by

$$F = \frac{Gm\,m_x}{x^2},$$

where m_x is mass of the sphere having radius x. It is given by

$$m_x = \frac{m_e}{(4/3)\,\pi R^3} \cdot \frac{4}{3}\pi x^3 = \frac{m_e \cdot x^3}{R^3}.$$

$$\therefore \qquad F = \frac{Gm}{x^2}\frac{m_e x^3}{R^3} = m\frac{Gm_e}{R^3}x.$$

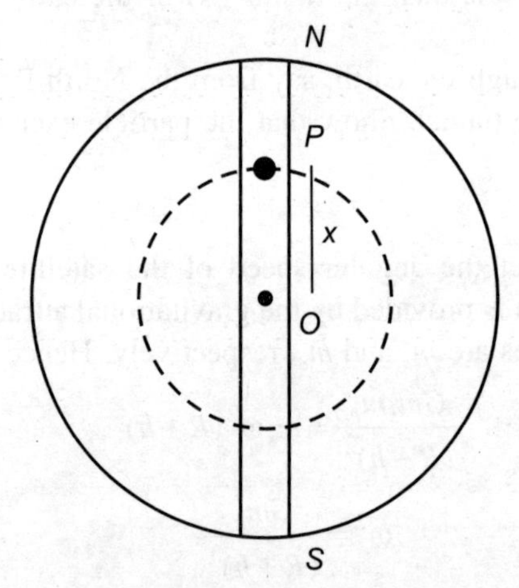

Fig. 2.20 A particle of mass m, dropped through the tunnel NS, at a distance x from the centre of the earth.

But $Gm_e = gR^2$. Hence this expression simplifies to

$$F = m\frac{g}{R}x. \tag{2.59}$$

From this result we note that the force experienced by the body of mass m is proportional to x and is directed towards the centre. So we can say that it executes SHM and the time period T_b is given by

$$T_b = 2\pi\sqrt{\frac{m}{mg/R}} = 2\pi\sqrt{\frac{R}{g}}. \tag{2.60}$$

From Eqs. (2.58) and (2.60), we find that

$$T_a = T_b. \tag{2.61}$$

In other words, the results of two apparently different problems are same. Is it a mere coincidence? Indeed, not. There is a missing link. The satellite revolves in a circular orbit of radius approximately equal to that of the earth. So the foot of the perpendicular on any diameter will execute SHM with a time period equal to the period of revolution of the satellite. The particle dropped through the tunnel indeed does that. Equation (2.61) re-establishes the fact that if a particle executes uniform circular motion, the foot of the perpendicular from its position on any diameter executes SHM with a time period equal to the period of rotation of the particle.

We now discuss the rotating vector representation of SHM.

2.10 ROTATING VECTOR REPRESENTATION OF SHM

Based on Section 2.9 we may define SHM as the projection of a uniform circular motion on any diameter. From Section 1.3 we recall that the complex quantity $Re^{j\omega t}$ represents a vector of magnitude R rotating anticlockwise at constant angular rate ω in the complex plane. We know that

$$\text{Re}(Re^{j\omega t}) = R\cos\omega t \tag{2.62}$$

and

$$\text{Im}(Re^{j\omega t}) = R\sin\omega t. \tag{2.63}$$

Thus, the real and imaginary parts of $Re^{j\omega t}$ are the two components of SHM which are mutually at right angles. The motion of the rotating vector starts from the x-axis, i.e., the real axis.

On the basis of Eq. (2.57), we can conclude that $Re^{j(\omega t+\delta)}$ is a rotating vector of same magnitude and angular speed as $Re^{j\omega t}$, but its motion starts from a position which makes an angle δ with the real axis. The rotating vector $Re^{-j\omega t}$ describes the circle clockwise. Figure 2.21 shows that the projection of a rotating vector plotted against time gives a sinusoidal function.

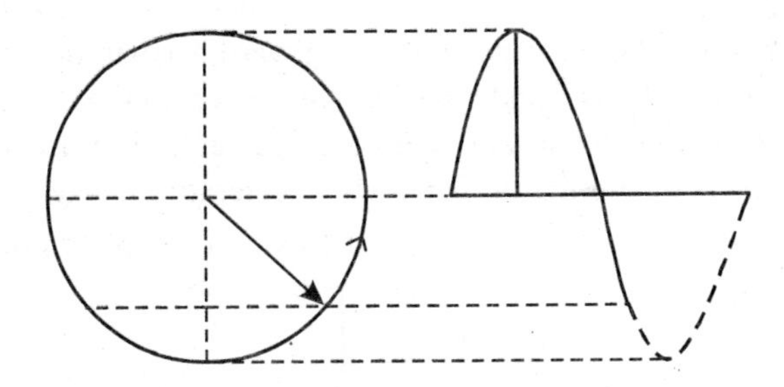

Fig. 2.21 The time variation of the projection of a rotating vector is sinusoidal.

REVIEW EXERCISES

2.1 The force constant of a spring shown in Fig. 2.22 is 50 Nm^{-1}. A block of mass 0.5 kg, attached to it is pulled through a distance of 0.01 m before being released. Calculate the (a) time period of oscillation, (b) frequency, (c) circular frequency, (d) velocity amplitude, (e) acceleration amplitude, (f) velocity and acceleration midway between the extreme and the mean position, (g) time required by the block to move half-way towards the centre from its initial position, and (h) total energy of the system.

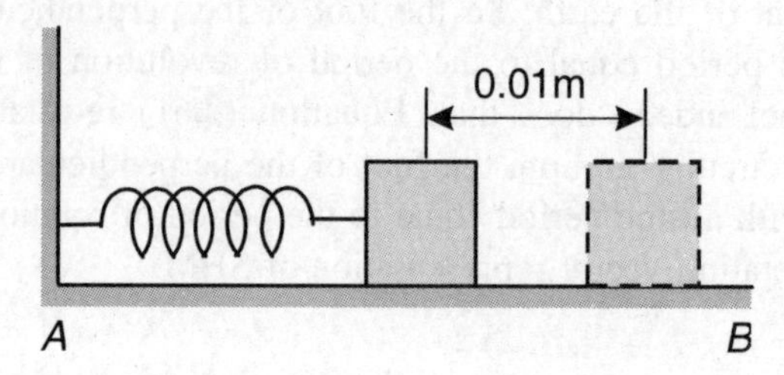

Fig. 2.22 The spring-mass system of Exercise 2.1; m = 0.5 kg. The surface AB is frictionless.

[*Ans.* (a) $\pi/5$ s; (b) $5/\pi\ \text{s}^{-1}$; (c) 10 s^{-1}; (d) 0.1 ms^{-1}; (e) 1 ms^{-2}; (f) $-0.0866\ \text{ms}^{-1}$, $-0.5\ \text{ms}^{-2}$; (g) $\pi/30$ s; (h) 1.25 J.]

2.2 If the spring of Exercise 2.1 has a circular frequency $\omega = 2\pi\ \text{rads}^{-1}$, then determine its displacement vs. time equation, given that at $t = 0$, the displacement is 0.2 m and velocity is $-(0.2)\ \pi\ \text{ms}^{-1}$.

$$\left[\textit{Ans.}\quad x = (0.2)\cos\left(2t + \frac{\pi}{6}\right)\text{m}\right]$$

2.3 If in Exercise 2.1, the block of mass m rests on a plate such that the coefficient of friction between the plate and the block is μ, calculate the maximum amplitude of oscillation that the spring and the plate-block system can have in terms of ω and μ if the block is not to slip on the plate.

[*Ans.* $\mu g/\omega^2$]

2.4 Refer to Fig. 2.23. A spring-mass system is attached to an elevator. Here m = 1 kg and k = 100 Nm^{-1}. The elevator is rising with an upward acceleration of $g/5$ and the spring is not vibrating. Take $g = 10\ \text{ms}^{-2}$ and calculate the (a) tension in the spring when the elevator is rising, (b) extension of the spring when the acceleration ceases, and (c) amplitude and the angular frequency of oscillation when the acceleration ceases.

[*Ans.* (a) 30 N; (b) 0.3 m; (c) 0.2 m, 10 s^{-1}]

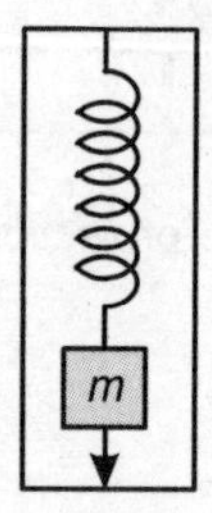

Fig. 2.23 A spring attached to the ceiling of an elevator.

2.5 A 2 kg mass is hanging in equilibrium from a vertical spring. An additional force further stretches the spring. When released from this position, it vibrates with a frequency 2 Hz and amplitude 2 cm. Calculate (a) the magnitude of the additional force applied; (b) the force exerted by the spring on the body when it is at the lowest, middle and highest point of the path and (c) the kinetic and the potential energies of the system when the body is respectively 2 cm and 1 cm below the middle of the path.

Ans. (a) 6.32 N; (b) 25.27 N, 18.95 N, 12.64 N; (c) At x = 2 cm, KE = ω, and PE = 0.0632 J; At x = 1 cm, KE = 0.9475 J, and PE = 0.0158 J.

2.6 Two particles of masses m_1 and m_2 lying on a smooth horizontal plane are connected by a light spring which is such that when m_1 is held fixed, m_2 oscillates harmonically in the line of the spring with period T_{12}. Show that if m_2 is held fixed, m_1 will oscillate harmonically with period T_{21}, such that $m_2 T_{21}^2 = m_1 T_{12}^2$.

2.7 (a) Refer to Figs. 2.24a and b. The horizontal surfaces are frictionless.

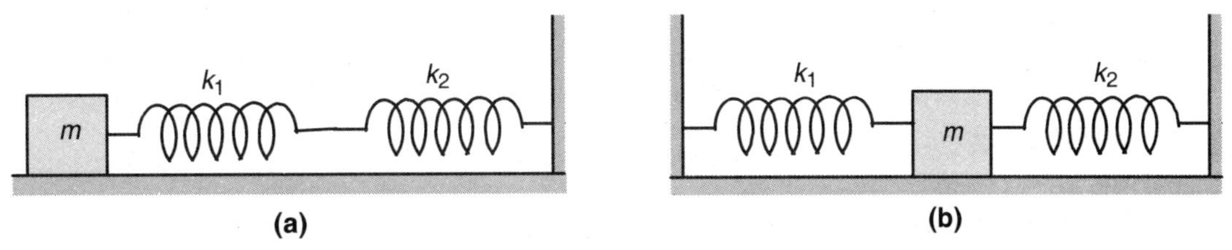

Fig. 2.24 Exercise 2.7.

Show that the time periods of oscillation in Figs. 2.24a and b respectively are

$$2\pi\left\{\frac{(k_1+k_2)m}{k_1 k_2}\right\}^{1/2} \text{ and } 2\pi\left(\frac{m}{k_1+k_2}\right)^{1/2}$$

(b) A uniform spring of length l has force constant k. The spring is cut into two pieces of unstressed lengths l_1 and l_2, where $(l_1/l_2) = (n_1/n_2)$. Express the force constants k_1 and k_2 of the two pieces in terms of k, n_1 and n_2. What happens when (i) $n_1 = n_2$ and (ii) $n_1/n_2 \to \infty$.

[***Ans.*** $k_1 = k\left(1+\frac{n_2}{n_1}\right)$; $k_2 = k\left(1+\frac{n_1}{n_2}\right)$; (i) $k_1 = k_2 = 2k$; (ii) $k_1 \to k$, $k_2 \to \infty$]

2.8 (a) A particle starts from rest and executes SHM in a straight line. Its displacement is p during the first second. It moves a further distance q in the same direction in the next second. Show that the amplitude of the motion is $2p^2/\{(3p+q)(p+q)\}^{1/2}$.

(b) In SHM, the distances of a particle from the middle point of its path at three consecutive seconds are respectively p, q and r. Show that frequency of oscillation is

$$\frac{1}{2\pi}\cos^{-1}\left(\frac{p+r}{2q}\right).$$

2.9 A particle is executing SHM with amplitude a about a mean position O. When it moves to a distance $\frac{a}{4}\sqrt{10}$ from O, it receives a push so that its velocity increases by 200%. What would be the amplitude of the resulting SHM?

[***Ans.*** $2a$]

2.10 A particle of mass m moves along x-axis under the influence of a force of magnitude k times its distance from a point C on the x-axis. The force is directed towards C, which moves along OC (O being the origin) with uniform acceleration f. If at $t = 0$, the point C was at rest at $x = 0$ and the particle at $x = a$ with velocity u, prove that the displacement vs. time equation of the particle is given by

$$x = \left(a + \frac{mf}{k}\right)\cos\left(\sqrt{\frac{k}{m}}t\right) + u\sqrt{\frac{m}{k}}\sin\left(\sqrt{\frac{k}{m}}t\right) - \frac{mf}{k} + \frac{1}{2}ft^2.$$

2.11 A particle is placed at the centre of a circular wire of radius R and mass per unit length λ. If the particle is displaced through a small distance r ($<< R$, such that $(r/R)^2 \to 0$) along a vertical line perpendicular to the plane of the wire and then released, show that it executes SHM with frequency $\frac{1}{R}\sqrt{\frac{\lambda G}{2\pi}}$, where G is the universal constant of gravitation.

2.12 A ball suspended by a thread 2 m long is deflected through an angle of 2° and then released. Assume that the subsequent motion is simple harmonic. Calculate the velocity of the ball when it passes through the mean position. Verify your result by applying energy considerations.

[**Ans.** $v = 0.154\ \text{ms}^{-1}$]

2.13 A simple pendulum of effective length l is suspended from the ceiling of a car. The car moves along a circular path of radius R with a uniform velocity v. If the pendulum is displaced slightly from its mean position, write down the equation of motion for its subsequent motion. Under what condition will the motion be simple harmonic? If the frequency of the SHM is f, show that $16\pi^4R^2l^2f^4 = g^2R^2 + f^4$.

2.14 A simple pendulum of length l is suspended from the ceiling of a lift. The lift starts rising with acceleration f and the pendulum is slightly displaced from its mean position. Write down the expression for the total mechanical energy of the pendulum at time t. Using the idea that the rate of change of energy is equal to force times velocity, obtain the equation of motion of the pendulum and show that for small oscillations, it executes SHM with period $2\pi\sqrt{\frac{l}{f+g}}$.

2.15 A hypothetical pendulum of infinite length is displaced slightly from its mean position. Write down its equation of motion and show that it executes SHM with a time period equal to that of an artificial satellite moving around the earth above the equator at a very small distance (compared to the radius of the earth) from the surface of earth.

3

Superposition of Harmonic Oscillations

Expected Learning Outcomes

In this Chapter, you will acquire capability to:

- state the principle of superposition;
- use the principle of superposition to obtain the resultant of two collinear harmonic oscillations of same as well as different frequencies;
- apply the principle of superposition to obtain the resultant of a number of collinear harmonic oscillations of same frequency;
- apply the principle of superposition to obtain the resultant of two mutually perpendicular harmonic oscillations having different amplitudes but same or different frequencies;
- depict formation of Lissajous' figures; and
- show equivalence of two mutually perpendicular harmonic oscillations with uniform circular motion.

3.1 INTRODUCTION

In the previous chapter, you learnt that under small oscillation approximation, motion of different physical systems is governed by a homogeneous second order differential equation. This corroborates an ideal situation and quite often, in situations of practical interest, we have to deal with a combination of two or more harmonic oscillations. For example, our ear drums receive a complex combination of harmonic oscillations. Similarly, you must have been enchanted while listening to a flute or any stringed instrument like guitar, sitar, and violin individually or in an orchestra. The resultant effect in such cases is obtained by the *principle of superposition*. In this chapter, you will learn about it for different combinations of frequencies, amplitudes or their relative orientations. As such, this principle finds wide applications in wave motion also. The phenomena of interference and diffraction are also caused by superposition of waves.

In Section 3.2, you will learn the principle of superposition and linearity of oscillations. In Section 3.3, you will learn that the sum of two collinear harmonic oscillations of same frequency is also a harmonic oscillation of that frequency, but the amplitude is different. However, superposition of collinear oscillations of different frequencies leads to an oscillation whose amplitude varies with time. That is, the resultant oscillation is periodic, but not simple harmonic. In Section 3.4, you will learn that when two mutually perpendicular harmonic oscillations of equal frequency but different amplitudes are superposed, the resultant motion is along a straight line or an ellipse. And if the frequencies are also not equal, we obtain Lissajous figures.

3.2 PRINCIPLE OF SUPERPOSITION AND LINEARITY

In the preceding chapter, you learnt about the rotating vector representation of SHM. Now, refer to Fig. 3.1. The vectors **A** and **B** represent two linear SHMs having a phase difference. These vectors rotate on the same plane with the same angular velocity about the centre of the circle. Their resultant **C** = **A** + **B** can be obtained by the parallelogram law of vectors. It also rotates with the same angular speed and their projection varies with time in a sinusoidal manner.

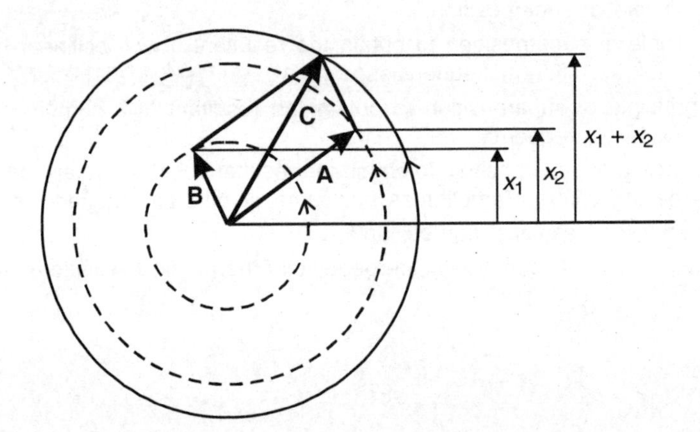

Fig. 3.1 Vector representation of two collinear SHMs having a phase difference.

Since vectors **A** and **B** follow the law of vector addition, the resultant motion can be obtained by adding the respective displacement versus time relations. In other words, if at a particular time the displacements of two SHMs are x_1 and x_2, the resultant displacement is obtained by their algebraic addition:

$$x = x_1 + x_2. \tag{3.1}$$

Note that this linear combination is tenable only for collinear oscillations.

3.3

SUPERPOSITION OF TWO COLLINEAR SIMPLE HARMONIC OSCILLATIONS OF SAME FREQUENCY

Let us begin by considering two collinear harmonic oscillations of same frequency but different amplitudes a_1 and a_2. Suppose that their initial phases differ by α. Then we can write

$$x_1 = a_1 \cos \omega t \tag{3.2}$$

and

$$x_2 = a_2 \cos (\omega t + \alpha). \tag{3.3}$$

Using the principle of superposition, we can write the resultant oscillation as

$$\begin{aligned} x &= x_1 + x_2 \\ &= a_1 \cos \omega t + a_2 \cos (\omega t + \alpha) \\ &= (a_1 + a_2 \cos \alpha) \cos \omega t - (a_2 \sin \alpha) \sin \omega t \end{aligned}$$

If we put $a_1 + a_2 \cos \alpha = A \cos \theta$, and $a_2 \sin \alpha = A \sin \theta$ (See Practice Exercise 1.2), we get

$$\begin{aligned} x &= A (\cos \omega t \cos \theta - \sin \omega t \sin \theta) \\ &= A \cos (\omega t + \theta). \end{aligned} \tag{3.4}$$

Note that Eq. (3.4) has the same form as that of the individual oscillations but different amplitude and phase. So we may conclude that the resultant of two harmonic oscillations of same frequency is also a harmonic oscillation of that frequency and is collinear with individual oscillations.

The amplitude of resultant oscillation is obtained by squaring the expressions for $A \cos \theta$ and $A \sin \theta$ and adding them:

$$\begin{aligned} A &= \sqrt{(A\cos\theta)^2 + (A\sin\theta)^2} \\ &= \sqrt{(a_1 + a_2\cos\alpha)^2 + (a_2\sin\alpha)^2} \\ &= \sqrt{a_1^2 + a_2^2 + 2a_1a_2\cos\alpha}. \end{aligned} \tag{3.5}$$

The phase of the resultant oscillation is given by

$$\tan \theta = \frac{A\sin\alpha}{A\cos\alpha} = \frac{a_2 \sin\alpha}{a_1 + a_2\cos\alpha}$$

or

$$\theta = \tan^{-1}\left(\frac{a_2 \sin\alpha}{a_1 + a_2\cos\alpha}\right). \tag{3.6}$$

The resemblance of Eqs. (3.5) and (3.6) with the expression of the magnitude and direction of the resultant of two vectors of magnitude a_1 and a_2 inclined at an angle α, corroborates the vector addition principle of obtaining linear combination of two SHMs. The amplitudes behave as vectors, and the phase difference is akin to the angle between the vectors.

Special cases

(i) When $\alpha = 0$, $\cos\alpha = 1$, $A = a_1 + a_2$, $\theta = 0$.
That is, when two oscillations are in phase, the amplitude of resultant oscillation is equal to the sum of the amplitudes of individual oscillations.

(ii) When $\alpha = \pi$, $\cos\alpha = -1$, $A = a_1 - a_2$, $\theta = 0$.
This result shows that when two oscillations are out of phase, the amplitude of resultant oscillation is equal to the difference of the amplitudes of individual oscillations.

(iii) When $\alpha = \dfrac{\pi}{2}$, $\cos\alpha = 0$, $A = \sqrt{a_1^2 + a_2^2}$, $\theta = \tan^{-1}\dfrac{a_2}{a_1}$.
You should now read Example 3.1 carefully.

EXAMPLE 3.1

Use the complex forms of representation of rotating vectors to obtain the value of $x = a_1 \cos \omega t + a_2 \cos(\omega t + \alpha)$.

Solution: The complex forms of rotating vectors for two SHMs are $a_1 e^{j\omega t}$ and $a_2 e^{j(\omega t + \alpha)}$. Then the result of superposition can be written as

$$x = \text{Real part of } \{a_1 e^{j\omega t} + a_2 e^{j(\omega t+\alpha)}\}$$

$$= \text{Re}\{e^{j\omega t}(a_1 + a_2 e^{j\alpha})\}$$

$$= \text{Re}(Z_1 Z_2), \text{ where } Z_1 = e^{j\omega t}, Z_2 = a_1 + a_2 e^{j\alpha}$$

From Eqs. (1.11a) and (1.11b), we recall that if $Z = Z_1 Z_2$, then $|Z| = |Z_1||Z_2|$ and Arg Z = Arg Z_1 + Arg Z_2. Proceeding further, we note that

$$|Z_1| = |e^{j\omega t}| = 1$$

$$|Z_2| = |a_1 + a_2 e^{j\alpha}| = |(a_1 + a_2\cos\alpha) + j(a_2 \sin\alpha)|$$

$$= \{(a_1 + a_2\cos\alpha)^2 + (a_2\sin\alpha)^2\}^{1/2}$$

$$= (a_1^2 + a_2^2 + 2a_1 a_2 \cos\alpha)^{1/2}$$

$$\therefore \quad |Z| = (a_1^2 + a_2^2 + 2a_1 a_2\cos\alpha)^{1/2} = A, \tag{3.7}$$

and

$$\text{Arg } Z_1 = \omega t$$

$$\text{Arg } Z_2 = \tan^{-1}\left(\frac{a_2\sin\alpha}{a_1 + a_2\cos\alpha}\right) = \theta, \text{ say}$$

$$\therefore \quad \text{Arg } Z = (\omega t + \theta) \tag{3.8}$$

and

$$x = \text{Re}\left\{|Z| e^{j(\text{Arg } Z)}\right\} = \text{Re}\left\{A e^{j(\omega t+\theta)}\right\}$$

Hence

$$x = A\cos(\omega t + \theta),$$

where A and θ are given by Eqs. (3.7) and (3.8).

You now know how to obtain the resultant of two collinear harmonic oscillations of equal frequencies but different amplitudes. We now extend the same principle to study superposition of a number of collinear harmonic oscillations of same frequency.

3.4 SUPERPOSITION OF *N* COLLINEAR HARMONIC OSCILLATIONS OF SAME FREQUENCY

Let us now consider n collinear harmonic oscillations, each of frequency ω but different amplitudes and initial phases. We represent these as

$$x_1 = a_1 \cos(\omega t + \alpha_1),$$
$$x_2 = a_2 \cos(\omega t + \alpha_2),$$
$$x_3 = a_3 \cos(\omega t + \alpha_3), \tag{3.9}$$
$$x_n = a_n \cos(\omega t + \alpha_n).$$

The resultant oscillation is obtained by their superposition:

$$x = x_1 + x_2 + x_3 + \cdots + x_n$$

or
$$x = a_1 \cos(\omega t + \alpha_1) + a_2 \cos(\omega t + \alpha_2) + a_3 \cos(\omega t + \alpha_3) + \cdots$$

As before, we use the concept of rotating vector representation to handle the superposition. Since these oscillations are collinear, the associated rotating vectors will be coplanar, rotating with constant angular velocity ω about the same origin.

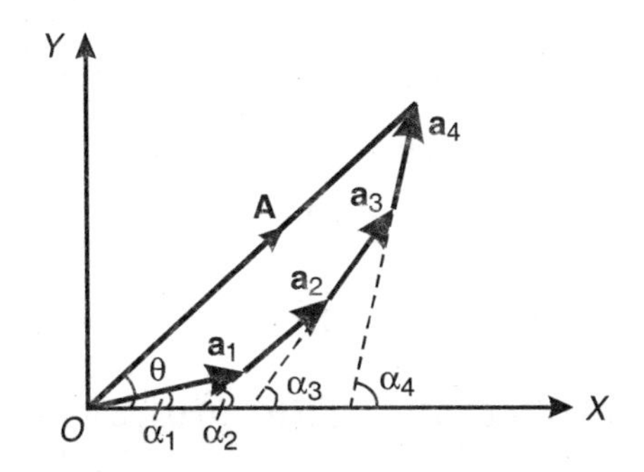

Fig. 3.2 Addition of n harmonic oscillations.

Refer to Fig. 3.2. It shows the initial orientation of rotating vectors $\mathbf{a}_1$, $\mathbf{a}_2$, $\mathbf{a}_3$, ... (at angles α_1, α_2, α_3, ..., respectively) with the x-axis. The resultant A makes an angle θ with the positive direction of the x-axis. By equating the x and y components of A with the algebraic sums of the said components of the vectors $\mathbf{a}_1$, $\mathbf{a}_2$, ..., we get

$$A \cos\theta = a_1 \cos\alpha_1 + a_2 \cos\alpha_2 + \cdots = \sum a \cos\alpha,$$

and
$$A \sin\theta = a_1 \sin\alpha_1 + a_2 \sin\alpha_2 + \cdots = \sum a \sin\alpha.$$

$\therefore$
$$A^2 = \left(\sum a \cos\alpha\right)^2 + \left(\sum a \sin\alpha\right)^2. \tag{3.9}$$

The resultant motion will be simple harmonic and the displacement versus time relation is given by

$$x(t) = A \cos(\omega t + \theta).$$

The amplitude A is defined by Eq. (3.9) and phase θ of the resultant oscillation is given by

$$\theta = \tan^{-1}\left(\frac{\sum A \sin \alpha}{\sum A \cos \alpha}\right). \tag{3.10}$$

Now, read the following examples carefully.

EXAMPLE 3.2

Calculate the amplitude and initial phase of the harmonic oscillation obtained by the superposition of identically directed oscillations corresponding to the equations: $x_1 = 0.02 \sin\left(5\pi t + \frac{\pi}{2}\right)$ m and $x_2 = 0.03 \sin\left(5\pi t + \frac{\pi}{4}\right)$ m.

Solution: We know that

$$x_1 = 0.02 \sin\left(5\pi t + \frac{\pi}{2}\right) = 0.02 \cos 5\pi t$$

and
$$x_2 = 0.03 \sin\left\{\left(5\pi t - \frac{\pi}{4}\right) + \frac{\pi}{2}\right\} = 0.03 \cos\left(5\pi t - \frac{\pi}{4}\right).$$

Thus, if we write $x_1 = a_1 \cos \omega t$, and $x_2 = a_2 \cos(\omega t + \delta)$, we get

$$a_1 = 0.02 \text{ m}, a_1 = 0.03 \text{ m}, \omega = 5\pi, \delta = -\frac{\pi}{4}.$$

$\therefore$ The resulting oscillation would be given by

$$x = a \cos(\omega t + \delta),$$

where

$$a^2 = a_1^2 + a_2^2 + 2a_1 a_2 \cos \delta$$

$$= (0.02)^2 + (0.03)^2 + 2(0.02)(0.03) \cos\left(-\frac{\pi}{4}\right)$$

$$= (4 + 9 + 6\sqrt{2}) \times 10^{-4} \text{m}^2$$

$$= (13 + 6\sqrt{2}) \times 10^{-4} \text{m}^2$$

$\therefore$
$$a = 4.6 \times 10^{-2} \text{ m}.$$

and
$$\tan \delta = \frac{a_2 \sin \delta}{a_1 + a_2 \cos \delta} = \frac{0.03 \sin(-\pi/4)}{0.02 + 0.03 \cos \pi/4}$$

$$= \frac{-3 \cdot (1/\sqrt{2})}{2 + 3 \cdot (1/\sqrt{2})} = \frac{-3}{2\sqrt{2} + 3}$$

$\therefore$
$$\delta = 152.8°.$$

EXAMPLE 3.3

n collinear harmonic oscillations of same amplitude and frequency but having a constant phase difference are superposed. Calculate the amplitude of the resultant, using the complex form of representation of SHM.

Solution: Let the complex form of the individual SHMs be

$$x_1 = ae^{j\omega t},\ x_2 = ae^{j(\omega t+\alpha)},\ x_3 = ae^{j(\omega t+2\alpha)}, \ldots, x_n = ae^{j(\omega t+\overline{n-1}\cdot\alpha)}$$

The resultant,

$$\begin{aligned} x &= x_1 + x_2 + \cdots + x_n \\ &= ae^{j\omega t}(1 + e^{j\alpha} + e^{2j\alpha} + \cdots + e^{(n-1)j\alpha}) \\ &= a\cdot\left(\frac{1-e^{jn\alpha}}{1-e^{j\alpha}}\right)e^{j\omega t} = \mathbf{A}e^{j\omega t} \end{aligned}$$

$$\begin{aligned} |\mathbf{A}|^2 &= AA^* = a^2\frac{(1-e^{jn\alpha})(1-e^{-jn\alpha})}{(1-e^{j\alpha})(1-e^{-j\alpha})} \\ &= a^2\frac{\{1-(e^{jn\alpha}+e^{-jn\alpha})+1\}}{\{1-(e^{j\alpha}+e^{-j\alpha})+1\}} \\ &= a^2\frac{(2-2\cos n\alpha)}{(2-2\cos\alpha)} = a^2\frac{1-\cos n\alpha}{1-\cos\alpha} = a^2\frac{\sin^2 n\alpha/2}{\sin^2\alpha/2} \end{aligned}$$

$\therefore$ Resultant amplitude $= a\dfrac{\sin n\alpha/2}{\sin\alpha/2}$.

3.5 SUPERPOSITION OF TWO COLLINEAR HARMONIC OSCILLATIONS OF NEARLY EQUAL FREQUENCIES

Let us consider two harmonic oscillations of nearly equal frequencies ν and $\nu + \nu'$. Then we can write

$$x_1(t) = a_1 \sin 2\pi\nu t = a_1 \sin ft \tag{3.11}$$

and

$$x_2(t) = a_2 \sin 2\pi(\nu+\nu')t = a_2 \sin(f+g)t, \tag{3.12}$$

where we have put $f = 2\pi\nu$ and $g = 2\pi\nu'$. We further assume that $a_1 > a_2$. On superposition, the instantaneous displacement of resultant oscillation is given by

$$\begin{aligned} x(t) &= x_1 + x_2 = a_1 \sin ft + a_2 \sin(f+g)t \\ &= (a_1 + a_2\cos gt)\sin ft + (a_2 \sin gt)\cos ft. \end{aligned}$$

To simplify this expression, we put

$$a_1 + a_2 \cos gt = R\cos\theta \tag{3.13}$$

and

$$a_2 \sin gt = R\sin\theta. \tag{3.14}$$

This leads to a compact expression for instantaneous displacement of resultant oscillation:

$$x(t) = R \sin (ft + \theta). \tag{3.15}$$

where amplitude of the resultant oscillation is given by

$$R = \sqrt{(R \cos \theta)^2 + (R \sin \theta)^2}$$

$$= \sqrt{a_1^2 + a_2^2 + 2a_1 a_2 \cos gt}. \tag{3.16}$$

The expression for phase has the form

$$\theta = \tan^{-1}\left(\frac{a_2 \sin gt}{a_1 + a_2 \cos gt}\right) \tag{3.17}$$

Let us pause for a while and ask: Does the resultant displacement given by Eq. (3.15) represent simple harmonic motion? By looking at it, you may be tempted to conclude that it is so. But we note that neither R nor θ is a constant; both vary with time. It means that the resultant of two harmonic oscillations of nearly equal frequencies does not lead to simple harmonic motion.

You should now read the following example carefully.

EXAMPLE 3.4

Calculate the maximum and minimum values of R in Eq. (3.15). What are the corresponding values of gt?

Solution: The maximum value of R is obtained when cos gt is +1:

$$R_{\text{max}} = a_1 + a_2 \text{ for } \cos gt = 1, \text{ i.e., } gt = 2n\pi.$$

The minimum value of R is obtained when cos gt is –1:

$$R_{\text{min}} = a_1 - a_2 \text{ for } \cos gt = -1, \text{ i.e., } gt = (2n + 1)\pi.$$

Here we note that the value of R fluctuates between its minimum and maximum values. The time between two successive maxima and minima is given by

$$gt = 2\pi$$

or

$$t = \frac{2\pi}{g} = \frac{1}{\nu'}. \tag{3.18}$$

In your school physics course, you have learnt that when two sound waves of slightly different frequencies are excited, say using two tuning forks, and made to superpose, the intensity fluctuates periodically between maximum and minimum; the periodicity being equal to the difference between the superimposed frequencies. This phenomenon of periodic waxing and waning of sound is known as *beat*. The difference of the superimposed frequencies is called the *beat frequency*. Equation (3.18) defines *beat period*. This result is depicted graphically in Fig. 3.3 when frequencies of the oscillations are in the ratio 4 : 5. Note that between two successive maxima (or minima), the resultant has four oscillations of A and five of B. We shall discuss it in detail in Chapter 8.

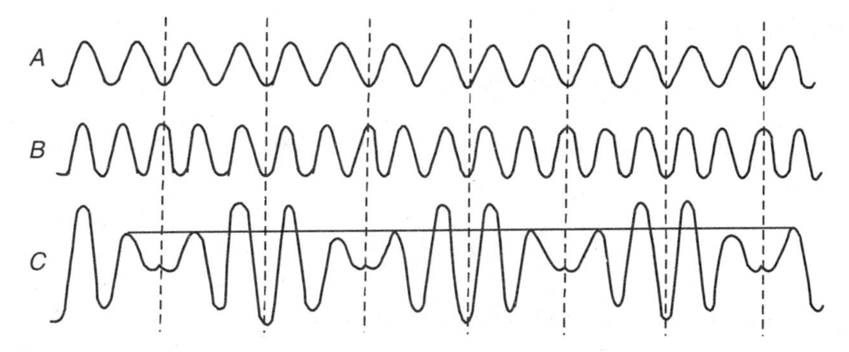

Fig. 3.3 Time variation of the resultant of two oscillations of nearly equal frequencies. *A* and *B* are component vibrations.

So far we have confined our discussion to collinear oscillations, i.e., we have been dealing with one-dimensional systems. But it is also possible to excite oscillatory motion in two-dimensions by allowing the bob of a simple pendulum to swing freely in *x-y* plane. We, therefore, consider superposition of two mutually perpendicular harmonic oscillations of same frequency.

3.6 SUPERPOSITION OF TWO MUTUALLY PERPENDICULAR HARMONIC OSCILLATIONS

3.6.1 Oscillations of Equal Frequencies

Let us consider two oscillations of same frequency acting at right angles; one along x-axis and the other along y-axis. If their amplitudes are a and b such that $b < a$, we can represent these by the equations

$$x = a \sin \omega t \tag{3.19}$$

and

$$y = b \sin (\omega t - \alpha). \tag{3.20}$$

Note that the initial phase of the oscillation along x-axis is zero and the phase of oscillation along y-axis is α. That is, the initial phase difference between two orthonormal oscillations is α.

We wish to know the nature of oscillations of a particle or path followed by it when it is subject simultaneously to two orthogonal oscillations. It implies that we wish to obtain a relation between displacements along x and y-axes. For this we have to eliminate terms containing t in Eqs. (3.19) and (3.20). From Eq. (3.19), we note that

$$\sin \omega t = \frac{x}{a}. \tag{3.21}$$

$$\therefore \qquad \cos \omega t = \sqrt{1 - \frac{x^2}{a^2}}. \tag{3.22}$$

We can rewrite Eq. (3.20) as

$$\frac{y}{b} = \sin \omega t \cos \alpha - \cos \omega t \sin \alpha.$$

On inserting the values of $\sin \theta$ and $\cos \theta$ respectively from Eqs. (3.21) and (3.22) in this expression, we get

$$\frac{y}{b} = \frac{x}{a} \cos \alpha - \sqrt{1 - \frac{x^2}{a^2}} \sin \alpha$$

On rearranging terms and squaring both sides, we can write

$$\left(1 - \frac{x^2}{a^2}\right) \sin^2 \alpha = \left(\frac{x}{a} \cos \alpha - \frac{y}{b}\right)^2$$

or

$$\left(1 - \frac{x^2}{a^2}\right) \sin^2 \alpha = \frac{x^2}{a^2} \cos^2 \alpha + \frac{y^2}{b^2} - \frac{2xy}{ab} \cos \alpha$$

Hence

$$\frac{x^2}{a^2} + \frac{y^2}{b^2} - \frac{2xy}{ab} \cos \alpha = \sin^2 \alpha. \tag{3.23}$$

Equation (3.23) represents an ellipse (Fig. 3.4) inscribed within a rectangle of sides $2a$ and $2b$. The major axis of the ellipse is inclined at an angle ϕ with the x-axis such that

$$\tan 2\phi = \frac{2ab \cos \alpha}{a^2 - b^2} \tag{3.24}$$

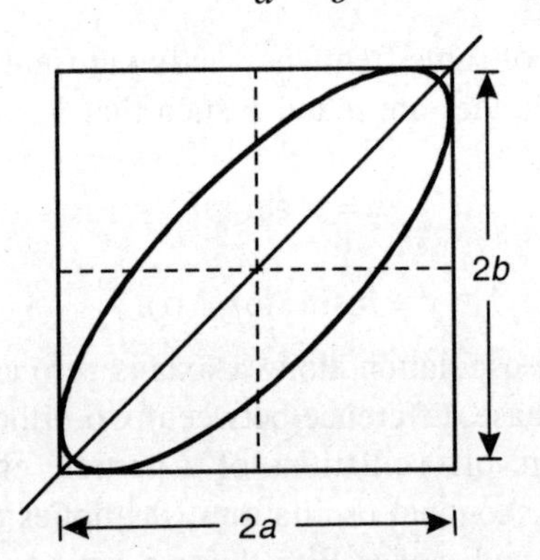

Fig. 3.4 Plot of Eq. (3.23) is an ellipse.

We may now conclude that when two mutually perpendicular linear harmonic oscillations of unequal amplitudes and having finite non-zero initial phase difference are superposed, the resultant motion is along an ellipse whose principal axes lie along x and y-axes. The semi-major and semi-minor axes of the ellipse are a and b.

Let us now consider a few special cases.

***Case* (i)** When the oscillations are in phase, i.e., $\alpha = 0$

From Eq. (3.23), we get

$$\left(\frac{x}{a}\right)^2 + \left(\frac{y}{b}\right)^2 - \frac{2xy}{ab} = 0$$

or

$$\left(\frac{x}{a} - \frac{y}{b}\right)^2 = 0.$$

That is,

$$y = \frac{b}{a}x.$$

It shows that motion is to and fro along a straight line passing through the origin (Fig. 3.5a).

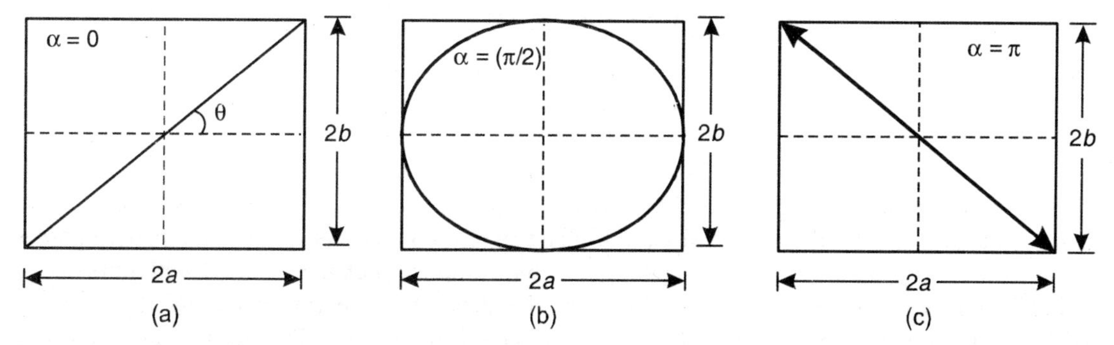

Fig. 3.5 Resultant of superposition of mutually perpendicular oscillations of same frequency when initial phase difference between them is (a) 0, (b) $\pi/2$ and (c) π.

***Case* (ii)** When $\alpha = \pi/2$,

From Eq. (3.23), we get

$$\frac{x^2}{a^2} + \frac{y^2}{b^2} = 1.$$

This represents an ellipse with semi major and minor axes respectively equal to a and b. From Eq. (3.24), we get $\phi = 0$, which means that the axes of the ellipse coincide with the perpendicular axes. It shows that when two orthogonal oscillations of unequal amplitude having initial phase difference of $\pi/2$ are superposed, the resultant oscillation is along an ellipse whose principal axes lie along x- and y-axes, as shown in Fig. 3.5b.

If we further have $a = b$, the equation of ellipse reduces to $x^2 + y^2 = a^2$, which represents a circle of radius a with centre at the origin. This corresponds to uniform circular motion with angular velocity ω.

***Case* (ii)** When $\alpha = \pi$, i.e., the oscillations are in opposite phase

From Eq. (3.23), we get

$$\left(\frac{x}{a}\right)^2 + \left(\frac{y}{b}\right)^2 + \frac{2xy}{ab} = 0$$

or

$$\left(\frac{x}{a}+\frac{y}{b}\right)^2 = 0$$

or

$$y = -\frac{b}{a}x.$$

That is, the oscillations are along a straight line passing through the origin (Fig. 3.5c).

3.6.2 Oscillations of Slightly Different Frequencies: Lissajous' Figures

Let us suppose that the displacements of the orthogonal oscillations are given by

$$x = a \sin \omega t$$

and

$$y = b \sin (\omega' t - \alpha).$$

The phase difference between them is given by

$$\delta = (\omega' t - \alpha) - \omega t$$

$$= (\omega' - \omega)t - \alpha.$$

Since the superposed orthogonal oscillations are of slightly different frequencies, one of them changes faster than the other and will gain in phase over the other. As a result, the pattern passes through different phases. The resultant pattern is quite complex; it changes slowly with time due to gradually changing phase difference. However, the general shape traced out by the resultant oscillation is similar to that obtained for the case of equal frequencies. That is, motion is confined within the rectangle of sides $2a$ and $2b$.

For different values of δ, the curves traced out the resultant of two mutually perpendicular oscillations are shown in Fig. 3.6. These are known as *Lissajous' figures*. As may be seen, for $\delta = 0$, the motion is along a straight line (Fig. 3.6a). This is analogous to that shown in Fig. 3.4a. For $\delta = \pi/4$, it opens out and takes the shape of an ellipse, i.e., the motion is along an ellipse whose major and minor axes are inclined with the co-ordinate axes (Fig. 3.6b). For $\delta = \pi/2$, the axes of

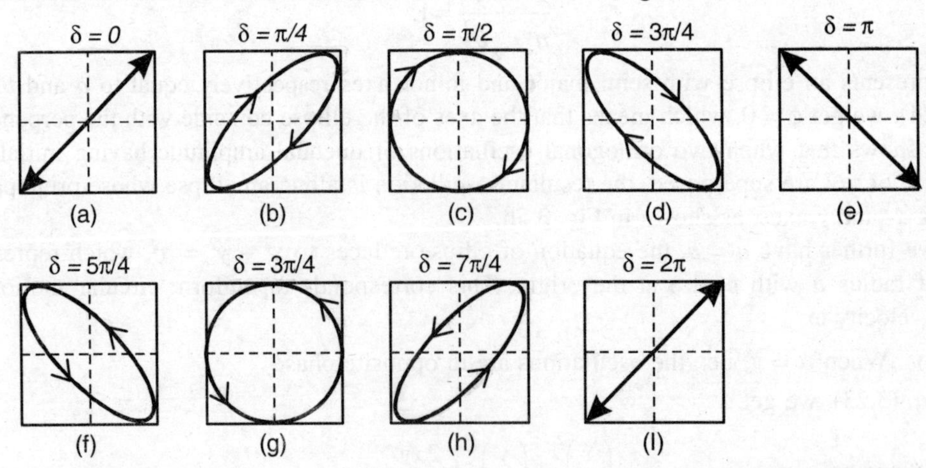

Fig. 3.6 Lissajous' figures for different values of phase difference between individual mutually perpendicular oscillations.

the ellipse coincide with the co-ordinate axes (Fig. 3.6c). For $\delta = 3\pi/4$, we get an ellipse as in the case of (b) but the major axis is in the second and the fourth quadrants (Fig. 3.6d). For $\delta = \pi$, we get a straight line passing through the origin but lying in the second and fourth quadrants (Fig. 3.6e). For $\delta > \pi$, the sequence of the figures is repeated but with opposite orientation. However, for $\delta = 2\pi$, we arrive at exactly the same position as that of Fig. 3.6a.

If the time taken between (a) and (i) is τ seconds, the frequency difference is $1/\tau$. If one of the frequencies is known, the other can be determined.

Let us now work out a simple example.

EXAMPLE 3.5

A point is under the influence of two simultaneous simple harmonic motions in mutually perpendicular directions given by $x = \cos \pi t$, $y = \cos \pi t/2$. Find the trajectory of the resulting motion of that point.

Solution: If we put $\pi t/2 = \theta$, we find that $x = \cos 2\theta$ and $y = \cos \theta$

$$\therefore \qquad x = 2\cos^2\theta - 1 = 2y^2 - 1.$$

i.e., $2y^2 = x + 1$, which is the equation of a parabola.

You may now like to solve a Practice Exercise.

***Practice Exercise* 3.1** A point participates simultaneously in two mutually perpendicular oscillations $x = \sin \pi t$ and $y = 2 \sin (\pi t + \pi/2)$. Draw its trajectory.

[***Ans.*** $4x^2 + y^2 = 4$]

Lissajous' figures corresponding to the frequencies whose ratio is an integer are particularly important. The final shape of the curve depends on this ratio as well as the position and phase relation at the initial instants. We would like you to convince yourself by working out the shape by taking the ratio as 2 and 3.

Lissajous figures can be seen by using a cathode ray oscilloscope (CRO). The different alternating sinusoidal voltages are applied at the *XX* and *YY* deflection plates of the CRO. We then get to see the trace of the resultant effect in the form of an electron beam on the fluorescent screen. When the applied voltages have the same frequency, we obtain various curves shown in Fig. 3.6 by adjusting the phases and the amplitudes. We now discuss the equivalence of two harmonic oscillations at right angles with uniform circular motion.

3.7 EQUIVALENCE OF UNIFORM CIRCULAR MOTION AND MUTUALLY PERPENDICULAR HARMONIC OSCILLATIONS

Let us assume that the mutually perpendicular harmonic oscillations have same frequency and amplitude but an initial phase difference of $\pi/2$. Then we can represent these as

$$x(t) = a \sin \omega t \qquad (3.25)$$

and

$$y(t) = a \sin \left(\omega t + \frac{\pi}{2}\right) = a \cos \omega t. \qquad (3.26)$$

We eliminate t between these equations. To do so, we square and add them. This leads to the relation

$$x^2 + y^2 = a^2. \tag{3.27}$$

Do you recognise this equation? It represents a circle of radius a (See Fig. 3.7). To determine the sense of rotation, we put $t = 0$ in Eqs. (3.25) and (3.26). This gives $x = 0$, $y = a$, which are the co-ordinates of point Q.

Again at $t = T/4$, $x = a \sin \frac{\omega T}{4} = a \sin \frac{\pi}{2} = a$ and $y = a \cos \frac{\omega T}{4} = a \cos \frac{\pi}{2} = 0$, which are the co-ordinates of point P. It means that the circle is being described in the clockwise sense.

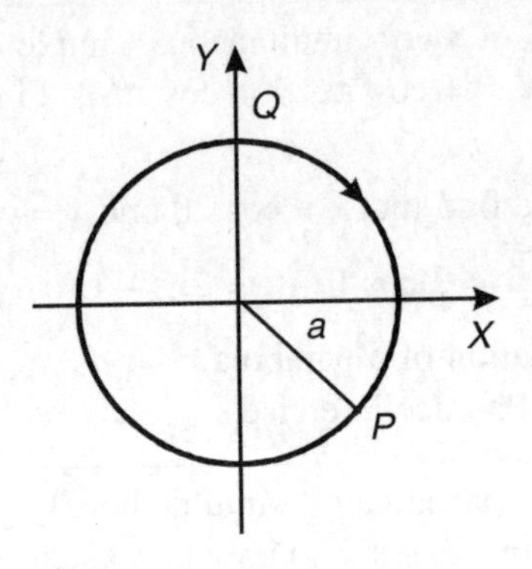

Fig. 3.7 Plot of Eq. (3.27) is a circle of radius a.

You may now like to answer a Practice Exercise.

***Practice Exercise* 3.2** Show that the oscillations represented by $x = a \sin \omega t$ and $y = -a \cos \omega t$ correspond to circular motion described in the anticlockwise sense.

REVIEW EXERCISES

3.1 Apply the method described in Example 3.1 to obtain the result of superposition of harmonic oscillations given by

$$y_1 = a_1 \sin \omega t$$

and

$$y_2 = a_2 \sin (\omega t + \alpha).$$

3.2 A particle of mass m moves along a trajectory given by

$$x = x_0 \cos \omega_1 t$$

and

$$y = y_0 \sin \omega_2 t.$$

(a) Calculate the x and y components of the applied force.
(b) Obtain an expression for PE as a function of x and y.
(c) Determine the KE of the particle.
(d) Show that the total energy of the particle is conserved.

[**Ans.**

(a) $F_x = -m\omega_1{}^2 x,\ F_y = -m\omega_1{}^2 y$

(b) $\text{PE} = \dfrac{m}{2}[x_0^2\omega_1^2 \cos^2 \omega_1 t + y_0^2\omega_2^2 \sin^2 \omega_2 t]$,

(c) $\text{KE} = \dfrac{m}{2}[x_0^2\omega_1^2 \sin^2 \omega_1 t + y_0^2\omega_2^2 \cos^2 \omega_2 t]$,

(d) Total energy $= \dfrac{m}{2}(x_0^2\omega_1^2 + y_0^2\omega_2^2)$ = constant]

3.3 Write the equation of the resultant oscillation obtained by superposition of two identical harmonic oscillations of same period (= 10 s) and amplitude 0.01 m. The difference of phase between them is $\pi/4$ and the epoch of one of them is zero.

$$\left[\textbf{Ans.}\quad x = \left(0.02\cos\frac{\pi}{8}\right)\cos\left(\frac{\pi t}{5} + \frac{\pi}{8}\right)\text{m}\right]$$

3.4 The resultant oscillation of two identical harmonic oscillations directed in the same direction has the same period and amplitude as that of the component SHMs. Calculate the difference in their initial phases.

$$\left[\textbf{Ans.}\quad \delta = \frac{2\pi}{3}\right]$$

3.5 Calculate the amplitude and epoch of the resultant of two identical harmonic oscillations given by $x_1 = 4 \sin \pi t$ and $x_2 = 3\sin\left(\pi t + \dfrac{\pi}{2}\right)$. The displacements are in cm and t is in seconds.

$$\left[\textbf{Ans.}\quad a = 5\text{ cm};\ \theta = \tan^{-1}\frac{3}{4}\right]$$

3.6 An oscillation is described by the equation $x = A \sin \omega_1 t$, where A is a function of time: $A = A_0 (1 + \cos \omega_2 t)$, A_0 being a constant. Determine the nature of the oscillation and plot it for $A_0 = 5$ cm, $\omega_1 = 4\pi$ rad s^{-1} and $\omega_2 = 2\pi$ rad s^{-1}.

[**Ans.** Modulated oscillations will have amplitudes A_0, $A_0/2$, $A_0/2$ and frequencies ω_1, $\omega_1 + \omega_2$, $\omega_1 - \omega_2$, respectively]

3.7 Determine the trajectory of motion of a point which participates simultaneously in two mutually perpendicular oscillations given by

(i) $x = 2 \sin \omega t$, $y = 2 \cos \omega t$;

(ii) $x = \sin \omega t$, $y = 4 \sin (\pi t + \pi)$.

Displacement and the time are given in metres and seconds.

[**Ans.** (i) $x^2 + y^2 = 4$; (ii) $y = -4x$]

3.8 Two rectangular harmonic oscillations of amplitudes a_1 and a_2 have frequencies 2ω and ω, respectively. If their initial phases are δ and zero, respectively, determine the resultant motion and draw diagrams corresponding to $\delta = 0$, $\pi/2$ and π.

$$\left[\textbf{Ans.}\quad 2ay^2 = b^2(x-a),\ \left(\frac{x}{a}\right)^2 = 4\left(1-\frac{y^2}{b^2}\right)\frac{y^2}{b^2},\ 2ay^2 = b^2(a-x)\right]$$

3.9 Show that $ae^{j\omega t}$ and $ae^{-j\omega t}$ represent anticlockwise and clockwise rotations respectively.

3.10 Use the identity, $\cos \omega t = \frac{1}{2}(e^{j\omega t} + e^{-j\omega t})$ to prove that a simple harmonic motion is equivalent to two oppositely directed circular motions of diameters equal to the amplitude of the SHM.

4

Damped Oscillations

EXPECTED LEARNING OUTCOMES

In this Chapter, you will acquire capability to:

- write the equation of motion of a damped oscillator and solve it;
- differentiate between weakly damped, critically damped and overdamped systems;
- critically discuss the effect of damping on amplitude, energy and period of oscillation of a damped oscillator; and
- analyse a weakly damped system in terms of logarithmic decrement and quality factor.

4.1 INTRODUCTION

In Chapter 2, you learnt the characteristics of simple harmonic motion (SHM). You may recall that the displacement-time graph of such a system exhibits a sine (or cosine) curve and the total energy of the oscillator does not change with time. It means that once such a system has been set in motion, it should continue to oscillate indefinitely. Such oscillations are called *free oscillations*. But in practice, we can not think of any real physical system which executes completely free oscillatory motion that persists indefinitely. You now also know that the superposition of two free collinear oscillations of different frequencies leads to formation of beats. However, superposition of orthogonal oscillations of different amplitudes results in Lissajous figures; their nature is determined by the phase difference between the superposing oscillations.

Recall the oscillations of a simple pendulum or vibrations of the prongs of a tuning fork when left to it. In these systems, amplitude of oscillations is seen to decrease gradually depending on the drag exerted by the medium, say air. It means that as time passes, every oscillating system loses energy due to the presence of frictional forces. Such a motion is called *damped motion*. You may now like to know: Where does this energy go? The energy of the oscillating system is spent in overcoming damping and is dissipated to the surrounding environment. Do you know that, in general, loss of energy due to damping is wasteful and we make every effort to minimise it? But in some engineering systems, we knowingly introduce damping. A familiar example is that of brakes in automobiles. When we apply brakes, we increase friction between the tyres and the road. This helps to reduce the speed of a vehicle in a short time. Otherwise, no vehicular traffic would have become possible. Similarly, dampers are introduced in a system for vibration control.

In this Chapter, you will learn about the salient features of damped harmonic motion and appreciate how these differ from those of free oscillations. In Section 4.2, you will learn about different types of damping forces. These, in general, are quite complicated in real systems. But while accounting for these, we model them depending on our experience, knowledge of the process(es) involved and mathematical convenience. In Section 4.3, we have discussed the equation of motion of a damped oscillator. You will learn to solve it for different types of damping in Section 4.4 and discover different behaviours. From Chapter 2, you may recall that the damped systems of widely different nature can be analysed by analogy to a mechanical system. In Section 4.5, we discuss non-mechanical damped systems, establish their equations of motion and write their solutions by analogy. While solving the equation of motion of a damped system, you will discover that weak damping leads to oscillatory motion whose amplitude diminishes gradually. We quantify weak damping in terms of logarithmic decrement, which is related to the observed behaviour of amplitude of oscillations, relaxation time and quality factor. These are also intimately related to energy of the system. In Section 4.6, you will learn to derive an expression for the energy of a weakly damped system, averaged over one cycle. In Section 4.7, we discuss logarithmic decrement and quality factor. This is followed by solved examples and practice exercises to help you get an idea about the numbers involved and fix your ideas.

4.2 TYPES OF DAMPING FORCES

You now know that every physical system experiences damping, which depends on the system under consideration and, in general, can be quite complex. In your earlier classes, you have learnt about some of these. A familiar example that you studied in Chapter 2 is a spring-mass system executing longitudinal oscillations on a horizontal surface. The oscillating mass experiences frictional force from the surface, which opposes its motion. It is referred to as *Coulomb friction force*. We normally model it as a force of constant magnitude.

In your school physics, you have learnt about Millikan's oil drop experiment. A charged oil drop experiences viscous drag when it is made to fall freely through an electric field. The damping force is proportional to velocity of the body (Stoke's law) and is referred to as the *viscous damping force*. The direction of the resistive force is opposite to that of velocity. When an aircraft begins its descend or a spacecraft enters the atmosphere of the earth, the magnitude of upward thrust, which acts as resistive force, can be very large. Usually, the space-craft experiences large stress which raises its outside body temperature. In fact, the disaster that hit space-craft Columbia and killed all astronauts on board, including an Indian astronaut, Kalpana Chawla, was partly due to the impact of viscous drag.

In a solid substance, some part of energy may be lost due to imperfections or dislocations. It is not easy to estimate or model what is known as *structural damping*. However, it is usually taken as independent of frequency and proportional to amplitude.

Note that damping is not a peculiar characteristic of mechanical systems only. It is encountered in an *LCR*-circuit, ballistic galvanometer, magnetometer, fluxmeter etc., where the damping is electromagnetic in nature. You will learn about these in this Chapter in detail a little later. In general, inclusion of damping force makes mathematical analysis somewhat difficult. But for simplicity, it is customary to model it by an equivalent viscous damping. In our discussion, we make the simplifying assumption that velocity of the moving part of the system is small so that the damping force can be taken to be linear in velocity. We find an analogous situation in case of electromagnetic oscillations.

4.3 EQUATION OF MOTION OF A 1-D DAMPED OSCILLATOR

To study the effect of damping on the motion of a 1-D oscillator such as a spring-mass system, refer to Fig. 4.1. The mass is made to move in a viscous medium, such as a lubricated cylinder or filled with some other viscous substance. (You can alternatively attach a damper between the mass and the vertical rigid support.) We have taken the X-axis to be along the length of the spring. We

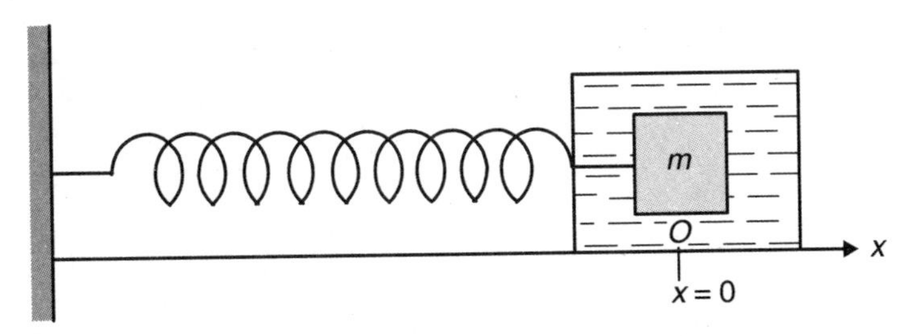

Fig. 4.1 A damped spring-mass system; the mass is immersed in a viscous medium.

mark a point O to mark the equilibrium position of mass (corresponding to its position of rest) and define the origin of the co-ordinate system by this point. That is, at equilibrium, O lies at $x = 0$. You may recall from Chapter 2 that the force law for an undamped spring-mass system executing SHM is given by

$$F = -kx.$$

Here F is the restoring force and x is the instantaneous displacement from the equilibrium position. In considering the motion of a damped oscillator, we need to ask: Is the restoring force in this case also linearly proportional to the displacement of mass from its equilibrium position? If not, what modification is necessary to describe the motion of the system under consideration quantitatively?

To discover answers to these questions, we note that as mass oscillates, it experiences a drag, which tends to damp motion. We denote its magnitude by F_d. You may like to know: How to predict the magnitude of this force? As mentioned earlier, *it is difficult to quantify it exactly*. However, for oscillations of sufficiently small amplitude, it is fairly reasonable to model the damping force on Stoke's law. That is, we take F_d to be proportional to the (magnitude of) velocity of the oscillating mass and write

$$F_d = -\gamma v. \tag{4.1}$$

The magnitude of force experienced by a spherical body of radius r falling freely in a viscous medium (like water or oil) is given by

$$F_d = 6\pi\eta rv.$$

Here $-\eta$ is the coefficient of viscosity of the medium and v is the velocity of the body experiencing free fall.

The negative sign signifies that the damping opposes motion. The constant of proportionality γ is known as *damping coefficient*. Numerically, it is equal to force per unit velocity and is measured in $\frac{\text{N}}{\text{ms}^{-1}} = \frac{\text{kg ms}^{-2}}{\text{ms}^{-1}} = \text{kg s}^{-1}$.

Let us now derive the equation which describes the longitudinal motion of a 1-D damped harmonic oscillator. Suppose that the mass (in the spring-mass system) is pulled along its length and then released. It will be displaced from its equilibrium position. At any instant, a restoring force and a damping force will act simultaneously on the (spring-mass) system. We can express these forces as:

- *Restoring force*: $-kx$, where k is spring constant, and
- *Damping force*: $-\gamma v$, where $v = (dx/dt)$ is magnitude of instantaneous velocity of the oscillator and γ is damping coefficient.

Note that the restoring force is proportional to displacement, whereas the damping force is proportional to velocity. This is because a faster moving body experiences greater opposition from the medium. From aforesaid discussion, it is clear that for a damped harmonic oscillator, the force law for free oscillations has to be modified to include the restoring force as well as the damping force. So we can write

$$F = -kx - \gamma v$$

$$= -kx - \gamma \frac{dx}{dt}.$$

On inserting this expression for force in Eq. (2.2), we can write the equation of motion for a damped oscillator as

$$m\frac{d^2x}{dt^2} = -kx - \gamma\frac{dx}{dt}. \tag{4.2}$$

On rearranging terms and dividing throughout by m, the equation of motion of a damped oscillator takes the form

$$\frac{d^2x}{dt^2} + 2b\frac{dx}{dt} + \omega_0^2 x = 0, \tag{4.3}$$

where $\omega_0^2 = k/m$ and $2b = \gamma/m$. You will learn that that the factor of 2 has been introduced in the damping term to help express the solution of this equation in a mathematically convenient form. This equation gives complete mathematical description of a damped oscillator.

Note that b has the dimension of s^{-1}, which is the same as that of ω_0:

$$\frac{\gamma}{m} = \frac{\text{Force}}{\text{Velocity} \times \text{Mass}} = \frac{\text{MLT}^{-2}}{\text{LT}^{-1}\text{M}} = \text{T}^{-1}.$$

Can you point out similarities between Eqs. (1.46b) and (4.3)? You will note that both differential equations are linear, of second order and homogeneous with constant coefficients. But before solving Eq. (4.3), it is worthwhile to consider two special cases which should give us an idea about the nature of the solution.

(i) Suppose that there is *no damping*. Mathematically speaking, the second term in Eq. (4.3) will drop out and solution of the resulting equation will represent free oscillations discussed in Chapter 2.

(ii) Suppose that there were *no restoring force*. Then, the third term on the left hand side of Eq. (4.3) will be zero, and you can easily verify that the solution of the resulting equation will be

$$x(t) = Ce^{-2bt} + D,$$

where C and D are constants. From this result, we note that $x(t)$ decreases exponentially with time, but it will never become zero. *This suggests that a system can not oscillate in the absence of restoring force.* This is expected because there will be no physical agency to bring the system back to its equilibrium position once it has been displaced from there.

> The solution of the differential equation
>
> $$\frac{d^2x}{dt^2} + 2b\frac{dx}{dt} = 0$$
>
> is given as
>
> $$x(t) = Ce^{-2bt} + D, \qquad \text{(i)}$$
>
> where C and D are constants. You can check the correctness of the solution as follows:
>
> 1. Calculate $\frac{dx}{dt}$ and $\frac{d^2x}{dt^2}$ with $x(t)$ given by (i).
> 2. Substitute for $\frac{dx}{dt}$ and $\frac{d^2x}{dt^2}$ in the given equation.
> 3. If LHS equals RHS, you can say that the solution is correct.

Can you now guess the nature of solution of Eq. (4.3) and write its form that we expect when damping as well as restoring force is present? *The general solution of Eq.* (4.3) *is expected to represent an oscillatory motion whose amplitude decreases with time and is expected to contain both exponential and harmonic terms.* Let us now discover how far this guess is valid.

4.4 SOLUTIONS OF THE EQUATION OF MOTION OF A 1-D DAMPED OSCILLATOR

To know how damping influences the amplitude of oscillation of a damped oscillator, we take a solution of the form

$$x(t) = a \exp(\alpha t), \qquad (4.4)$$

where a and α are unknown constants. Obviously a has dimension of inverse time.

On differentiating Eq. (4.4) twice with respect to time, we have

$$\frac{dx}{dt} = a\alpha \exp(\alpha t)$$

and

$$\frac{d^2x}{dt^2} = a\alpha^2 \exp(\alpha t).$$

> The justification for presuming the exponential form in Eq. (4.4) is the fact that x, $\dot{x}$, $\ddot{x}$, as per Eq. (4.3) are linearly dependent. This is possible only when x vs. t is exponential.

On substituting these expressions in Eq. (4.3), we get

$$(\alpha^2 + 2b\alpha + \omega_0^2)\, a \exp(\alpha t) = 0. \qquad (4.5)$$

For this result to hold at all times, we have two possibilities:

- The amplitude of oscillations is zero ($a = 0$). But this does not make any sense and corresponds to a trivial condition.
- The term within the parentheses vanishes identically:

$$\alpha^2 + 2b\alpha + \omega_0^2 = 0. \qquad (4.6)$$

> The roots of a quadratic equation of the form
> $$ax^2 + bx + c = 0,$$
> are given by
> $$x = \frac{-b \pm (b^2 - 4ac)^{1/2}}{2a}$$

This equation is quadratic in α and has two roots. Let us denote these by α_1 and α_2:

$$\alpha_1 = -b + (b^2 - \omega_0^2)^{1/2} \tag{4.7a}$$

and

$$\alpha_2 = -b - (b^2 - \omega_0^2)^{1/2}. \tag{4.7b}$$

These roots determine the motion of a damped oscillator. On substituting these values of a in Eq. (4.4), we obtain two possible solutions of Eq. (4.3):

$$x_1(t) = a_1 \exp [-\{b + (b^2 - \omega_0^2)^{1/2})\} t]$$

and

$$x_2(t) = a_2 \exp [-\{b - (b^2 - \omega_0^2)^{1/2}\} t]. \tag{4.8}$$

Since Eq. (4.3) is linear, its general solution is obtained by superposing the individual solutions:

$$x(t) = \exp (-bt) [a_1 \exp \{(b^2 - \omega_0^2)^{1/2}\}t] + a_2 \exp \{-(b^2 - \omega_0^2)^{1/2}\}t]. \tag{4.9}$$

This result specifies the instantaneous displacement of a damped oscillator. From this we can easily determine the nature of oscillations. As expected, these will be governed by the interplay of mutually opposed damping force (represented by the constant b) and the restoring force (represented by ω_0^2). The quantity $(b^2 - \omega_0^2)$ in Eq. (4.9) can be positive, zero or negative, depending respectively on whether b is greater than, equal to or less than ω_0. These conditions give rise to three different kinds of motion:

- When $b > \omega_0$, we say that damping is heavy. The system is then said to be *heavily damped* or *overdamped*. This damping does not allow oscillatory motion to take place. The system takes very long time to regain its equilibrium position once it has been displaced from there. Can you name a heavily damped system of practical interest? Springs joining wagons of a train constitute one of the most important heavily damped systems. If a pendulum is made to oscillate in a viscous medium, such as thick oil, it will constitute a heavily damped system.
- When $b = \omega_0$, we say that the system is *critically damped*. A critically damped system also does not execute oscillatory motion. However, when such a system is displaced from its equilibrium position, it attains the equilibrium position in minimum time. That is why door closing mechanism in a door closer and shock absorbers in a car are usually designed to experience critical damping. This helps the door to close quickly but gently. Similarly, shock absorbers quickly damp the bounce (oscillation) that a car may experience immediately after hitting a road bump. Some other examples of critically damped systems are the indicator needles in electrical and electronic instruments and level indicators on tape recorders.
- When $b < \omega_0$, we say that the system is *weakly damped*. When a weakly damped system is displaced from the equilibrium position and released, it executes oscillatory motion with gradually decreasing amplitude. This is of maximum interest in physics and we shall discuss it in some detail.

Note that the nature of relationship between b and ω_0, as listed above, influences Eq. (4.9) and leads to different solutions. Each of these solutions describes a particular behaviour of a damped oscillator.

To fix these ideas, we will like you to answer the following Practice Exercise.

***Practice Exercise* 4.1** Consider the following equations and state the kind of motion they represent:

(i) $4\dfrac{d^2x}{dt^2}+16\dfrac{dx}{dt}+x=0.$

(ii) $\dfrac{d^2x}{dt^2}+10\dfrac{dx}{dt}+25x=0.$

(iii) $2\dfrac{d^2x}{dt^2}+4\dfrac{dx}{dt}+9x=0.$

Ans. (i) heavily damped, (ii) critically damped, (iii) weakly damped

Let us now discuss the different forms that Eq. (4.9) takes for different values of damping.

4.4.1 Heavy Damping

When resistance to motion is very high, the system is said to be heavily damped. For $b > \omega_0$, the quantity $(b^2 - \omega_0^2)$ will be positive and definite. For mathematical convenience and compactness, we introduce a new variable β through the relation

$$\beta = \sqrt{b^2 - \omega_0^2}.$$

Then Eq. (4.9) takes the following compact form:

$$x(t) = \exp(-bt)\,[a_1 \exp(\beta t) + a_2 \exp(-\beta t)]. \tag{4.10}$$

To know how displacement varies with time, we have to determine constants a_1 and a_2. For this, we use initial conditions. Let us suppose that

(i) the oscillator is at its equilibrium position, i.e., $x = 0$ at $t = 0$, and

(ii) the system is suddenly disturbed, i.e., it experiences a sudden impulse (kick) and acquires a speed v_0, i.e., $v = v_0$ at $t = 0$.

On applying the first condition, Eq. (4.10) leads to the result

$$a_1 + a_2 = 0.$$

This result leads to the relation $a_1 = -a_2$.

To apply condition (ii), we have to differentiate Eq. (4.10) with respect to time (see margin). You can now convince yourself that the condition on initial speed leads us to the relation

$$\frac{dx(t)}{dt} = -b \exp(-bt)\,[a_1 \exp(\beta t) + a_2 \exp(-\beta t)] + \exp(-\beta t)\,[a_1\beta \exp(\beta t) - a_2\beta \exp(-\beta t)]$$

$$-b(a_1 + a_2) + \beta(a_1 - a_2) = v_0.$$

On inserting the result $a_1 = -a_2$ in this equation, you can easily solve it to obtain

$$a_1 = -a_2 = \frac{v_0}{2\beta}.$$

$\sinh x = \dfrac{\exp(x) - \exp(-x)}{2}$ is a hyperbolic sine function.

On substituting this result in Eq. (4.10), the solution for a heavily damped system takes a compact form:

$$x(t) = \frac{v_0}{2\beta} \exp(-bt) [\exp(\beta t) - \exp(-\beta t)].$$

$$= \frac{v_0}{\beta} \exp(-bt) \sinh \beta t. \tag{4.11}$$

Note that $x(t)$ is a product of two terms: an increasing hyperbolic function $\phi(t) = \sinh \beta t$ and a decaying exponential function $\psi(t) = \exp(-bt)$. The interplay of these terms determines the form of resultant solution.

Refer to Fig. 4.2(a), where we have plotted these two terms separately. Note that β and b are constants and these do not in any way influence the time variation of the functions $\phi(t)$ and $\psi(t)$, respectively. In other words, $\phi(t)$ and $\psi(t)$ vary with time essentially as sine hyperbolic function and exponential function, respectively. Figure 4.2b shows the displacement-time graph for a heavily damped system when it is suddenly disturbed from its equilibrium position. As may be noted, initially the displacement increases with time but the exponential (decaying) term begins to dominate soon thereafter resulting in gradual decrease in displacement. However, the motion is non-oscillatory; $x(t)$ does not attain a negative value. Such a motion is called *dead-beat*.

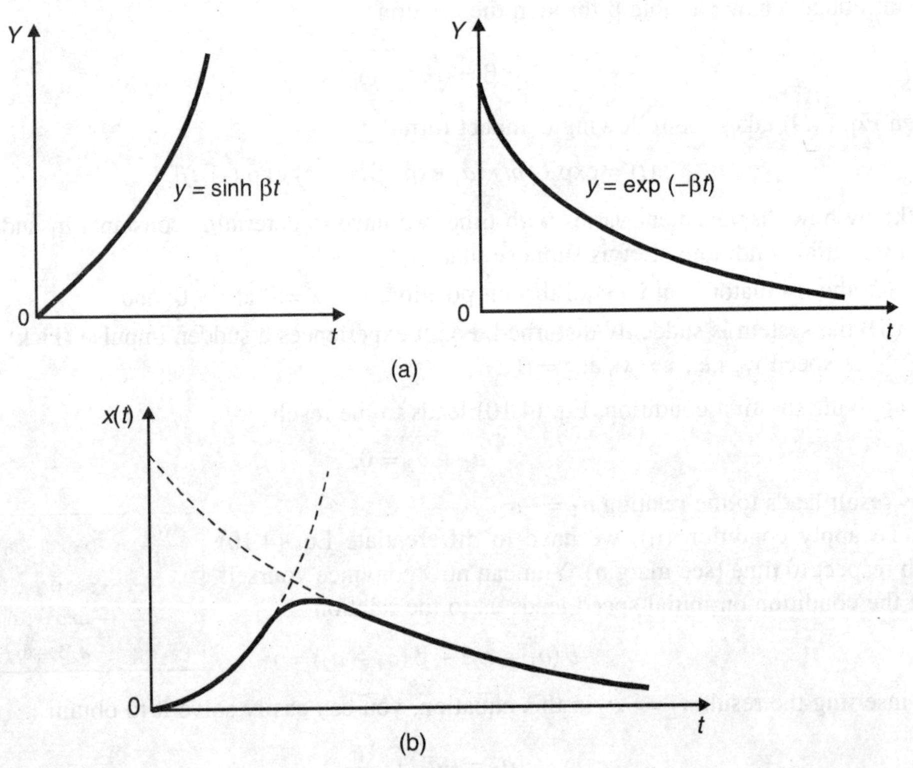

Fig. 4.2 (a) Time variation of sinh βt and exp $(-bt)$; and (b) displacement of a heavily damped system.

4.4.2 Critical Damping

Mathematically, we say that a system is critically damped if b is equal to the natural frequency ω_0, of the system. This means that $b^2 - \omega_0^2 = 0$, so that Eq. (4.9) takes a very compact form:

$$x(t) = (a_1 + a_2) \exp(-bt)$$
$$= a \exp(-bt), \tag{4.12}$$

where $a = a_1 + a_2$.

You may recall from Chapter 2 that the solution of the differential equation for SHM, which is of second order, involves two arbitrary constants. These are fixed by specifying the initial conditions. But Eq. (4.12) has only one constant. It means that this does not give a complete solution of Eq. (4.9) for critical damping. To show this, we differentiate Eq. (4.12) with time. This gives

$$\frac{dx}{dt} = -ba \exp(-bt).$$

If the system experiences sudden jerk, i.e., $v = v_0$ at $t = 0$, we get

$$v_0 = \left.\frac{dx}{dt}\right|_{\substack{x=0 \\ t=0}} = -ba.$$

This relation suggests that $v_0 = 0$, since $a = 0$ at $t = 0$. But this result contradicts our initial condition that at $t = 0$, $v = v_0$. You may ask: How has this fallacy crept in? It only means that Eq. (4.12) does not specify a complete solution of Eq. (4.3). You can easily convince yourself by solving Review Exercise 4.1 that the general solution of Eq. (4.3) for a critically damped oscillator is given by

$$x(t) = (p + qt) \exp(-bt), \tag{4.13}$$

where p and q are constant. Note that p has the dimension of length and q has the dimensions of speed. As before, these constants have to be determined by specifying two initial conditions.

Now, refer to Fig. 4.3, which shows displacement-time variation of a critically damped system. Note that instantaneous displacement does not take negative value at any time and attains maximum value when $t = b^{-1}$. Physically, it means that a critically damped system stops just before reaching the equilibrium position, i.e., it neither overshoots nor oscillates about the equilibrium position. Moreover, the time taken by the oscillator to come back to the equilibrium position is minimum for critical damping.

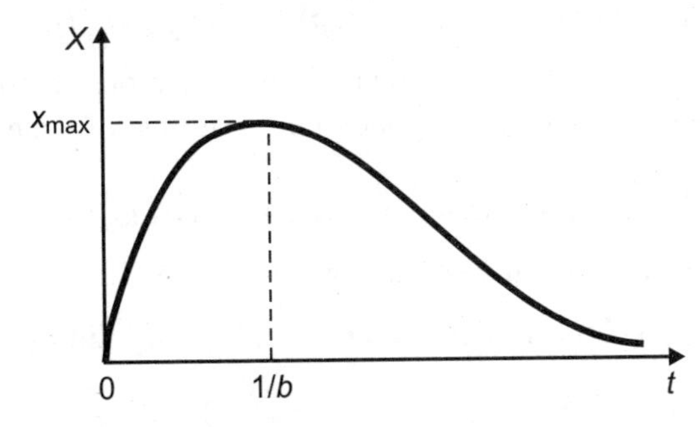

Fig. 4.3 Displacement-time graph for a critically damped system.

You must have observed that when a door closer is new, the motion of the door is very smooth. Similarly, when the shock absorbers in an automobile are new, we do not experience jerks even when it hits a road bump. But after some time, the door may begin to slam with a bang and the scooter/car/bus may bounce up and down several times on hitting a road bump and cause discomfort to the occupants. Do you know the reason for such a drastically different behaviour in two situations? In physical terms, this is because the springs in the door closer as also the shock absorbers in an automobile *wear out with time*. Mathematically speaking, this change in behaviour in a critically damped system arises due to the time dependence of the constant q in Eq. (4.13).

You may now ask: What will happen if indicator needles in electrical and electronic instruments and level indicators on tape recorders are not critically damped. Underdamped indicator needle will swing back and forth excessively before coming to the required/exact position/value. As a result, rapid changes in the signal, as in a recording level indicator, will go undetected.

In Physics, overdamped and critically damped oscillations have very limited use. It is, therefore, more instructive to study the behaviour of a weakly damped system. This forms the subject of discussion of Section 4.4.3.

4.4.3 Weak Damping

When $b < \omega_0$, the system is said to be weakly damped and this gives rise to *oscillatory damped harmonic motion.* To mathematically analyse such a system, we note that the quantity $(b^2 - \omega_0^2)$ will be negative and $(b^2 - \omega_0^2)^{1/2}$ becomes an imaginary number. Therefore, we rewrite it as

It may be noted that both 'i' and 'j' are used to represent $\sqrt{-1}$.

$$(b^2 - \omega_0^2)^{1/2} = \sqrt{-1}\,(\omega_0^2 - b^2)^{1/2}$$
$$= i\omega_d,$$

where $i = \sqrt{-1}$ and

$$\omega_d = (\omega_0^2 - b^2)^{1/2} = \left[\frac{k}{m} - \frac{\gamma^2}{4m^2}\right]^{1/2} \tag{4.14}$$

is a real positive quantity. Note that for free oscillations, i.e., when there is no damping ($b = 0$), $\omega_d = \omega_0$, the natural frequency of the oscillator.

On combining Eqs. (4.9) and (4.14), the expression for displacement of a weakly damped harmonic oscillator takes the form

$$x(t) = \exp(-bt)\,[a_1 \exp(i\omega_d t) + a_2 \exp(-i\omega_d t)]. \tag{4.15}$$

$\exp(\pm ix) = \cos x \pm i \sin x$

To put this expression in a more familiar form, we write the complex exponential in terms of sine and cosine functions. This gives

$$x(t) = \exp(-bt)\,[a_1(\cos\omega_d t + i\sin\omega_d t) + a_2(\cos\omega_d t - i\sin\omega_d t)]. \tag{4.16}$$

On collecting coefficients of $\cos\omega_d t$ and $\sin\omega_d t$, we can write

$$x(t) = \exp(-bt)\,[(a_1 + a_2)\cos\omega_d t + i(a_1 - a_2)\sin\omega_d t)]. \tag{4.17}$$

The constants a_1 and a_2 are, in general, complex numbers. It means that either the real or the imaginary part of Eq. (4.17) will correspond to a physically acceptable solution. Therefore, we introduce a change of variable by putting

$$a_1 + a_2 = a_0 \cos \phi$$

and

$$i\,(a_1 - a_2) = a_0 \sin \phi. \tag{4.18}$$

On inserting these in Eq. (4.17), we find that the general solution of Eq. (4.3) for a weakly damped oscillator ($b < \omega_0$) exhibits sinusoidal character:

$$\begin{aligned} x(t) &= a_0 \exp(-bt)\,[\cos \phi \cos \omega_d t + \sin \phi \sin \omega_d t] \\ &= a_0 \exp(-bt) \cos(\omega_d t - \phi), \end{aligned} \tag{4.19}$$

where ω_d is given by Eq. (4.14), $a_0 = \sqrt{4a_1a_2}$ and $\tan \phi = i\left(\dfrac{a_1 - a_2}{a_1 + a_2}\right)$.

> Strictly speaking, we are not quite justified in using the terms amplitude and frequency for an aperiodic motion. But for small damping, these may be used with some caution.

Note that Eq. (4.19) describes sinusoidal motion of frequency ω_d ($< \omega_0$). And the amplitude ($= a_0 \exp(-bt)$) has been modified vis-à-vis SHM; it decreases exponentially with time at a rate governed by the magnitude of damping. You may now ask: Does a weakly damped system execute SHM? The answer to this question is: *The motion of a weakly damped system is oscillatory but not simple harmonic*. In fact, the motion is not even periodic, since it does not repeat itself identically; each successive swing has smaller amplitude than the preceding one.

The displacement-time graph for a weakly damped oscillator, described by Eq. (4.19), is shown in Fig. 4.4 for $\phi = 0$. You may recall that the sine and cosine functions vary between +1 and –1. For this reason, the displacement-time curve lies between $a_0 \exp(-bt)$ and $-a_0 \exp(-bt)$. Thus, we conclude that *damping influences amplitude as well as angular frequency adversely*. Further, for $bt = 1$, the amplitude of oscillation drops to e^{-1} (= 0.368) of its initial value. The time a system takes to attain this value is referred to as *relaxation time*: $\tau = 1/b$.

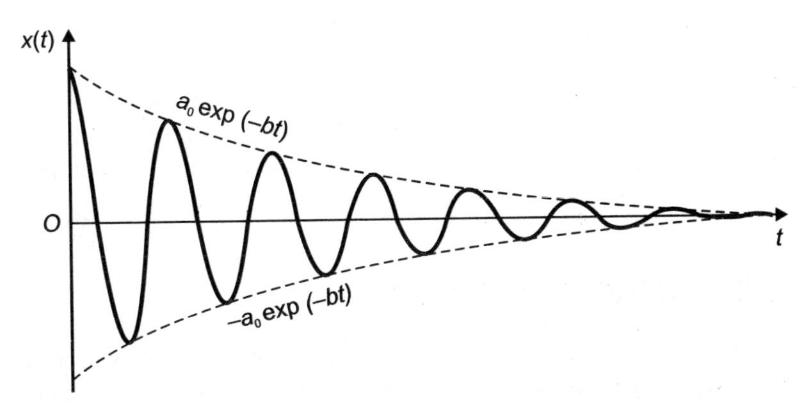

Fig. 4.4 Displacement-time graph for a weakly damped oscillator.

Let us now examine how damping influences the period of oscillation of a weakly damped oscillator. From Eq. (4.14), we note that if $b > 0$, $\omega_d < \omega_0$. It means that period of oscillation of a weakly damped oscillator increases with damping, i.e., it is more than that of an ideal oscillator. Did you not expect this result since damping resists motion?

Mathematically, we express the period of oscillation of a weakly damped oscillator as

$$T = \frac{2\pi}{\omega_d} = \frac{2\pi}{(\omega_0^2 - b^2)^{1/2}} = \frac{2\pi}{\sqrt{(k/m) - (\gamma^2/4m^2)}} \tag{4.20}$$

To give you an idea about the magnitude of time period for a weakly damped system, let us consider Eq. (iii) given in Practice Exercise 4.1:

$$\frac{d^2x}{dt^2} + 0.2\frac{dx}{dt} + 36x = 0.$$

You can easily identify that in this case, $b = 0.1$ and $\omega_0 = 6$. On using these values in Eq. (4.20), we get

$$T = \frac{2\pi}{\omega_d} \cong \frac{\pi}{3}\text{ s}.$$

You now know that the motion of heavily damped and critically damped systems is non-oscillatory. However, the motion of a weakly damped system is oscillatory but not periodic. Moreover, its frequency is lower than that of a free oscillator.

You may now like to answer a Practice Exercise.

***Practice Exercise* 4.2** (a) The amplitude of vibration of a damped spring-mass system decreases from 20 cm to 4.0 cm in 80 s. If this oscillator completes 40 oscillations in this time, calculate the time period of damped oscillator.

(b) A one kg mass executes one dimensional motion. It experiences a restoring force which is proportional to displacement and a resisting force characterised by damping constant of 0.6 Ns m^{-1} (i) For what value of force constant of the restoring force will motion be critically damped? (ii) If the force constant is changed to 2 Nm^{-1}, for what mass will the given forces make the motion critically damped.

Ans: (a) 2.01×10^{-3} s; (b) (i) 0.09 Nm^{-1}, (ii) 45 g

We have discussed solutions of the equation of motion for a heavily, critically and weakly damped mechanical oscillator such as the spring-mass system. From Chapter 2, you may recall that mechanical oscillations of an undamped spring-mass system are analogous to the oscillations of charge in an *LC* circuit. It is, therefore, interesting to discover the nature and effect of damping on oscillations in systems of practical utility or of frequent use in Physics Laboratory.

4.5 NON-MECHANICAL DAMPED SYSTEMS

A series *LCR* circuit is of particular interest as it finds important applications in radio transmission. In your Physics laboratory, you must have seen or worked with ballistic galvanometers to study charging and discharging of a capacitor, measure self-inductance or mutual inductance. Similarly, a fluxmeter is used to determine the field intensity. We, therefore, begin this discussion by considering a series *LCR* circuit.

4.5.1 A Series *LCR* Circuit

In Chapter 2 you have learnt that in an *LC* circuit, charge oscillates between inductor and capacitor periodically. Do you expect any change in this behaviour when a resistor *R* is added in the circuit? To discover answer to this question, let us consider a series *LCR* circuit shown in Fig. 4.5.

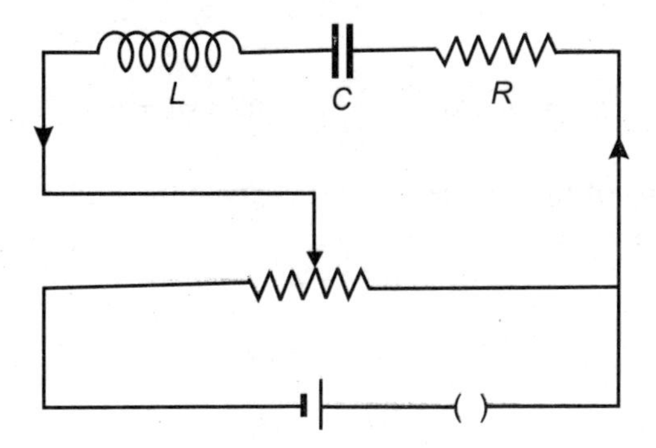

Fig. 4.5 A series *LCR* circuit.

The capacitor *C* in the circuit is charged using an external battery. Thereafter, the battery connection is removed and the charged capacitor is connected in series with an inductor *L* and resistor *R* as shown in Fig. 4.5. As the capacitor begins to discharge through the inductor and resistor, let us assume that at any instant of time *t*, current *I* flows in the circuit and charge *q* resides on the capacitor.

> We know that
> $$I(t) = \frac{dq}{dt}(t)$$
> and
> $$\frac{dI(t)}{dt} = \frac{d^2q(t)}{dt^2}.$$

Then the voltage drop across the resistor will be *RI*. And the equation of motion of charge, using Kirchhoff's loop rule, can be written as

$$\frac{q(t)}{C} = -L\frac{dI(t)}{dt} - R\frac{dq(t)}{dt}.$$

This may be rewritten as

$$L\frac{d^2q(t)}{dt^2} + R\frac{dq(t)}{dt} + \frac{q(t)}{C} = 0. \tag{4.21}$$

On comparing Eq. (4.21) with Eq. (4.2), we discover that *L*, *R* and $1/C$ are respectively analogous to *m*, γ and *k*. This means that the effect of a resistor in an electric circuit is *exactly* analogous to that of the viscous force in a mechanical system.

To proceed further, we divide Eq. (4.21) throughout by *L*. This gives

$$\frac{d^2q(t)}{dt^2} + \frac{R}{L}\frac{dq(t)}{dt} + \frac{1}{LC}q(t) = 0. \tag{4.22}$$

This equation is analogous to Eq. (4.3), if we identify

$$\omega_0^2 = \frac{1}{LC} \tag{4.23}$$

and

$$b = \frac{R}{2L}. \tag{4.24}$$

Equation (4.24) suggests that damping in a series *LCR* circuit is determined by resistance and inductance. You may recall that b has dimensions of time inverse. It means that R/L also has the unit of s^{-1}, which is same as that of ω_0. It implies that $\omega_0 L$ will be measured in ohm.

With these analogies, all the results of section 4.4.3 apply to Eq. (4.22). Therefore, for a weakly damped *LCR* circuit, the instantaneous value of charge on the capacitor plates can be expressed as

$$q(t) = q_0 \exp\left(-\frac{R}{2L}t\right) \cos(\omega_d t + \phi), \tag{4.25}$$

where angular frequency ω_d of the damped circuit is given by

$$\omega_d = \sqrt{\omega_0^2 - b^2}$$

$$= \sqrt{\frac{1}{LC} - \frac{R^2}{4L^2}}. \tag{4.26}$$

Equation (4.25) shows that the rate of decay of charge amplitude, $q_0 \exp\left(-\frac{R}{2L}t\right)$ depends on resistance and inductance. However, only resistance acts as the dissipative element in a series *LCR* circuit; an increase in R increases the rate of decay of charge and decreases the frequency of oscillations.

When $1/LC >> R^2/4L$, Eq. (4.26) gives

$$\omega_d^2 \cong \omega_0^2 = \frac{1}{LC}$$

or

$$\omega_0 L = \frac{1}{\omega_0 C}. \tag{4.27}$$

Since $\omega_0 L$ is measured in ohm, $1/\omega_0 C$ will also be measured in ohm. You may recall from your school Physics classes that these are respectively referred to as *inductive reactance* and *capacitive reactance*.

For $R = 0$, $\omega_d = \omega_0$ and Eq. (4.25) reduces to $q(t) = q_0 \exp\left(-\frac{R}{2L}t\right)$.

You should now answer the following Practice Exercise.

***Practice Exercise* 4.3** In a series *LCR* circuit, $L = 1$ mH and $C = 5$ μF. Will discharge be oscillatory for $R = 10\ \Omega$? If so calculate the frequency of oscillation.

Ans: 2.11 kHz

4.5.2 A Ballistic Galvanometer

A ballistic galvanometer is a suspension type galvanometer in oscillatory mode. You may recall that it consists of a coil suspended in a magnetic field produced by a horse-shoe magnet. It carries a small mirror which is used to observe deflection produced in the coil when it is disturbed from its equilibrium position on passing a small charge through it. An iron cylinder suspended between the poles of the magnet ensures that the field is uniform. When we pass charge through the galvanometer coil, it rotates through a small angle θ, say. Since the coil is akin to a torsional pendulum, it is subjected to a restoring couple and a damping couple. Note that damping arises partly due to air dragged by the coil as it rotates (air resistance) and partly due to electromagnetic damping (Lenz's law of electromagnetic induction).

The equation of motion of the galvanometer coil can be written as

$$I\frac{d^2\theta}{dt^2} = -k\theta - \gamma\frac{d\theta}{dt},$$

where I is moment of inertia of the coil about its axis of suspension. You can rewrite it as

$$\frac{d^2\theta}{dt^2} + 2b_1\frac{d\theta}{dt} + \omega_1^2\theta = 0, \tag{4.28}$$

where $\omega_1^2 = k/I$ and $2b_1 = (\gamma/I)$; I being the moment of inertia of the galvanometer coil about the axis of suspension. Note that this equation is identical to Eq. (4.3) and for low damping, we can write the solution as analogous to Eq. (4.19):

$$\theta = a_0 \exp(-b_1 t)\cos(\omega_d t - \phi), \tag{4.29}$$

where $a_0 \exp(-b_1 t)$ is amplitude of oscillation of the coil. This equation represents oscillatory motion with time period T given by

$$T = \frac{2\pi}{\omega_d} = \frac{2\pi}{(\omega_1^2 - b_1^2)^{1/2}} = \frac{2\pi}{\sqrt{(k/I) - (\gamma^2/4I^2)}}. \tag{4.30}$$

Note that damping will be small if moment of inertia is large and/or γ is small. To reduce γ, we must minimise induced emf. For his reason, the coil is wound on a non-conducting frame of bamboo or ivory.

You now know that a weakly damped oscillator involves a lot of good Physics and finds far greater applications in a variety of physical systems. Therefore, it is important to quantify/characterise weak damping. You will learn about it now in some detail.

4.6 ENERGY OF A WEAKLY DAMPED SYSTEM

In Chapter 2 you learnt that the average energy E_0 associated with a free oscillator is given by

$$E_0 = \frac{1}{2}ka^2.$$

From this expression we note that larger the amplitude of oscillation, greater is the average energy of the system. To get an idea about the energy dissipation in a weakly damped system, refer to Fig. 4.4 again. You will note that the amplitude decreases gradually with time, i.e., the oscillation dies down as time passes. It means that a damped oscillating system loses energy with time due to the work done in overcoming opposition of damping force to motion. And the rate of loss of energy is governed by the magnitude of damping present.

From Section 4.4 you will recall that instantaneous amplitude $a(t)$ of a damped harmonic oscillation is given by $a_0 \exp(-bt)$. But we expect the average energy of the damped oscillator, $< E >$, to be proportional to the square of decaying amplitude. So we can write

$$< E > = Da^2 = Da_0^2 \exp(-2bt),$$

where D is constant of proportionality. Hence, the average energy of a damped oscillator can be written in terms of the average energy of an undamped oscillator as

$$< E > = E_0 \exp(-2bt). \tag{4.31}$$

From this equation we can say that average energy of a weakly damped oscillator decreases exponentially with time and its rate of decrease is faster [exp (–2*bt*)] than the rate of decay of amplitude [exp (–*bt*)]. This is shown in Fig. 4.6.

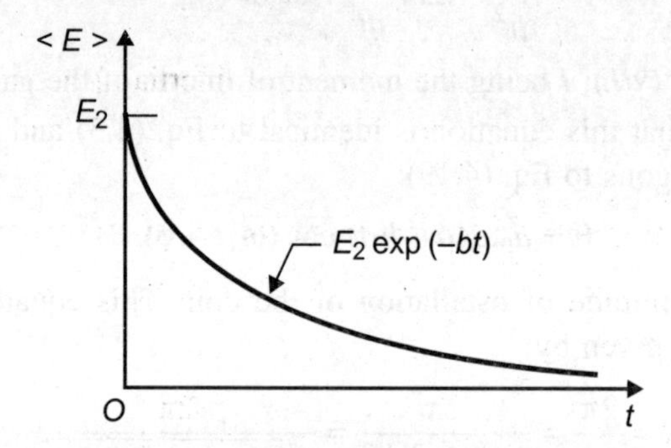

Fig. 4.6 Time variation of average energy for a weakly damped system.

You can also derive the result contained in Eq. (4.31) from first principle, i.e., by writing energy as a sum of kinetic and potential energies, inserting the expression for displacement and its first time derivative using Eq. (4.19) and averaging it over one complete cycle. We leave it as an Practice Exercise for you.

***Practice Exercise* 4.4** Derive Eq. (4.31) from the first principle.

4.7 CHARACTERISING WEAK DAMPING

To characterise weak damping, we use three methods. Each of these methods involves determination of a separate parameter defined in terms of ω_0 and b based on the behaviour of amplitude and energy: logarithmic decrement λ, relaxation time τ and quality factor Q. Depending on the system, one or more of these parameters may suitably be used to quantify damping. You have already learnt about relaxation time. It refers to the time in which the amplitude of a weakly damped system reduces to e^{-1} times (0.368 of) its initial value. Let us now take a closer look at the other two.

4.7.1 Logarithmic Decrement

The most convenient way to know the magnitude of damping present in a system is to measure the rate at which its amplitude of oscillation dies down with time. Suppose that a weakly damped system, initially at rest at its equilibrium position, is given an impulse suddenly, i.e., at $t = 0$, $x = 0$ and $v = v_0$. You can easily show that its instantaneous displacement is given by

$$x(t) = a(t)\sin\left(\frac{2\pi t}{T}\right),$$

where $a(t) = a_0 \exp(-bt)$ and a_0 is the amplitude in the absence of damping ($b = 0$). The displacement-time graph of this motion is shown in Fig. 4.4. Note that at $t = T/4$, the displacement rises to its first maximum value, say a_1,

$$x\left(\frac{T}{4}\right) = a_1 = a_0 \exp\left(-\frac{bT}{4}\right)\sin\left(\frac{2\pi t}{T}\cdot\frac{T}{4}\right)$$

$$= a_0 \exp\left(-\frac{bT}{4}\right). \tag{4.32a}$$

At $t = 3T/4$, the amplitude becomes maximum again (but on the opposite side). Let us denote it by a_2. The amplitude a_2 is given by

$$a_2 = a_0 \exp\left(-\frac{3bT}{4}\right). \tag{4.32b}$$

The next maximum occurs at $t = 5T/4$ and we can write

$$a_3 = a_0 \exp\left(-\frac{5bT}{4}\right). \tag{4.32c}$$

Similarly

$$a_4 = a_0 \exp\left(-\frac{7bT}{4}\right), \tag{4.32d}$$

and so on.

You can easily calculate the ratio of successive amplitudes separated by half-a-period:

$$\frac{a_1}{a_2} = \frac{a_2}{a_3} = \frac{a_3}{a_4} = \cdots = \frac{a_{n-1}}{a_n} = \exp\left(\frac{bT}{2}\right) = d \tag{4.33}$$

That is, d is constant and is called *decrement* of motion.

The logarithmic decrement is defined as the natural logarithm of the ratio of successive amplitudes of oscillation separated by half a period; the larger amplitude being in the numerator. We will denote logarithmic decrement by λ and write

$$\lambda = \ln d = \ln\left[\exp\left(\frac{bT}{2}\right)\right]$$

$$= \frac{bT}{2} = \frac{\gamma T}{4m} \tag{4.34}$$

$$e^{-x} = 1 - x + \frac{x^2}{2!} - \frac{x^3}{3!} + \cdots$$

For $x << 1$,

$$e^{-x} \cong 1 - x$$

Note that λ depends on the damping factor b as well as the time period T of the oscillating system. On combining Eqs. (4.33) and (4.34), we obtain

$$\frac{a_1}{a_2} = d = \exp \lambda, \tag{4.35a}$$

$$\frac{a_1}{a_3} = \frac{a_1}{a_2} \cdot \frac{a_2}{a_3} = d^2 = \exp (2\lambda), \tag{4.35b}$$

and so on. Therefore, we can write

$$\frac{a_1}{a_n} = \frac{a_1}{a_2} \cdot \frac{a_2}{a_3} \cdot \frac{a_3}{a_4} \cdots \frac{a_{n-1}}{a_n} = \exp [(n-1)\lambda]$$

or
$$\lambda = \frac{1}{(n-1)} \ln \left(\frac{a_1}{a_n} \right). \tag{4.36}$$

This result shows that the logarithmic decrement can be measured by observing several successive amplitudes on both sides of the equilibrium position.

We will now like you to go through Example 4.1.

EXAMPLE 4.1

A harmonic oscillator is represented by the equation

$$m\frac{d^2 x}{dt^2} + \gamma \frac{dx}{dt} + kx = 0$$

with m = 0.25 kg, γ = 0.070 kg s^{-1} and k = 85 Nm^{-1}. Calculate (i) the period of oscillation, and (ii) the number of oscillations in which its amplitude will become half of its original value.

Solution: For the given damped harmonic oscillator, m = 250 g = 0.25 kg,

$$\gamma = 0.070 \text{ kg s}^{-1} \text{ and } k = 85 \text{ Nm}^{-1}.$$

(i) The period of oscillation

$$T = \frac{2\pi}{\sqrt{\dfrac{k}{m} - \left(\dfrac{\gamma}{2m}\right)^2}} = \frac{2\pi}{\sqrt{\dfrac{85 \text{ Nm}^{-1}}{0.25 \text{ kg}} - \left(\dfrac{0.070 \text{ kg s}^{-1}}{2 \times 0.25 \text{ kg}}\right)^2}}$$

$$= \frac{2\pi}{\sqrt{\dfrac{85 \text{ Nm}^{-1}}{0.25 \text{ kg}} - \left(\dfrac{0.070 \text{ kg s}^{-1}}{2 \times 0.25 \text{ kg}}\right)^2}}$$

$$= \frac{2\pi}{\sqrt{(340 - 0.14) \text{ s}^{-2}}}$$

$$= \frac{2\pi}{18.44 \text{ s}^{-1}}$$

$$= 0.34 \text{ s}.$$

(ii) We know that amplitude of a damped oscillator is given by

$$a(t) = a_0 e^{-\gamma t/2m}.$$

As per the problem,

$$\frac{a(t)}{a_0} = \frac{1}{2} = e^{-\gamma t/2m}.$$

On taking natural logarithm of both sides and rearranging terms, we get

$$t = \frac{2m \ln 2}{\gamma} = \frac{2 \times (0.25\,\text{kg}) \times 0.692}{0.070\,\text{kg}\,\text{s}^{-1}}$$

$$= 4.95\ \text{s}.$$

Since the period of the oscillator is 0.34 s, the amplitude will reduce to half in 4.95 s/0.34 s $\approx$ 15 oscillations.

4.7.2 Quality Factor

Another way to obtain a quantitative measure of damping is to calculate the quality factor Q of the system. The quality factor represents the rate of decay of energy of the system. Suppose E_0 is the initial energy of an undamped oscillating system. Due to damping, howsoever small, the system gradually loses energy. *The number of radians through which a weakly damped system oscillates as its average energy decays to* e^{-1} *times its initial energy, i.e., to* $E_0\, e^{-1}$, *gives a measure of the quality factor* Q.

From Eq. (4.31) you will note that $\langle E \rangle$ reaches the value $E_0 e^{-1}$ in a time t given by

$$t = \frac{1}{2b} = \frac{m}{\gamma}. \tag{4.37}$$

Since the number of radians covered during this time is given by $\omega_d \times t$, we find that quality factor

$$Q = \omega_d t = \frac{\omega_d}{2b} = \frac{\omega_d \tau}{2}, \tag{4.38}$$

where relaxation time, $\tau = 1/b$.

From Eq. (4.38), we note that the quality factor is directly proportional to relaxation time or inversely proportional to damping. However, it is a pure number because it is ratio of two energies. For a weakly damped oscillator, we can assume that

$$\omega_d \approx \omega_0 = \sqrt{\frac{k}{m}},$$

so that an approximate expression for quality factor is

$$Q = \frac{\omega_d}{2b} = \frac{m}{\gamma}\sqrt{\frac{k}{m}} = \sqrt{\frac{km}{\gamma^2}}. \tag{4.39}$$

Note that Q is directly proportional to the square root of the spring constant k and inversely proportional to the damping coefficient γ. For a weakly damped oscillator, γ is very small and Q is very large. For example, Q of a tuning fork is about one thousand. On the other hand, for a rubber band, Q is nearly equal to 10. What is the value of Q for an undamped system? On the basis of

Eq. (4.39) you can say that Q is infinite for an undamped ($\gamma = 0$) oscillator. The quality factor Q of a weakly damped series LCR circuit is given by

$$Q = \frac{\omega_d}{2b} \cong \omega_0 \frac{L}{R} = \frac{1}{R}\sqrt{\frac{L}{C}}. \quad (4.40)$$

This equation shows that for a purely inductive circuit ($R = 0$), quality factor will be infinite.

In Chapter 5, you will learn about the quality factor from a different perspective.

Now, study the following solved examples carefully.

EXAMPLE 4.2

For the harmonic oscillator given in Example 4.1, calculate (i) the number of oscillations in which its mechanical energy will drop to one-half of its initial value. Also calculate its quality factor.

Solution: For the given damped harmonic oscillator, $m = 250$ g $= 0.25$ kg, $\gamma = 0.070$ kg s^{-1} and $k = 85$ Nm^{-1}.

(i) From Eq. (4.31), we recall that average energy associated with a damped harmonic oscillator is given by

$$\langle E \rangle = E_0 e^{-2bt}$$
$$= E_0 e^{-\gamma t/m}.$$

$$\therefore \quad \frac{\langle E \rangle}{E_0} = e^{-\gamma t/m}.$$

For $\langle E \rangle / E_0 = 1/2$, we have

$$\frac{1}{2} = e^{-\gamma t/m}.$$

Taking natural logarithm on both sides and rearranging the terms, we can rewrite it as

$$t = \frac{m \ln 2}{\gamma} = \frac{(0.25\ \text{kg}) \times 0.693}{0.070\ \text{kg s}^{-1}}$$
$$= 2.48\ \text{s}.$$

Since the period of the oscillator is $T = 0.34$ s, we find that the energy of the oscillator will drop to half of its initial value in about 2.48 s/0.34 s $\approx$ 7 oscillations.

(ii) From Eq. (4.37), we recall that quality factor of a damped harmonic oscillator, Q is given by

$$Q = \frac{\omega_d \tau}{2} \approx \frac{\omega_0 m}{\gamma}$$

since $\omega_d \approx \omega_0$ and $\tau = \dfrac{1}{b} = \dfrac{2m}{\gamma}$. On substituting the values of various quantities, we get

$$Q = \frac{(18.43\ \text{s}^{-1}) \times (0.25\ \text{kg})}{0.070\ \text{kg s}^{-1}} \approx 66.$$

EXAMPLE 4.3

The quality factor of the wire of a musical instrument is 3000. The wire vibrates at a frequency of 300 Hz. Calculate the time in which its amplitude will decrease to half of its initial value.

Solution: Since the quality factor Q of a weakly damped system is given by

$$Q = \frac{\omega_0 \tau}{2},$$

we can write the relaxation time τ as

$$\tau = \frac{2Q}{\omega_0}$$

$$= \frac{2 \times 3000}{2\pi \times 300 \text{ s}^{-1}}$$

$$= 3.18 \text{ s}.$$

Now, the amplitude a of a damped system at time t is given by

$$a = a_0 e^{-bt} = a_0 e^{-t/\tau}$$

$$\therefore \quad t = \tau \ln \frac{a_0}{a}$$

$$= (3.18 \text{ s}) \times \ln 2$$

$$= 2.194 \text{ s}.$$

EXAMPLE 4.4

A small pan of mass 0.2 kg is attached to one end of a spring whose other end is fixed to a rigid support. When a mass of 0.9 kg is placed on the pan, the system performs 240 oscillations per minute and the amplitude falls from 2 cm to 1 cm in 60 s. Calculate (i) the force constant, (ii) the relaxation time, and (iii) Q-factor.

Solution:

(i) Here $v = 240/60 = 4 \text{ s}^{-1}$.
Hence $\omega_0 = 2\pi v = 2 \times (3.14 \text{ rad}) \times 4 \text{ s}^{-1} = 25 \text{ rad s}^{-1}$.
We also know that $m = (0.1 + 0.9) \text{ kg} = 1 \text{ kg}$ and

$$k = m\omega_0^2$$

$$= (1 \text{ kg}) \times (25 \text{ s}) = 625 \text{ Nm}^{-1}$$

(ii) $a = a_0 e^{-bt}$
or,

$$0.01 \text{ m} = (0.02 \text{ m}) e^{-60b}.$$

or

$$\exp(60b) = 2$$

and

$$b = \frac{\ln 2}{60}$$

$$= \frac{2.303 \times \log_{10} 2}{60}$$

$$= 1.15 \times 10^{-2}\ \text{s}^{-1}.$$

Hence, relaxation time

$$\tau = \frac{1}{b} = \frac{1}{1.15 \times 10^{-2}\ \text{s}^{-1}} = 87\ \text{s}.$$

(iii) For a weakly damped system,

$$Q = \frac{\omega_0 \tau}{2}$$

$$= 25\ \text{s}^{-1} \times 87\ \text{s}$$

$$= 2175.$$

EXAMPLE 4.5

In a series LCR circuit, $L = 10$ mH, $C = 1$ μF and $R = 0.4\ \Omega$. Will the discharge be oscillatory? If so, calculate the frequency and quality factor of the circuit. How long does the charge oscillation take to decay to half of its initial value? What value of R will make the discharge just non-oscillatory?

Solution: Here

$$\frac{1}{LC} = \frac{1}{(10 \times 10^{-3}\ \text{H}) \times (1 \times 10^{-6}\ \text{F})} = 10^{8}\ \text{s}^{-2}$$

and

$$\frac{R^2}{4L^2} = \frac{(0.4)^2\ \Omega^2}{4 \times (10 \times 10^{-3}\ \text{H})^2} = 400\ \text{s}^{-2}.$$

Since $\frac{1}{LC} > \frac{R^2}{4L^2}$, the discharge is oscillatory and has frequency

$$f = \frac{1}{2\pi}\sqrt{\frac{1}{LC} - \frac{R^2}{4L^2}}$$

$$= 1.59 \times 10^{3}\ \text{Hz}.$$

The quality factor of the circuit is

$$Q = \frac{\omega_0 L}{R}$$

$$= \frac{2\pi \times (1.59 \times 10^{3}\ \text{s}^{-1}) \times (10 \times 10^{-3}\ \text{H})}{0.4\ \Omega}$$

$$= 250.$$

The time taken in decay of change oscillation to half of its initial value q_0 can be written as

$$t = \frac{R}{2L}\ln\left(\frac{q_0}{q}\right)$$

$$= \frac{0.4\ \Omega}{2\times(10\times10^{-3}\ \text{H})}\ln 2$$

$$= 14\ \text{s}.$$

The discharge will be just non-oscillatory when

$$\frac{1}{LC} = \frac{R^2}{4L^2}$$

or
$$R^2 = \frac{4L}{C}$$

$$= \frac{4\times(10\times10^{-3}\ \text{H})}{1\times10^{-6}\ \text{F}}$$

$$= 4\times10^4\ \text{HF}^{-1} = 4\times10\ \Omega^2$$

so that $R = 200\ \Omega$.

You may now like to answer a Practice Exercise.

***Practice Exercise* 4.5** The upper end of a mass-less spring is fixed to a rigid support. It carries a horizontal disc of mass 200 g at the lower end. It is observed that the system oscillates with frequency 10 Hz and the amplitude of the damped oscillations reduces to half its undamped value in one minute. Calculate (i) the damping force constant, (ii) the relaxation time of the system, (iii) its quality factor, and (iv) the force constant of the spring.

[***Ans:*** (i) 0.0046 Nsm^{-1}, (ii) 86.6 s, (iii) 273, (iv) 788.7 Nm^{-1}]

REVIEW EXERCISES

4.1 An object of mass m is subject to restoring and frictional forces of magnitudes kx and $\gamma\,(dx/dt)$, respectively. It oscillates with a frequency of 0.5 Hz. Its amplitude is known to reduce to half in 2 s. Calculate its damping coefficient γ and spring constant k in terms of mass. Also write the differential equation of motion.

[***Ans.*** $\gamma = 0.693$; $k = 9.98$; $\dfrac{d^2x}{dt^2} + 0.693\dfrac{dx}{dt} + 9.98x = 0$]

4.2 The period of a simple pendulum is 2 s and its amplitude is 5°. After 20 complete oscillations, its amplitude is reduced to 4°. Calculate the damping constant and relaxation time.

[***Ans.*** $b = 5.57\times10^{-3}\ \text{s}^{-1}$; $\tau = 179.5$ s]

4.3 The quality factor of a tuning fork of frequency 512 Hz is 6×10^4. Calculate the time in which its energy drops to E_0e^{-1}. How many oscillations will the tuning fork make in this time?

[***Ans.*** $t = 18.7$ s; $n = 9570$]

4.4 A steady force of 60 N is required to lift a mass of 1 kg vertically through a viscous liquid at a constant speed of 5 ms^{-1}. Assuming that the effect of viscosity can be taken to be proportional to velocity, calculate the proportionality constant. The mass is suspended in the same liquid by a spring of force constant 50 Nm^{-1}. Calculate the equilibrium extension of the spring. The mass is pulled down and released from rest. It executes oscillatory motion of continuously decaying amplitude. Calculate damping coefficient and period of free oscillations. Take $g = 10$ ms^{-2}.

[*Ans.* $C = 10$ Nsm^{-1}; Extension = 0.2 m; $b = 5$ s^{-1}; $\omega_0 = 7.07$ s^{-1}]

4.5 For a system executing damped oscillation the time period is 4s, the logarithmic decrement is 1.6 and the epoch is zero. The instantaneous displacement of the oscillation at 1s is 4.5 cm. Write down the expression for displacement of the oscillator.

[*Ans.* $x = 6.7e^{-0.4t \sin \pi t/2}$]

4.6 The displacement-time equation of a damped oscillation is given by $x = e^{-0.1t} \sin(\pi t/4)$, where x is in metres and t is in seconds. Calculate the velocity of an oscillation point at $t = 0, T, 2T, 3T, 4T$.

[*Ans.* $v = 0.785$ ms^{-1}, 0.353 ms^{-1}, 0.158 ms^{-1}, 0.071 ms^{-1}, 0.032 ms^{-1}]

4.7 (a) A 24.7 cm long compound pendulum executes damped oscillations. In what time will the energy of the oscillation become 10% of the initial energy if the logarithmic decrement factor is (i) 0.01, and (ii) 0.1.
(b) A compound pendulum performs damped oscillations with a logarithmic decrement factor equal to 0.2. By how many times will the acceleration of the pendulum decrease in its extreme position during one oscillation?

[*Ans.* (a) (i) 120 s, (ii) 1.22 s; (b) 1.22]

4.8 A vertically hanging spring is extended by 9.8 cm when a weight is suspended from it. The weight is then pulled down and released to make it oscillate. For what value of the damping coefficient will (i) the amplitude of oscillation become 1% of its initial value in 10 s, (ii) the weight return to equilibrium aperiodically?
Ans. (i) 0.46 s^{-1}; (ii) 10 s^{-1}

4.9 A particle of mass m is constrained to move along a straight line under the influence of a restoring force $m\mu x$ and resistive force mkx^2. If at $t = 0$, $x = a$ and $\dot{x} = 0$, prove that it will come to rest again at $x = b$, where $\frac{1}{2}\log_e\left(\frac{1+2ak}{1-2bk}\right) = (a+b)k$.

4.10 A particle is executive SHM under the influence of an attractive force $\mu x/a$, where a is its amplitude. If a small force Fx^3/a^3 is further made to act on it, show that the time period, to a first approximation, decreases in ratio $\left(1-\frac{3F}{8\mu}\right):1$.

5

Forced Oscillations

EXPECTED LEARNING OUTCOMES

In this Chapter, you will acquire capability to:

- list differences between free, forced and resonant oscillations;
- establish the equation of motion of a one dimensional (1-D) weakly damped system driven by a harmonic force and solve it to obtain steady-state solution;
- calculate power absorbed, resonance width and quality factor;
- analyse the frequency response of a forced system; and
- write the equation of motion for charge in a forced LCR circuit and solve it by analogy with a mechanical system.

5.1 INTRODUCTION

In the previous Chapter, you have learnt that an oscillating system loses energy due to damping and amplitude of its oscillations decreases gradually with time. However, you must have seen a wall clock and noted that its pendulum continues to oscillate. That is, its motion does not seem to die out. Is this because of absence of damping? The answer to this question is: No, It is not so. You may, therefore, like to know as to what might be the cause behind maintaining the oscillations of the pendulum of a wall clock and similar other systems. Since damping causes loss of energy of a system, we can compensate for it by feeding energy from outside using an external source like a dry pencil cell or such other agent/device.

In general, the frequencies of the oscillating *driven* system to which energy is supplied and the external *driving* force may not be the same. But, irrespective of its natural frequency, the driven system ultimately begins to oscillate with the frequency of the *driver*. Such oscillations are called *forced oscillations.* The amplitude of such forced oscillations is in general small. In everyday life, we come across several systems which execute forced oscillations under the action of a periodic external force. The diaphragms in our ears vibrate under the influence of the vibrations of a musical instrument. Note that the diaphragm executes forced vibrations and transfer of energy is unidirectional, i.e., from the instrument to the ear. Moreover, the driver is not affected in any way whatsoever by the oscillations of the driven system.

When the frequency of the driver exactly matches the natural frequency of the driven (system), i.e., the driver and driven systems are in unison, we observe a spectacular effect - the amplitude of forced oscillations becomes very large. This phenomenon is known as *resonance*. Resonances are desirable in transmission and reception of radio and TV signals, structural engineering, and various other mechanical and molecular phenomena. But these can cause disaster also; literally break a system apart. For instance, if the natural frequency of some part of a car, say engine, gear levers or glass windows equals the frequency of the piston, then that part begins to execute resonant vibrations and rattle vigorously. If the frequency of oscillations is large, it could cause considerable inconvenience and discomfort. In a moving bus, the noise of glass windows is also due to the occurrence of resonance. Fast blowing wind can make a suspension bridge to oscillate. If the frequency of the force generated by the wind matches the natural frequency of the bridge, it gains in amplitude and may ultimately collapse. In 1940, the Tacoma Narrows Bridge in Washington State, USA collapsed within four months of its opening. In your school physics, you must have learnt that for this reason, army is made to break steps rather than march in rhythm while crossing a suspension bridge.

In this Chapter, you will learn how a system responds when it is driven by an external harmonic force. In Section 5.2, you will recapitulate and refresh your knowledge about free, forced and resonant oscillations. Various typical examples of resonance phenomenon observed in actual physical systems are also discussed in this section. In Section 5.3, we have discussed the equation of motion of a weakly damped 1-D oscillator driven by a harmonic force. In Section 5.4, you will learn to solve it and investigate steady-state behaviour. Since forced oscillations involve transfer of energy from the driver to the driven system, and if the natural frequency of the driver and the driven happen to match, these begin to oscillate in unison leading to resonance. This forms the subject matter of discussion of Section 5.5. In Section 5.6, we have derived an expression for the power absorbed by a forced oscillator. You will learn to quantify resonance in terms of quality factor in Section 5.7. An *LCR* circuit as an example of forced electrical oscillations under the influence of a harmonic e.m.f and the condition for obtaining resonance in this circuit are discussed in Section 5.8.

5.2
FREE AND FORCED OSCILLATIONS: RESONANCE

You now know that when an undamped mechanical system is made to oscillate by displacing it from its equilibrium position, it executes SHM; i.e., its displacement-time graph exhibits a sine (or cosine) curve and total energy of the oscillator does not change with time. It means that once such a system has been set in motion, it would continue to oscillate indefinitely. Such oscillations are called *free oscillations*. But do you know of any real physical system which executes completely free oscillatory motion that goes on and on indefinitely? We are sure that your answer to this question will be: The amplitude and energy of oscillations of every oscillating system decreases gradually due to the drag exerted by the medium, say air, which acts as a frictional dissipative force. Recall the oscillations of a swing, a simple pendulum and a spring-mass system left to it. In all these systems, oscillations decrease gradually, depending on the magnitude of damping. It means that every freely oscillating system loses energy in overcoming damping as time passes.

As mentioned earlier, we can compensate for loss of energy and maintain the oscillations of a system by feeding energy to it from outside using an external source/device. The system to which energy is supplied from an external source is said to be *driven* and the external force is known as *driver*. In general, the natural frequencies of the driver and driven are not same. However, as time passes, irrespective of its natural frequency, the driven system ultimately begins to oscillate with the frequency of the driver. Such oscillations are called *forced oscillations.*

When the frequency of the driver exactly matches the natural frequency of the driven (system), we observe that amplitude of forced oscillations becomes very large. This leads to a spectacular effect—the phenomenon is known as *resonance*. Resonances are of two types: *Amplitude resonance* and *velocity resonance*. You will learn about these in this chapter. For the first time, correct theoretical explanation as to how and why resonances occur was given by Galileo.

Forced vibrations and resonance are commonly experienced in everyday life. These are not confined only to mechanical systems; these are also observed in structural engineering, acoustics, electromagnetism and optics. We now discuss some of these.

5.2.1 Examples of Forced Vibrations and Resonance

1. While travelling in a bus, you may have heard the noise of glass windows or some other parts vibrating vigorously. Actually each of its parts, say engine, gear levers, glass windows, etc. have their own natural frequencies. As the piston moves, it imparts a periodic force whose frequency is directly proportional to the speed of the bus. As the speed changes, the frequency of the piston may match the natural frequency of some part of the bus. This will result in the occurrence of resonance, making that part to rattle vigorously. If the frequency of oscillations is large, it could be quite inconvenient, even uncomfortable, to occupants.
2. Resonant vibrations are extremely important in structural engineering. If a heavy machine is operated in a high rise building, the vibrations of the machine induce forced oscillations in its structure. If the frequency of machine is close to the natural frequency of the building, the resulting resonant vibrations will produce large stresses and may result in cracks. That is why in the design of such structures, engineers make allowance for the effect of external periodic forces. This is particularly true while designing an earthquake resistant building in earthquake prone regions such as Japan and Hawaii, among others. You must have read about the earthquake on 26 January, 2001 in Gujarat, India or 2008 earthquake in Sandung region in China which caused huge loss of life and property. The energy released in the quake makes the buildings to oscillate and inflict damage in them.
3. You may have visited Rishikesh sometime and walked over *Lakshman Jhula* and *Ram Jhula* bridges to cross over the Ganges. Did you experience the swing of these suspension bridges caused by the wind? If the frequency of the wind matches the frequency of the bridge, it may gain in amplitude and collapse. In fact, disaster struck Tacoma Narrows Bridge in Washington State in 1940 within six months of its opening. Such oscillations can also arise when we move on such a bridge. That is why army is advised to break steps while marching over a suspension bridge.

4. In stringed instruments like guitar, violin, sitar, etc., strings are stretched on a wooden board covering a hollow box. When the strings are made to vibrate, the air molecules in the box begin to oscillate. This significantly intensifies sound emitted by the strings. In your school physics, you must have learnt about resonance in air columns and worked with them to determine the frequency of a tuning fork. A sonometer also uses the condition of resonance to determine the frequency of a tuning fork.
5. The principle of resonance operates in radio and TV transmission and reception. When you tune into a radio station, which corresponds to a particular wavelength of radio waves, e.g. 400 metre band, you change the capacitance in the tank circuit of your radio. (It contains inductance L and capacitance C joined in parallel.) You may recall that the natural frequency of oscillations of such a circuit is given by $f = (1/2\pi)\sqrt{LC}$. When you turn the knob, you change the natural frequency of the circuit. And when it equals the frequency of radio-waves from the radio station you want to tune in, the periodic electric field of the incoming radio waves sets up resonant electrical oscillations in the radio circuit. These electrical oscillations, after suitable amplification by other electronic circuits, make the membrane of the loudspeaker to vibrate. That is how you hear the desired radio programme.

You may now ask: Do we, in real world, observe infinitely large amplitudes for forced oscillators under the resonance condition? The amplitude significantly increases but it remains finite since one or other form of damping is always present in every system. To understand the behaviour of such systems, we now consider a weakly damped forced oscillator.

5.3 FORCED OSCILLATIONS OF A 1-D WEAKLY DAMPED OSCILLATOR

To discuss the effect of damping on a forced oscillator, refer to Fig. 5.1. Note that the spring-mass system considered in Chapter 4 is now also subject to an external driving force, $F(t) = F_0 \cos \omega t$, where F_0 is the amplitude of driving force. That is, the damped system is pushed back and forth periodically at a frequency ω.

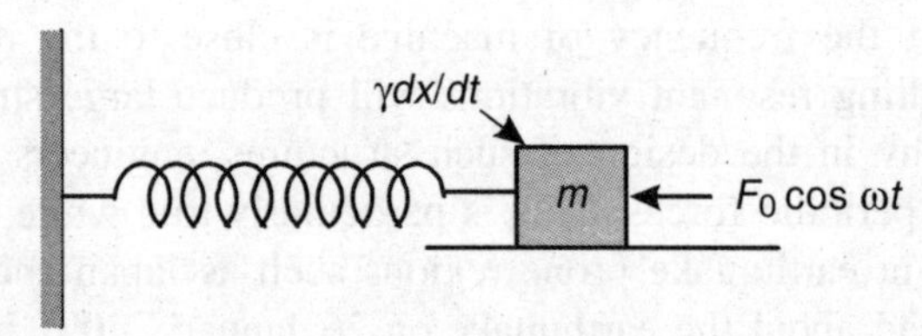

Fig. 5.1 A weakly damped forced spring-mass system. A driving force $F_0 \cos \omega t$ provides energy to the system continuously.

Suppose that the mass m is displaced from its equilibrium position through a distance x and then released. You will agree that at any time t, it is subject to a (i) driving force, $F_0 \cos \omega t$, (ii) restoring force, $-kx$, and (ii) damping force, $-\gamma(dx/dt)$. So for a damped forced oscillator, Eq. (4.2) modifies to

$$m\frac{d^2x}{dt^2} = -kx - \gamma\frac{dx}{dt} + F_0 \cos \omega t. \tag{5.1}$$

On dividing by m and rearranging terms, you will get the differential equation governing the motion of a damped forced oscillator:

$$\frac{d^2x}{dt^2} + 2b\frac{dx}{dt} + \omega_0^2 x = f_0 \cos \omega t, \tag{5.2}$$

where $2b = \gamma/m$, $\omega_0^2 = k/m$ and $f_0 = F_0/m$.

Note that we derived Eq. (5.2) for a mass on a spring. You may now ask: Does Eq. (5.2) apply to any damped oscillator? The answer to this question is: It applies to any damped oscillator – mechanical or electrical—whose natural frequency is different from the frequency of an external harmonic force applied on the system.

To proceed further, we note that Eq. (5.2) is inhomogeneous, of second order and linear with constant coefficients. We should solve it to discover how a weakly damped oscillator behaves under the influence of an external periodic force. But before we do so, it is worthwhile to analyse the situation physically. From Chapter 4, you may recall that a weakly damped ($b < \omega_0$) free system oscillates with an angular frequency, ω_d (= $\sqrt{\omega_0^2 - b^2}$) which is less than the natural frequency of the oscillator. Moreover, the amplitude of oscillations decreases continuously due to loss of energy in overcoming damping. Therefore, when a driving force is applied, it tends to compensate for the loss of energy. In the process, the oscillator begins to acquire the frequency of the applied force and we expect that *initially the actual motion of a weakly damped forced oscillator will arise from superposition of damped oscillations (of frequency ω_d) and those of the driving force (of frequency ω).*

For $\omega \neq \omega_0$, the general solution of Eq. (5.2) is written as

$$x(t) = x_1(t) + x_2(t), \tag{5.3}$$

where $x_1(t)$ is a solution of its homogeneous part. It corresponds to the solution for a weakly damped oscillator discussed in Chapter 4. You may recall that such a system exhibits oscillatory motion and at any instant, the displacement is given by

> The transient state persists from the moment the driving force is applied to the time oscillator acquires completely the frequency of driving force.

$$x_1(t) = a_0 e^{-bt} \cos (\omega_d t + \phi). \tag{5.4}$$

Since $x_1(t)$ decays exponentially, after some time, it will cease to exist, that is, it will disappear. For this reason, it is also referred to as the *transient solution.* The transient motion corresponds to frequency ω_d, which is different from the natural frequency of the oscillator as well as the frequency of the driving force. However, as the driver is feeding energy continuously, the system continues to oscillate. Mathematically, we say that the displacement arrived through the general solution of Eq. (5.2) will not decay with time. Physically, it means *that a forced system gets energy from external source continuously and will ultimately oscillate with the frequency of the driving force.* The system is then said to be in *steady-state.* (The transient part has no role once steady-state has been reached.) We now discuss the steady-state solution of Eq. (5.2).

5.4 STEADY STATE BEHAVIOUR OF A 1-D WEAKLY DAMPED FORCED OSCILLATOR

To obtain the steady-state solution of Eq. (5.2), we combine it with Eq. (5.3). Since $x_1(t)$, given by Eq. (5.4), specifies the solution of homogenous part of Eq. (5.2), only $x_2(t)$ is relevant in steady-state and we can write

$$\frac{d^2x_2}{dt^2} + 2b\frac{dx_2}{dt} + \omega_0^2 x_2(t) = f_0 \cos \omega t. \tag{5.5}$$

Let us now suppose that $x_2(t)$, which specifies displacement of the forced oscillator, lags behind the driving force through an angle θ in the steady-state. Then we can write

$$x_2(t) = a \cos (\omega t - \theta), \tag{5.6}$$

where a and θ are unknown constants. To determine these, we differentiate $x_2(t)$ twice with respect to time. This gives

$$\frac{dx_2(t)}{dt} = -a\omega \sin (\omega t - \theta)$$

and

$$\frac{d^2x_2(t)}{dt^2} = -a\omega^2 \cos (\omega t - \theta).$$

On substituting these expressions in Eq. (5.5) and collecting coefficients of cos $(\omega t - \theta)$ and sin $(\omega t - \theta)$, we get

$$(\omega_0^2 - \omega^2)\, a \cos (\omega t - \theta) - 2ab\omega \sin (\omega t - \theta) = f_0 \cos \omega t.$$

To put this expression in a more mathematically compact form, we use the trigonometric formulae $\cos (\omega t - \theta) = \cos \omega t \cos \theta + \sin \omega t \sin \theta$ and $\sin (\omega t - \theta) = \sin \omega t \cos \theta - \cos \omega t \sin \theta$ and collect the coefficients of $\cos \omega t$ and $\sin \omega t$. This gives

$$[(\omega_0^2 - \omega^2)\, a \cos \theta + 2\, ab\omega \sin \theta - f_0] \cos \omega t$$
$$+ [(\omega_0^2 - \omega^2)\, a \sin \theta - 2\, ab\omega \cos \theta] \sin \omega t = 0. \tag{5.7}$$

From elementary trigonometry, we know that sine and cosine functions never become zero simultaneously; when one increases, the other decreases. It means that the above equation will be satisfied only when both terms within the square brackets—one associated with $\cos \omega t$ and the other associated with $\sin \omega t$—become zero separately:

$$(\omega_0^2 - \omega^2)\, a \cos \theta + 2\, ab\omega \sin \theta = f_0 \tag{5.8}$$

or

$$(\omega_0^2 - \omega^2)\, a \sin \theta - 2\, ab\omega \cos \theta = 0. \tag{5.9}$$

The phase by which the driving force, $F_0 \cos \omega_0 t$ leads the displacement of steady state solution, $x_2(t)$, is easily obtained from Eq. (5.9):

$$\theta = \tan^{-1} \frac{2b\omega}{\omega_0^2 - \omega^2}. \tag{5.10}$$

To obtain the expression for the amplitude of steady-state oscillations, we have to substitute values of $\sin \theta$ and $\cos \theta$ in Eq. (5.8). You can easily calculate these values from Eq. (5.10) by constructing the so-called *impedance triangle*. Obviously, the measure of perpendicular side is $2b\omega$ and the measure of base is $(\omega_0^2 - \omega^2)$. Using Pythagoras theorem, you can readily get the expression for the hypotenuse. These are shown in Fig. 5.2.

From the acoustic impedance triangle, we can write

$$\sin \theta = \frac{2b\omega}{[(\omega_0^2 - \omega^2)^2 + 4b^2\omega^2]^{1/2}}$$

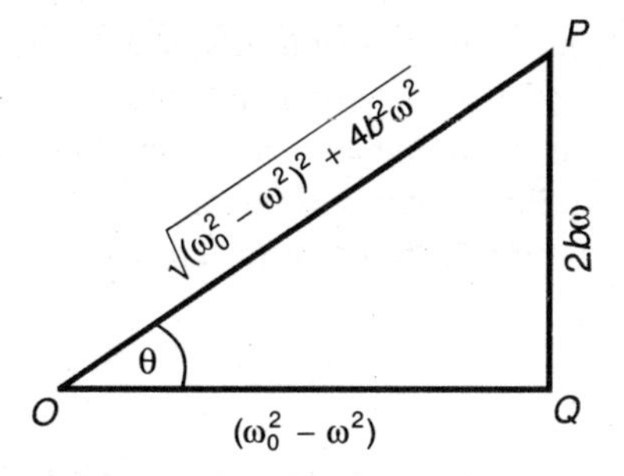

Fig. 5.2 An impedance triangle.

The impedance triangle is essentially a mathematical tool. Now, why the adjective impedance before triangle? We shall make reference to this again in Section 5.5 in connection with Eq. (5.19d) and also in Section 5.8 while working out Example 5.6. We expect you to apply your mind to seek the answer to this question.

and

$$\cos\theta = \frac{(\omega_0^2 - \omega^2)}{[(\omega_0^2 - \omega^2)^2 + 4b^2\omega^2]^{1/2}}.$$

On substituting these values of sin θ and cos θ in Eq. (5.8) and simplifying the resultant expression, we obtain the following expression for the amplitude in steady-state:

$$a(\omega) = \frac{f_0}{[(\omega_0^2 - \omega^2)^2 + 4b^2\omega^2]^{1/2}}$$

$$= \frac{F_0}{m[(\omega_0^2 - \omega^2)^2 + 4b^2\omega^2]^{1/2}}. \quad (5.11)$$

Thus, we find that the steady-state amplitude of a weakly damped forced oscillator depends on

- the amplitude and angular frequency of the driving force;
- the mass and natural angular frequency of the forced system; and
- the damping present in the system.

The steady-state solution of Eq. (5.5) is obtained on substituting this expression for $a(\omega)$ in Eq. (5.6):

$$x_2(t) = \frac{F_0}{m[(\omega_0^2 - \omega^2)^2 + 4b^2\omega^2]^{1/2}} \cos(\omega t - \theta). \quad (5.12)$$

Let us pause for a while and ask: What have we achieved so far? You will note that:

- The frequency of oscillations of a forced weakly damped oscillator in steady-state corresponds to the frequency ω of the driving force rather than its natural frequency ω_0. This is because the oscillator gradually loses its initial energy.
- The steady-state displacement lags the driving force in phase by θ. Moreover, since θ is completely defined by Eq. (5.10), we can say that steady-state displacement does not depend on the initial conditions. In other words, the motion of a forced weakly damped system in steady-state is not dependent on the way it began to oscillate.

The complete solution of Eq. (5.2) can now be written by combining Eqs. (5.4) and (5.12):

$$x(t) = x_1(t) + x_2(t)$$

$$= a_0 e^{-bt}\cos(\omega_d t + \phi) + \frac{F_0 \cos(\omega t - \theta)}{m[(\omega_0^2 - \omega^2)^2 + 4b^2\omega^2]^{1/2}}. \quad (5.13)$$

Now, refer to Fig. 5.3. It shows the time variation of the transient displacement, the steady-state displacement and the instantaneous displacement of a weakly damped forced oscillator. You will note that in the initial stages of motion, oscillations of frequencies ω_d as well as ω are present and the amplitude of resultant oscillation builds up and dies down periodically. These are referred to as *transient beats*. They occur as long as the contribution from the transient part remains significant (curve a).

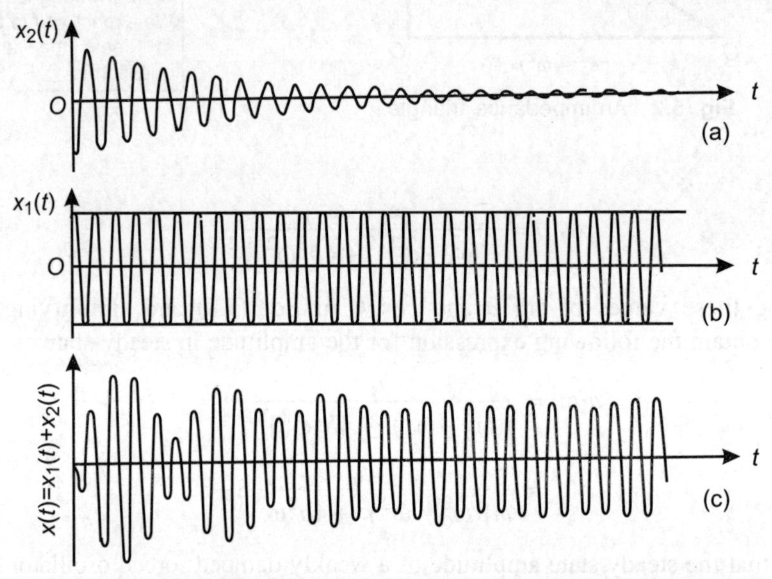

Fig. 5.3 Time variation of the transient displacement (curve a), steady-state displacement (curve b) and instantaneous displacement for a weakly damped forced system (curve c).

You should now go through Example 5.1 carefully to fix the concepts discussed in this section.

EXAMPLE 5.1

An object of mass 10 kg is suspended from a vertical spring of force constant 170 Nm^{-1}. A damping force 20v (given numerically in newton) is also present. (Here v is the instantaneous speed of the object in ms^{-1}). The initial displacement and speed are both equal to zero. An external periodic force $F(t)$ = 150 sin 4t (in newton) is applied. Determine the instantaneous position of the object. Also calculate the amplitude and period of the steady-state oscillations of the object.

Solution: (a) The oscillating object described in the problem is a damped forced oscillator. For such a system, the displacement of the oscillating mass at any instant t is given by Eq. (5.13):

$$x(t) = a_0 e^{-bt} \cos(\omega_d t + \phi) + \frac{F_0 \cos(\omega t - \theta)}{m[(\omega_0^2 - \omega^2)^2 + 4b^2\omega^2]^{1/2}} \qquad \text{(i)}$$

with
$$\theta = \tan^{-1} \frac{2b\omega}{\omega_0^2 - \omega^2}.$$

For the given system, k = 170 Nm^{-1} and m = 10 kg. Therefore,

$$\omega_0^2 = \frac{k}{m} = \frac{170\ \text{Nm}^{-1}}{10\ \text{kg}} = 17\ \text{s}^{-2}$$

Hence,

$$\omega_0 = 4.12 \text{ s}^{-1}. \tag{ii}$$

Further, since damping force constant $\gamma = 20$ Nm^{-1} s, we note that

$$b = \frac{\gamma}{2m} = \frac{20 \text{ Nm}^{-1}\text{ s}}{2 \times 10 \text{ kg}} = 1 \text{ s}^{-1}$$

Further, from the expression for the external periodic force, we note that $\omega = 4$ s^{-1}. Thus, we have, $\omega_0^2 - \omega^2 = 1$.

Substituting these values of b, ω, and $(\omega_0^2 - \omega^2)$ in the expression for phase θ of the steady-state oscillation, we get

$$\theta = \tan^{-1}(2 \times 4) = 0.46\ \pi.$$

For a weakly damped oscillator, $b << \omega_0$ and you can take $\omega_d = (\omega_0^2 - b^2)^{1/2} \approx \omega_0$. Hence, instantaneous displacement (Eq. (i) above) of the object is given by

$$x(t) = a_0 e^{-t} \cos(4.12t + \phi) + \frac{150 \cos(4t - 0.46\pi)}{10\,[1 + 4 \times 1 \times 16]^{1/2}}$$

$$= a_0 e^{-t} \cos(4.12t + \phi) + 1.86 \cos(4t - 0.46\pi) \tag{iii}$$

To determine a_0 and ϕ, we use the initial conditions: $x = 0 = dx/dt$ at $t = 0$. Using the first of these conditions in Eq. (iii), you will obtain

$$0 = a_0 \cos\phi + 1.86 \cos(0.46\pi),$$

which gives

$$a_0 \cos\phi = -0.2344. \tag{iv}$$

To apply the initial condition on velocity, we note from Eq. (iii) that

$$\frac{dx}{dt} = e^{-t} a_0\,[-\cos(4.12t + \phi) - 4.12 \sin(4.12t + \phi)] - (1.86 \times 4) \sin(4t - 0.46\pi).$$

Therefore,

$$\left.\frac{dx}{dt}\right|_{t=0} = -a_0(\cos\phi + 4.12 \sin\phi) + 7.44 \sin(0.46\pi)$$

$$= 0.2344 + 7.38 - 4.12 a_0 \sin\phi.$$

This gives

$$a_0 \sin\phi = \frac{7.61}{4.12} = 1.84. \tag{v}$$

By combining Eqs. (iv) and (v), we obtain

$$\tan\phi = -\left(\frac{1.84}{0.2344}\right) = -7.85$$

so that $\phi = +0.459\ \pi$.

Use this result in Eq. (v) to obtain amplitude of undamped oscillations:

$$a_0 = 1.85.$$

Hence, you can express the instantaneous displacement of the damped forced oscillator in steady state as

$$x(t) = 1.85 e^{-t} \cos(4.12t + 0.459\pi) + 1.86 \cos(4t - 0.46\pi). \tag{vi}$$

The amplitude of the steady-state damped forced oscillations = 1.86 m.

The frequency of the steady-state damped forced oscillations = $\dfrac{\omega}{2\pi} = \dfrac{2}{\pi}$.

The period of the steady-state oscillations = $\dfrac{\pi}{2}$.

EXAMPLE 5.2

A spring is stretched by 5 cm by a force of 5 N. A mass of 1 kg is placed on the lower end of the spring. After equilibrium has been attained, the upper end of the spring is moved up and down so that the external force acting on the mass is given by $F(t) = 20 \cos \omega t$. (a) Obtain expression for instantaneous displacement, and (b) the resonance frequency of the system.

Solution: $$k = \frac{5\text{ N}}{5 \times 10^{-2}\text{ m}} = 10^2\text{ Nm}^{-1}.$$

Since m = 1 kg, the natural angular frequency of the spring-mass system is

$$\omega_0 = \sqrt{\frac{k}{m}} = \sqrt{\frac{10^2\text{ Nm}^{-1}}{1\text{ kg}}} = 10\text{ s}^{-1}.$$

Since the given configuration corresponds to an undamped forced oscillator, we can write the complete solution as:

$$x(t) = A \cos \omega_0 t + B \sin \omega_0 t + \frac{F_0 \cos \omega t}{m(\omega_0^2 - \omega^2)}.$$

To determine A and B, we use initial conditions:

$$x(t = 0) = 0 = A + \frac{20}{(100 - \omega^2)}$$

$$\Rightarrow \qquad A = \frac{20}{\omega^2 - 100} \qquad \text{(i)}$$

and

$$\left.\frac{dx}{dt}\right|_{t=0} = 0 = \omega_0 \text{B}$$

$$\Rightarrow \qquad B = 0. \qquad \text{(ii)}$$

Hence

$$x(t) = \frac{20 \cos 10t}{\omega^2 - 100} + \frac{20 \cos \omega t}{(100 - \omega^2)}$$

$$= \frac{20}{(\omega^2 - 100)}(\cos 10t - \cos \omega t).$$

The steady-state amplitude will become infinite at $\omega = \omega_0 = 10\text{ s}^{-1}$. Hence resonance frequency

$$f_r = \frac{\omega_0}{2\pi} = \frac{5}{\pi}\text{ Hz}.$$

EXAMPLE 5.3

A body of mass 0.2 kg is suspended from a spring of force constant 80 Nm^{-1}. A damping force is acting on the system for which $\gamma = 4$ Nsm^{-1}. Write down the differential equation of motion of the system and calculate the period of free oscillations. Now, a harmonic force $F = 10 \cos 10t$ is applied. Calculate A and δ when the steady-state response is given by $A \cos(\omega t - \delta)$.

Solution: For free oscillations, the differential equation of motion is

$$m\frac{d^2x}{dt^2} = -kx - \gamma\frac{dx}{dt}$$

Substituting $m = 0.2$ kg, $k = 80$ Nm^{-1} and $\gamma = 4$ Nsm^{-1}, this equation takes the form:

$$0.2\frac{d^2x}{dt^2} = -80x - 4\frac{dx}{dt}.$$

We can rewrite it as

$$\frac{d^2x}{dt^2} + 20\frac{dx}{dt} + 400x = 0.$$

From this equation, we note that $2b = 20$ s^{-1} and $\omega_0^2 = 400$ s^{-2} so that $b = 10$ s^{-1} and $\omega_0 = 20$ s^{-1}. Since the system is a damped oscillator with no driving force acting on it, we can write its period T as

$$T = \frac{2\pi}{\omega_d}$$

$$= \frac{2\pi}{[\omega_0^2 - b^2]^{1/2}} = \frac{2\pi}{[400 - 100]^{1/2}}.$$

When a periodic driving force, $F = 10 \cos 10t$ is applied on the system, the equation of motion becomes

$$\frac{d^2x}{dt^2} + 20\frac{dx}{dt} + 400x = 10 \cos 10t.$$

From Eq. (5.11), we recall that amplitude of the forced oscillations is given by

$$A = \frac{f_0}{\left[(\omega_0^2 - \omega^2)^2 + 4b^2\omega^2\right]^{1/2}}$$

$$= \frac{f_0}{[(\omega_0^2 - \omega^2)^2 + 4b^2\omega^2]^{1/2}}$$

$$= 7.7 \times 10^{-3} \text{ m}.$$

The phase of the forced oscillations is given by Eq. (5.10):

$$\tan\delta = \frac{2b\omega}{\omega_0^2 - \omega^2} = \frac{2 \times 10 \times 20}{400 - 100}$$

$$= \frac{4}{3}$$

so that

$$\delta = \tan^{-1}\left(\frac{4}{3}\right).$$

From Chapter 4 you may recall that the time rate of decrease of amplitude of a weakly damped oscillation is quantified in terms of relaxation time τ. Since $\tau = b^{-1}$, small damping implies that transient state will persist longer. On the other hand, heavy damping (large b) leads to fast disappearance of transient state and the beat effect. That is, the existence and duration of beats depend on the nature of damping experienced by the oscillating system. It means that for $t >> \tau$, a damped forced oscillator virtually attains the steady state (This is illustrated in Fig. 5.3 curve b).

Before we discuss steady-state response of a forced weakly damped oscillator to changes in frequency of the applied periodic force of constant amplitude, you should solve Practice Exercise 5.1.

***Practice Exercise* 5.1** A spring-mass system with mass $m = 0.01$ kg, spring factor $k = 25$ Nm^{-1} and $\gamma = 0.2$ kg s^{-1} is subjected to a harmonic driving force of amplitude 2.14 N. Calculate the amplitude and the phase by which the driving force leads the instantaneous displacement in steady state for $\omega = 20$ s^{-1}.

Ans. 0.1 m; $\theta = \tan^{-1}(0.2)$

5.5 AMPLITUDE AND VELOCITY RESONANCE

You now know that in the steady-state, amplitude $a(\omega)$ of a weakly damped forced oscillator is a function of frequency of the (external) driving force. On examining Eq. (5.11) more closely, we note that

(i) When the driving frequency is low ($\omega << \omega_0$) or high ($\omega >> \omega_0$) compared to the natural frequency of the system, the amplitude of steady-state oscillation is small because the term $(\omega_0^2 - \omega^2)^2$ in the denominator will be large.

(ii) As the driving frequency approaches the natural frequency of the system, the amplitude of oscillations begins to increase gradually and attains a very high value when ω is very close to ω_0. The condition in which amplitude acquires maximum value is known as *resonance*. The corresponding value of frequency is called *resonant frequency* and we normally denote it by ω_r.

In the language of differential calculus, we say that the frequency at which the first frequency-derivative of amplitude is zero and the second derivative is negative, will correspond to resonance frequency. We, therefore, calculate the first derivative of $a(\omega)$ with respect to ω using Eq. (5.11) and equate it to zero. This gives

$$\frac{da(\omega)}{d\omega} = \frac{d}{d\omega}\left[\frac{f_0}{[(\omega_0^2 - \omega^2)^2 + 4b^2\omega^2]^{1/2}}\right]$$

$$= \frac{f_0[-4\omega(\omega_0^2 - \omega^2) + 8b^2\omega]}{2[(\omega_0^2 - \omega^2)^2 + 4b^2\omega^2]^{3/2}}. \tag{5.14}$$

For $\omega = \omega_r$, $da(\omega)/d\omega = 0$. Therefore, we can write

$$-4\omega_r(\omega_0^2 - \omega_r^2) + 8b^2\omega_r = 0$$

or

$$4\omega_r[\omega_r^2 - \omega_0^2 + 2b^2] = 0. \tag{5.15}$$

This equation will hold in two situations. We ignore the root $\omega_r = 0$, which is trivial and consider the other condition:

$$\omega_r^2 - \omega_0^2 + 2b^2 = 0.$$

This equation is quadratic in ω_r and its acceptable root is

$$\omega_r = (\omega_0^2 - 2b^2)^{1/2}. \tag{5.16}$$

The root corresponding to the negative sign is physically unacceptable since it is meaningless and has been ignored. You can easily verify that at this value of ω_r, $\left.\dfrac{d^2a}{d\omega^2}\right|_{\omega=\omega_r}$ is negative.

From Eq. (5.16), you can write

$$\omega_0^2 - \omega_r^2 = 2b^2$$

$$\therefore \quad (\omega_0^2 - \omega_r^2)^2 = 4b^4$$

and $\quad \omega_r^2 = \omega_0^2 - 2b^2.$

$$\therefore \quad 4b^2\omega_r^2 = 4b^2(\omega_0^2 - 2b^2).$$

Hence denominator of Eq. (5.11) takes the form

$$\sqrt{(\omega_0^2 - \omega_r^2)^2 + 4b^2\omega_r^2}$$

$$= \sqrt{4b^4 + 4b^2(\omega_0^2 - 2b^2)}$$

$$= 2b\sqrt{b^2 + (\omega_0^2 - 2b^2)}$$

$$= 2b\sqrt{(\omega_0^2 - b^2)}\,.$$

We pause for a while and examine the implications of the result contained in Eq. (5.16). We note that resonance frequency is less than ω_0 as well as ω_d. That is, $\omega_r < \omega_d < \omega_0$. Physically it means that the peak value of amplitude is attained at a frequency below the natural frequency of free oscillator, ω_0, as well as the frequency of weakly damped oscillator, ω_d (= $\sqrt{\omega_0^2 - b^2}$). And the shift in the value of the resonance frequency with respect to the natural frequency ω_0 is governed by damping.

To obtain an expression for the peak value of steady-state amplitude, we note that denominator of Eq. (5.11) takes to a very compact form (= $2b\sqrt{\omega_0^2 - b^2}$) so that

$$a_{\max} = \frac{f_0}{2b\sqrt{\omega_0^2 - b^2}}. \tag{5.17}$$

This equation gives the maximum value of amplitude of the driven system. Then we say that *amplitude resonance* has occurred at frequency $\omega = \omega_r$. Note that the coefficient of damping influences the steady state amplitude.

Before proceeding further, you should work out Practice Exercise 5.2.

***Practice Exercise* 5.2** For the spring-mass system specified in Practice Exercise 5.1, calculate the (i) frequency corresponding to amplitude resonance, (ii) difference between the resonance frequency and the natural frequency of the system, and (iii) amplitude at resonance.

Ans. (i) ω_r = 48 Hz; (ii) 2 Hz; (iii) 0.22 m

The expression for instantaneous velocity is obtained by differentiating Eq. (5.12) with respect to time. This gives

$$v(t) = \frac{dx_2(t)}{dt} = -\frac{F_0\omega}{m\ [(\omega_0^2 - \omega^2)^2 + 4b^2\omega^2]^{1/2}} \sin(\omega t - \theta)$$

$$= -v_0 \sin(\omega t - \theta)$$

$$= v_0 \cos\left(\omega t - \theta + \frac{\pi}{2}\right)$$

$$= v_0 \cos(\omega t + \phi), \tag{5.18}$$

where

$$v_0 = \frac{F_0 \omega}{m[(\omega_0^2 - \omega^2)^2 + 4b^2\omega^2]^{1/2}} \tag{5.19a}$$

defines the velocity amplitude and

$$\phi = \frac{\pi}{2} - \theta \tag{5.19b}$$

is the phase angle by which velocity leads the driving force. Note that like steady-state amplitude, the velocity amplitude of a damped forced oscillator is a function of the frequency of driving force. For high ($\omega >> \omega_0$) as well as low ($\omega << \omega_0$) frequencies, its value is low. However, at $\omega = \omega_0$,

$$(v_0)_{\max} = \frac{F_0}{2mb} = \frac{F_0}{\gamma}. \tag{5.19c}$$

This corresponds to *velocity resonance*.
Now, Eq. (5.19a) can be expressed as

$$v_0 = \frac{F_0}{\{[(m\omega_0^2/\omega) - m\omega]^2 + 4b^2m^2\}^{1/2}}$$

$$= \frac{F_0}{\{[(k/\omega) - m\omega]^2 + \gamma^2\}^{1/2}}$$

$$\therefore \qquad v_0 = \frac{F_0}{\sqrt{X_m^2 + R_m^2}} = \frac{F_0}{Z_m} \tag{5.19d}$$

where

$X_m = (k/\omega) - m\omega$ = Mechanical reactance

$R_m = \gamma$ = Mechanical resistance

Z_m = Mechanical impedance

With the above knowledge you may try to redraw the triangle of Fig. 5.2.

Let us pause for a while and reflect as to what we have learnt in this section. We have learnt about two types of resonances: Amplitude resonance and Velocity resonance. You may now ask: Which one of these is more fundamental? To discover answer to this question, we recall that resonance occurs when frequency of the driver equals the natural frequency of the driven system. This condition is applicable for Eq. (5.19c) rather than Eq. (5.16). So we can say that velocity resonance is more fundamental than amplitude resonance. Let us examine it further.

Since the oscillator absorbs energy/power from the external source, it is only natural to expect that its amplitude will depend on the power drawn by it. Can you list the factors on which the power absorbed by the oscillator depends? To get a feel of these, you may pause for a while and reflect on how children maintain oscillations while enjoying a swing ride. You may recall that they push the ground with their feet periodically in the direction of displacement. Alternatively, another child (or a parent) may give a push when the swing is moving away from them. In this process, energy is imparted to the swing and amplitude of oscillations is maintained when push is in the direction of increasing velocity. You may now ask: What will happen if the swing were pushed when it came towards them? In this case, the push will be directed against the velocity. As a result, the amplitude of swing will decrease and the swing will gradually come to a stop. This means that power drawn by a weakly damped forced system depends on the velocity and phase between the driving force and velocity rather than between the driving force and displacement. We hope that now you are convinced why velocity resonance is more fundamental.

You can view this situation from another angle. The energy given by the driver to the driven system manifests in the form of kinetic and potential energies. You will recall that KE is proportional to the square of velocity, whereas PE depends on the square of displacement. It means that kinetic energy imparts greater drive to the system making velocity resonance more important.

For weak damping ($b << \omega_0$), it follows from Eq. (5.17) that the amplitude at resonance is given by

$$a_{\max} \cong \frac{f_0}{2b\omega_0} = \frac{F_0}{2mb\omega_0} = \frac{F_0}{\gamma\omega_0}. \tag{5.20a}$$

Similarly, from Eq. (5.16) we note that for $b << \omega_0$:

$$\omega_r \approx \omega_0. \tag{5.20b}$$

That is, for weak damping, the amplitude of the forced oscillator is maximum when frequency of the driving force is practically equal to the natural frequency of the oscillator.

From Eq. (5.10), we note that for $\omega = \omega_r \approx \omega_0$, $\tan\theta \to \infty$, which implies that

$$\theta \to \frac{\pi}{2}. \tag{5.21}$$

This result shows that at resonance, the driving force and displacement of the oscillating mass are out of phase by $\pi/2$.

On substituting Eqs. (5.20a) and (5.21) in Eq. (5.12), the instantaneous displacement of a weakly damped forced oscillator at resonance ($\omega = \omega_r \approx \omega_0$) takes a very compact form:

$$x_2(t) = \frac{F_0}{\gamma\omega_0}\cos\left(\omega_0 t - \frac{\pi}{2}\right) = \frac{F_0}{\gamma\omega_0}\sin\omega_0 t. \tag{5.22}$$

Note that even though instantaneous displacement of forced oscillations is out of phase with the driving force, the amplitude of oscillations will be maximum at resonance.

Let us now examine the case of velocity resonance in a weakly damped system.
The velocity of a weakly damped oscillator at resonance ($\omega_r \approx \omega_0$) is obtained by differentiating Eq. (5.22) with respect to time:

$$\frac{dx_2(t)}{dt} = v = \frac{F_0}{\gamma}\cos\omega_0 t. \tag{5.23}$$

Note that velocity is in phase with the applied force ($F = F_0 \cos\omega_0 t$) at resonance.

Figure 5.4 shows the variation of amplitude $a(\omega)$ and phase $\theta(\omega)$ with the frequency of the driving force applied on a weakly damped forced oscillator.

We will now discuss the response of an oscillator in the limiting cases when the frequency of the driving force is significantly lower or higher than the natural frequency ω_0 of the oscillator.

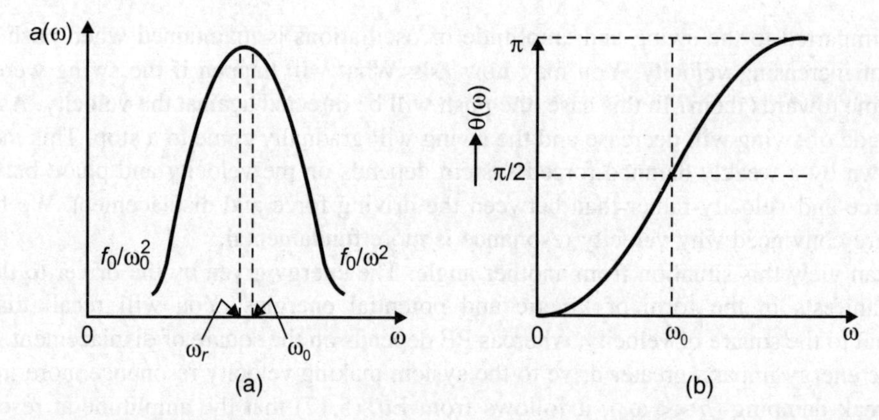

Fig. 5.4 Frequency variation of (a) steady-state amplitude; and (b) phase of a forced oscillator.

5.5.1 Low Driving Frequency

To know how amplitude of a weakly damped oscillator varies at low driving frequencies ($\omega << \omega_0$), we rewrite Eq. (5.11) as

$$a(\omega) = \frac{f_0}{\omega_0^2\left[[1-(\omega^2/\omega_0^2)]^2 + (4b^2\omega^2/\omega_0^4)\right]^{1/2}}.$$

For $\omega << \omega_0$, the ratio ω^2/ω_0^2 will be very small and we can ignore terms of second or higher order in (ω/ω_0). Then expression in the square bracket reduces to unity. This leads us to a compact expression for $a(\omega)$:

$$a(\omega) = \frac{f_0}{\omega_0^2} = \frac{F_0}{m\omega_0^2} = \frac{F_0}{k}. \tag{5.24}$$

That is, *at very low driving frequencies, stiffness constant of the spring and magnitude of the driving force determine the steady-state amplitude of a weakly damped forced oscillator.* Under this condition, Eq. (5.10) implies that

$$\tan\theta = \frac{2b\omega}{\omega_0^2 - \omega^2} \to 0. \tag{5.25}$$

That is, for $\dfrac{\omega}{\omega_0} << 1,\ \theta \to 0.$

Substituting for $a(\omega)$ from Eq. (5.24) and taking $\theta = 0$ in Eq. (5.13), we note that for $\omega << \omega_0$, the displacement of a weakly damped forced oscillator in steady-state is given by

$$x_2(t) = \frac{F_0}{k}\cos\omega t. \tag{5.26}$$

That is, *the driving force and steady-state displacement are in the same phase when the angular frequency of the driver is much less than the natural angular frequency of the system.* This is unlike the case of resonance.

Let us now examine as to what happens to the velocity of the oscillator. Differentiating Eq. (5.26) with time, we obtain

$$v = \frac{dx_2(t)}{dt} = -\frac{\omega F_0}{k}\sin\omega t = \frac{\omega F_0}{k}\cos\left(\omega t + \frac{\pi}{2}\right). \tag{5.27}$$

From this expression we note that velocity of the oscillator leads the driving force by 90°. It means that the force is directed against the velocity for half of the time period and directed along it for the other half. Thus, on an average, the energy transferred from an external driving agency to the oscillator is not significant during one period of oscillation. This explains why the amplitude of oscillations is very small in the limit of low driving frequency.

5.5.2 High Driving Frequency

When frequency of driving force is high ($\omega >> \omega_0$), we rewrite Eq. (5.11) as

$$a(\omega) = \frac{f_0}{\omega^2\{[(\omega_0^2/\omega^2) - 1]^2 + (4b^2/\omega^2)\}^{1/2}}$$

We can neglect terms containing second or higher powers of $(\omega_0/\omega)^2$ as well as $(2b/\omega)^2$ in comparison to one. So when driving frequency is high, the expression for the amplitude of resultant oscillation of a weakly damped forced oscillator takes a very compact form:

$$a(\omega) = \frac{f_0}{\omega^2}. \tag{5.28}$$

This result shows that *at high frequencies, the amplitude is inversely proportional to square of driving frequency.*

For $\omega >> \omega_0$, Eq. (5.10) for the phase of oscillation reduces to

$$\tan\theta = \frac{2b\omega}{(\omega_0^2 - \omega^2)} = -\frac{2b}{\omega} \xrightarrow[\omega \to \infty]{} 0 \tag{5.29}$$

so that $\theta = \pi$. Physically, this implies that *at high frequencies, the driving force and displacement are in opposite phases.*

You may now like to check your understanding and derive a relation connecting velocity and driving force.

***Practice Exercise* 5.3** By combining Eqs. (5.22), (5.28) and (5.29), discuss the phase relationship between velocity and driving force.

You now know that oscillations of a forced damped oscillator are maintained by the energy supplied by an external agency. In the steady-state, the average power spent in doing work against the damping force equals the average power supplied by the external source. This means that energy absorbed by a forced oscillator in one cycle is completely spent in overcoming dissipation of energy due to damping. You may, therefore, like to know the average rate at which energy (average power) must be supplied to the oscillator to sustain its steady-state oscillations. This forms the subject matter of discussion of Section 5.6.

5.6 POWER ABSORBED BY A WEAKLY DAMPED FORCED OSCILLATOR

From your elementary Physics classes, you will recall that power is defined as the work done per unit time. So we can write the expression for instantaneous power as:

$$\text{Power} = \frac{\text{Work}}{\text{Time}} = \frac{\text{Force} \times \text{Displacement}}{\text{Time}}$$

so that

$$P(t) = \text{Force} \times \text{Velocity}$$
$$= F(t) \times v(t). \tag{5.30}$$

On substituting $F(t) = F_0 \cos \omega t$ and using Eq. (5.18) in Eq. (5.30), the expression for instantaneous power absorbed by the oscillator takes the form

$$P(t) = F_0 v_0 \cos \omega t \cos (\omega t + \phi).$$

Using the trigonometric formula $\cos (\omega t + \phi) = \cos \omega t \cos \phi - \sin \omega t \sin \phi$ in the above relation, we can rewrite the expression for instantaneous power as

$$P(t) = F_0 v_0 [\cos^2 \omega t \cos \phi - \cos \omega t \sin \omega t \sin \phi].$$

From this you can easily calculate the average power absorbed in one cycle:

$$\langle P \rangle = F_0 v_0 \cos \phi \langle \cos^2 \omega t \rangle - \frac{1}{2} F_0 v_0 \sin \phi \langle \sin 2 \omega t \rangle. \tag{5.31}$$

From Chapter 1, we recall that $\langle \sin 2 \omega t \rangle = 0$ and $\langle \cos^2 \omega t \rangle = 1/2$. Though these results were obtained as applications of integral calculus, these provide significant insights in understanding the oscillatory behaviour of physical systems. The fact that average value of $\sin 2\omega t$ is zero over a complete cycle essentially implies that there is perfect balance in energy exchange between mechanical reactance and the driver. Similarly, the result $\langle \cos^2 \omega t \rangle = 1/2$ is responsible for dissipation of energy through the mechanical resistor.

On using these results in Eq. (5.31) and inserting the value of ϕ from Eq. (5.19b), we obtain the desired expression for average power absorbed over one cycle:

$$\langle P \rangle = \frac{1}{2} F_0 v_0 \cos \phi = \frac{1}{2} F_0 v_0 \sin \theta. \tag{5.32}$$

Note that the average power absorbed by a forced oscillator in a cycle will be maximum when $\sin \theta = 1 = \cos \phi$, i.e., $\theta = \pi/2$ or $\phi = 0$. For this reason, $\cos \phi$ is referred to as *power factor*.

We can also express this condition in terms of frequency. To do so, we substitute the value of $\sin \theta$ obtained from impedance triangle (Fig. 5.2) and v_0 from Eq. (5.19a) in Eq. (5.32). This gives

$$\langle P \rangle = \frac{1}{2} F_0 \frac{F_0 \omega}{m[(\omega_0^2 - \omega^2)^2 + 4b^2\omega^2]^{1/2}} \times \frac{2b\omega}{[(\omega_0^2 - \omega^2)^2 + 4b^2\omega^2]^{1/2}}$$

$$= \left(\frac{bF_0^2}{m} \right) \frac{\omega^2}{[(\omega_0^2 - \omega^2)^2 + 4b^2\omega^2]}. \tag{5.33}$$

From this result you will note that the average power absorbed by the driven system will be maximum at $\omega = \omega_0$. It means that $\cos\phi = 1$ corresponds to the condition of velocity resonance. The average power absorbed at resonance is

$$<P>_{\max} = \frac{1}{4bm} F_0^2. \tag{5.34}$$

That is, *maximum power is absorbed at resonance and its magnitude is controlled by damping in the system and the amplitude of driving force.*

As discussed in Section 5.2, the concept of forced oscillations and associated phenomenon of resonance play a significant role in designing mechanical, electrical and optical systems. In particular, engineers and scientists involved in the design and development of electronic and optical communication and transmission systems take into account several parameters associated with forced oscillations. One such useful parameter is quality factor of a forced oscillatory system. You will learn about this now.

5.7 QUALITY FACTOR: SHARPNESS OF RESONANCE

In Chapter 4 you learnt to characterise damping in terms of quality factor. This parameter can also be related to power absorbed in a circuit and obtain useful information about the efficiency and effectiveness of a forced oscillator. In fact, it also tells us the rate at which power decreases as a function of the difference of driving frequency and the resonance frequency. This parameter is particularly useful in the study of electrical and electronic circuits.

To obtain a relation between average power absorbed by a forced oscillatory system and its quality factor, we take out $4b^2\omega^2$ as the common factor from the denominator of Eq. (5.33) and substitute Q for $\omega_0/2b$ as per the expression obtained for quality factor in Chapter 4. Then, we can rewrite Eq. (5.33) as

$$<P> = \frac{F_0^2}{4bm\,[Q^2[(\omega_0/\omega) - (\omega/\omega_0)]^2 + 1]} = \frac{<P>_{\max}}{[Q^2[(\omega_0/\omega) - (\omega/\omega_0)]^2 + 1]}. \tag{5.35}$$

Figure 5.5 shows a plot of Eq. (5.35) for average power ($<P>$) absorbed by a weakly damped forced oscillator as a function of frequency of the driver for different values of quality factor. We note that

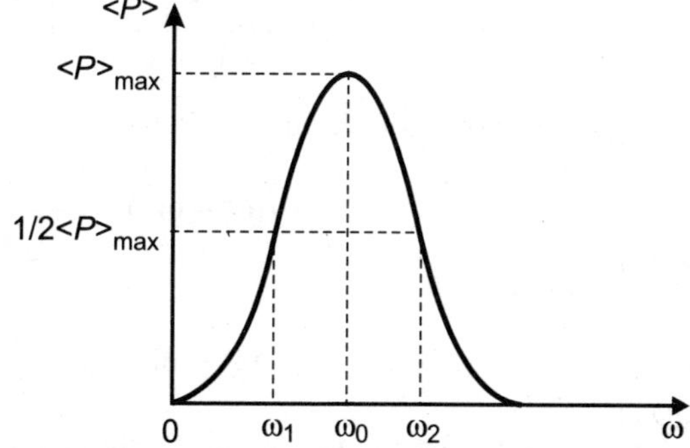

Fig. 5.5 Frequency variation of average power absorbed by a forced damped oscillator from the external source. $\omega = \omega_0$ denotes the position of resonance and ω_1 and ω_2 correspond to half-power points.

(i) The maximum value of average power is obtained at $\omega = \omega_0$ and its value drops fairly sharply around the resonance frequency.

(ii) At ω_1 ($<\omega_0$) and ω_2 ($> \omega_0$), the power absorbed by the oscillator drops to half-the maximum value. For this reason, these frequencies are known as *half-power frequencies* and are more or less evenly distributed about ω_0. $(\omega_2 - \omega_1)$ is called the *bandwidth or full width at half-power points.*

(iii) For large Q values (weak damping), the curve will have a higher maxima. Resonance in this case is said to be *sharp*.

(iv) For small Q (heavy damping), the maxima will be comparatively small value and drop gradually around ω_0. In this case, resonance is said to be *flat or broad*.

Thus, we may conclude that the quality factor controls not only the maximum average power absorbed by an oscillator but also the rapidity with which the absorbed power decreases as the driving frequency deviates away from the resonance frequency.

Before we proceed further, you should answer the following Practice Exercise.

Practice Exercise 5.4 The work done against a typical damping force is $2bmv^2$. Show that the average power dissipated by a damped forced oscillator in steady-state is equal to the average power supplied by the driver.

In Fig. 5.5, curve A corresponds to a sharp resonance. It means that its band width $(\omega_2 - \omega_1)$ is small. Since large Q value signifies a sharp resonance, we can say that smaller the band width, larger is the Q factor and vice-versa. It suggests a connection between quality factor and bandwidth. Let us now discover how these are related.

From Eq. (5.35), we note that half-power frequencies will satisfy the equation

$$\langle P\rangle = \frac{\langle P\rangle_{max}}{\{Q^2[(\omega_0/\omega) - (\omega/\omega_0)]^2 + 1\}} = \frac{\langle P\rangle_{max}}{2}.$$

We can rewrite it as

$$Q^2\left(\frac{\omega_0}{\omega} - \frac{\omega}{\omega_0}\right)^2 = 1.$$

On re-arranging terms, we can write

$$(\omega_0^2 - \omega^2)^2 = \frac{\omega^2\omega_0^2}{Q^2},$$

so that

$$\omega_0^2 - \omega^2 = \pm\frac{\omega\omega_0}{Q}. \tag{5.36}$$

Equation (5.36) has 4 roots. Of these, two roots correspond to negative frequencies and are physically unacceptable. The two acceptable roots are

We first rewrite Eq. (5.36) by considering the positive sign:

$$\omega^2 + \frac{\omega_0}{Q}\omega - \omega_0^2 = 0. \qquad \text{(i)}$$

The two roots of (i) are

$$\omega = \frac{-(\omega_0/Q) \pm \sqrt{(\omega_0^2/Q^2) + 4\omega_0^2}}{2}$$

$$= -(\omega_0/2Q) \pm \omega_0\sqrt{1 + (1/4Q^2)}. \qquad \text{(ii)}$$

Of these roots,

$$-\frac{\omega_0}{2Q} - \omega_0\sqrt{1 + \frac{1}{4Q^2}}$$

corresponds to a negative frequency and is physically absurd. Next, by considering the negative sign in Eq. (5.36), you will obtain the equation

$$\omega^2 - \frac{\omega_0}{Q}\omega - \omega_0^2 = 0,$$

Its roots are

$$\omega = \frac{(\omega_0/Q) \pm \sqrt{(\omega_0^2/Q^2) + 4\omega_0^2}}{2}.$$

$$= \frac{\omega_0}{2Q} \pm \omega_0\sqrt{1 + \frac{1}{4Q^2}}. \qquad \text{(iii)}$$

Of these, the root

$$\frac{\omega_0}{2Q} - \omega_0\left(1 + \frac{1}{4Q^2}\right)^{1/2}$$

will correspond to negative frequency.

Hence physically meaningful solutions correspond to those given in Eq. (5.37 a, b).

$$\omega_1 = -\frac{\omega_0}{2Q} + \omega_0 \left(1 + \frac{1}{4Q^2}\right)^{1/2} \tag{5.37a}$$

and

$$\omega_2 = \frac{\omega_0}{2Q} + \omega_0 \left(1 + \frac{1}{4Q^2}\right)^{1/2}. \tag{5.37b}$$

Obviously, $\omega_2 > \omega_0 > \omega_1$. This is shown in Fig. 5.5.

The frequency interval between two half-power points is given by

$$\omega_2 - \omega_1 = \frac{\omega_0}{Q}.$$

On re-arranging terms, we can express quality factor in terms of bandwidth:

$$Q = \frac{\omega_0}{\omega_2 - \omega_1} \tag{5.38}$$

This result shows that the quality factor is equal to the ratio of frequency at which power resonance occurs to the full width at half power points. Note that this expression for quality factor is in conformity with the fact that a high Q system has small bandwidth and a low Q system has large bandwidth.

There are other convenient and physically more meaningful definitions of quality factor also. Since $Q = \omega_0/2b$ and the relaxation time $\tau = 1/b$, we can write

$$Q = \frac{\omega_0 \tau}{2}. \tag{5.39}$$

In Section 5.3, you learnt that relaxation time is a measure of the time taken by a damped forced system to attain steady-state after the application of the periodic force. Therefore, from Eq. (5.39), we note that larger value of τ implies higher value of quality factor. And when Q is large, resonance is sharp, implying high *selectivity*. This runs counter to the requirement of high *fidelity* because faithful reproduction of impressed frequency demands quick attainment of steady-state and disappearance of transient beats (small Q and small τ). *Therefore, we can say that fidelity and selectivity are competitive phenomena and these have to be balanced in radio reception.*

Before proceeding further, you should read Example 5.4.

EXAMPLE 5.4

An oscillator of mass 0.01 kg draws maximum power at a frequency of 96 Hz with half-power points at 93 Hz and 99 Hz. If the maximum average power drawn by the oscillator is 20 W, calculate (i) ω_0, (ii) Q-factor, (iii) the damping factors γ and b, (iv) the amplitude of the driving force (F_0), and (v) the amplitude of the oscillator at $\omega = \omega_0$.

Solution: We note that resonance frequency $f_0 = 96$ Hz. Therefore, resonance angular frequency

$$\omega_0 = 2\pi f_0 = 2\pi \times 96 \text{ Hz} = 603.18 \text{ s}^{-1}.$$

Also, two half-power frequencies are

$$f_2 = 99 \text{ Hz}, \quad \text{i.e., } \omega_2 = 2\pi f_2 = 2\pi \times 99 \text{ Hz}$$

and

$$f_1 = 93 \text{ Hz}, \quad \text{i.e., } \omega_1 = 2\pi f_1 = 2\pi \times 93 \text{ Hz}.$$

$$\therefore \qquad Q = \frac{\omega_0}{(\omega_2 - \omega_1)} = \frac{f_0}{f_2 - f_1}$$

$$= \frac{96}{99 - 93} = 16.$$

We also know that $Q = \omega_0/2b$. Therefore,

$$b = \frac{\omega_0}{2Q} = \frac{2\pi \times 96}{2 \times 16} = 18.85 \text{ s}^{-1}.$$

Since mass of the oscillator (m) = 0.01 kg, the damping factor

$$\gamma = 2bm = 2 \times 18.85 \times 0.01 = 0.377 \text{ kg s}^{-1}.$$

Since $\langle P \rangle_{max} = 20$ W, we recall that

$$\langle P \rangle_{max} = \frac{F_0^2}{4bm}.$$

On substituting the given values, we get

$$20 = \frac{F_0^2}{4bm}.$$

Hence

$$F_0^2 = 20 \times 4bm = 80 \times 18.85 \times 0.01 = 15.08 \text{ N}^2$$

so that

$$F_0 = 3.88 \text{ N}.$$

At $\omega = \omega_0$, $a(\omega)$ is given by

$$a(\omega_0) = \frac{f_0}{2b\omega_0} = \frac{F_0}{2mb\omega_0}$$

$$= \frac{3.88 \text{ N}}{2 \times (0.01 \text{ kg}) \times (18.85 \text{ s}^{-1}) \times (603.18 \text{ s}^{-1})}$$

$$= 0.017 \text{ m}.$$

In Chapter 4, we discussed the oscillations of charge in an *LCR*-circuit. But if an external e.m.f is impressed on this circuit, we get a resonant *LCR* circuit, which finds use in signal reception (radio or TV) It forms subject matter of Section 5.8. For simplicity, we will discuss the behaviour of this system by drawing similarities with a mechanical system.

5.8

A RESONANT *LCR* CIRCUIT

From Chapter 4, you will recall that in a damped *LCR* circuit, oscillations of charge die out because of power losses in the resistance. What changes should we expect when a source of alternating e.m.f having frequency ω is introduced in the circuit? To answer this question, refer to Fig. 5.6. If I is the current in the circuit at a given time, the applied alternating e.m.f (= $E_0 \cos \omega t$) will be equal to the sum of the potential differences across the capacitor, resistor and the inductor. Then we can write

$$\frac{q(t)}{C} + RI + L\frac{dI}{dt} = E_0 \cos \omega t.$$

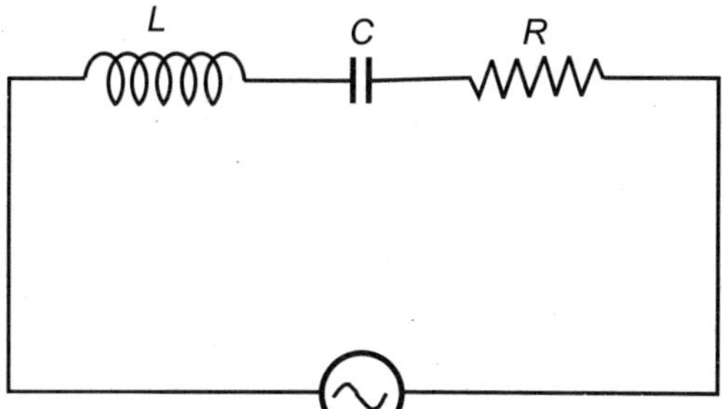

L C R

$E_0 \cos \omega t$

Fig. 5.6 A *LCR* circuit driven by source of alternating e.m.f.

Since $I = dq/dt$, we can rewrite the equation of motion of charge as

$$L\frac{d^2q(t)}{dt^2} + R\frac{dq(t)}{dt} + \frac{q(t)}{C} = E_0 \cos \omega t.$$

Dividing throughout by L, we get

$$\frac{d^2q(t)}{dt^2} + \frac{R}{L}\frac{dq(t)}{dt} + \frac{q(t)}{LC} = \frac{E_0}{L} \cos \omega t. \tag{5.40}$$

In this form, this equation is similar to Eq. (5.2). Hence its steady-state solution can be written by analogy with $2b = R/L$, $\omega_0^2 = 1/LC$ and $f_0 = E_0/L$. Thus, for a weakly damped *LCR* circuit, the charge on capacitor plates at any instant of time will be given by an expression analogous to Eq. (5.12):

$$q(t) = \frac{E_0/L}{\{[(1/LC) - \omega^2]^2 + (\omega R/L)^2\}^{1/2}} \cos(\omega t - \theta), \tag{5.41}$$

As you have learnt in the previous section, the resonant frequency, ω_r of a weakly damped forced oscillator is equal to the frequency, ω of the driver and is given by

$$\omega \approx \omega_r = (\omega_0^2 - 2b^2)^{1/2}$$

Substituting the values of ω_0 and $2b$ for a forced *LCR* circuit, we can write the expression for angular frequency of charge oscillations as:

$$\omega = \sqrt{\frac{1}{LC} - \frac{R^2}{4L^2}}. \tag{5.42}$$

Further, the expression for $q(t)$ can be rewritten as

$$q(t) = \frac{E_0}{\omega\{R^2 + [\omega L - (1/\omega C)]^2\}^{1/2}} \cos(\omega t - \theta)$$

$$= \frac{E_0}{\omega Z} \cos(\omega t - \theta), \tag{5.43}$$

where

$$Z = \left[R^2 + \left(\omega L - \frac{1}{\omega C}\right)^2\right]^{1/2}. \tag{5.44}$$

Similarly, using Eq. (5.10) and the value of $2b$ and ω_0, we can write the phase of the oscillating charge in the circuit as

$$\tan\theta = \frac{(\omega R/L)}{1/LC - \omega^2} = \frac{R}{(1/\omega C) - \omega L}. \tag{5.45}$$

The current in the circuit at any instant is obtained by differentiating Eq. (5.43) with respect to t. This gives

$$I = -\frac{E_0}{Z} \sin(\omega t - \theta)$$

$$= \frac{E_0}{Z} \cos(\omega t - \phi), \tag{5.46}$$

where $\phi = \theta + (\pi/2)$ defines the phase difference between impressed e.m.f and instantaneous current.

From Eq. (5.46), we note that current in a *LCR* circuit is a function of frequency. The relation between ϕ and θ gives $\tan\phi = -\cot\theta$. Thus, using Eq. (5.45), we can write

$$\tan\phi = \frac{\omega L - (1/\omega C)}{R}$$

so that

$$\phi = \tan^{-1} \frac{\omega L - (1/\omega C)}{R}. \tag{5.47}$$

So far we have obtained expressions for the instantaneous values of charge and current in a driven *LCR* circuit by drawing analogies with the damped forced mechanical oscillator. We also wrote expression for the phase difference between the current in the circuit and the applied e.m.f. You may now like to know: Under what conditions is maximum power absorbed by an *LCR* circuit (which is an electrical analogue of damped harmonic oscillator) from the driver? To discover answer to this question, we study the effect of frequency of the driver on the current-e.m.f phase relation. When frequency ω of the driver is low, we can write

$$\omega L << \frac{1}{\omega C}.$$

In this condition, the circuit is capacitive in nature and we can write

$$\left(\omega L - \frac{1}{\omega C}\right)^2 \simeq \frac{1}{\omega^2 C^2}.$$

Thus, when we are working at low frequencies and R is also small (weak damping), the current amplitude will be small. (Refer to Eq. (5.46) and the expression for Z.) What will be its magnitude for $\omega \to 0$? In this limit, $I \to 0$. This is depicted in Fig. 5.7a for different values of R. Further, from Eq. (5.47) we note that for $\omega \to 0$, $\tan \phi \to \infty$ and $\phi = \pi/2$. This means that current leads the applied emf by $\pi/2$. This is shown in Fig. 5.7b.

As the driving frequency increases, the reactance $\omega L - (1/\omega C)$ decreases and current amplitude increases. When

$$\omega L = \frac{1}{\omega C}$$

or

$$\omega = \frac{1}{\sqrt{LC}}, \tag{5.48}$$

the circuit will be purely resistive and impedance becomes minimum; equal to R . Then the current attains its peak value ($= E_0/R$) and the circuit is said to resonate with angular frequency ω_r:

$$\omega_r = \frac{1}{\sqrt{LC}}. \tag{5.49}$$

At resonance, the current and applied e.m.f are in phase ($\phi = 0$).

When the driving frequency is high $\omega L >> (1/\omega C)$, the circuit will be inductive. For very high frequencies, $(1/\omega C) \to 0$ and $\omega L \to \infty$. In this condition, current amplitude will again decrease and Eq. (5.47) shows that the current will lag behind e.m.f by $\pi/2$.

For different values of R (for a fixed L), the frequency variations of peak current and phase are shown in Fig. 5.7. You will observe that lower the resistance (higher Q), higher is the peak value of current and sharper is the resonance.

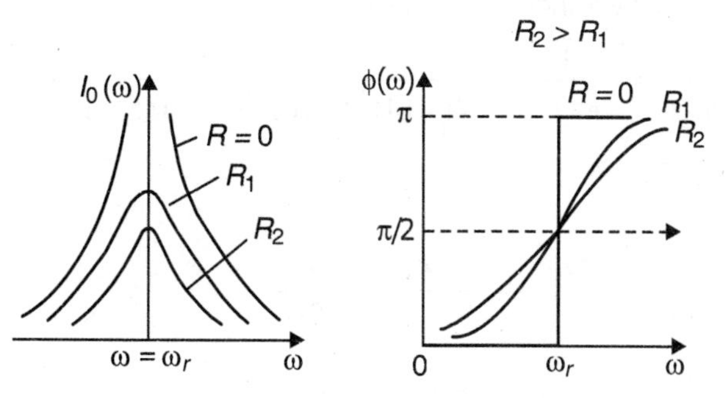

Fig. 5.7 Variation of the peak value of current I_0 and its phase ϕ with the frequency ω of applied voltage. $\omega = \omega_r$ is the resonant condition.

From Chapter 4 you may recall that quality-factor of an *LCR* circuit is given by

$$Q = \frac{\omega_0 L}{R},$$

where $\omega_0 = 1/\sqrt{LC}$. Note that Q-factor of a highly resistive *LCR* circuit will be low.

Equation (5.46) can be rewritten in terms of the Q-factor of the circuit as

$$I(t) = \frac{(E_0/R)}{\sqrt{Q^2[(\omega_0/\omega) - (\omega/\omega_0)]^2 + 1}} \cos(\omega t - \phi). \tag{5.50}$$

You can easily see that the maximum value of current at resonance is obtained at $\omega = \omega_r$:

$$I_{max} = \frac{E}{R}.$$

When driver frequency ω is slightly lower or higher than the natural frequency of the forced system (ω_0), the current decreases from the maximum value (E_0/R) at $\omega = \omega_r$. The rapidity with which the current drops with ω is determined by the quality factor; higher Q induces more rapid fall.

You can convince yourself that at the resonant frequency ω_r, the charge in the circuit will not have the maximum value, rather the current will be maximum. By referring to Section 2.8, you will recall that current is analogous to velocity and charge is analogous to displacement and now if you read Section 5.5 again you will perhaps have no doubt that velocity resonance is more fundamental than amplitude resonance.

The quality factor of a circuit determines its ability to select a narrow band of frequencies from a wide range of input frequencies. The radio receivers operating in the MHz region have Q values of the order of 10^2 to 10^3. This explains why oscillatory circuits in radio sets have low resistance. You may now like to study some Solved Examples.

EXAMPLE 5.5

An alternating potential of frequency 100 kHz and peak value 1.0 V is applied to a series *LCR* circuit. If $L = 0.5$ mH and $R = 20\ \Omega$, calculate the value of the capacitance C to obtain condition of resonance. Also calculate the rms value of current.

Solution: At resonance, capacitance is given by

$$C = \frac{1}{\omega_0^2 L} = \frac{1}{4\pi^2 f_r^2 L} = \frac{1}{4(3.14)^2 \times (10)^{10} \text{s}^{-2} \times 0.5 \times 10^{-3}\ \text{H}}$$

$$\approx 5 \times 10^{-9}\ \text{F}.$$

Hence

$$i_{rms} = \frac{E_{rms}}{R} = \frac{E_0}{R\sqrt{2}}$$

$$= \frac{1.0\ \text{V}}{20\sqrt{2}\ \Omega}$$

$$= 0.035\ \text{A}.$$

EXAMPLE 5.6

Use Eq. (5.38) to show that quality factor $Q = \omega_0 L/R$ for an LCR-circuit.

Solution: In analogy with Eq. (5.32), we can write

$$P = \frac{1}{2} V_0 I_0 \cos \phi$$

where R, X and Z are respectively the resistance, reactance and impedance of the circuit. By choosing R, X and Z as base, normal and hypotenuse and redraw the impedance triangle (Fig. 5.3), we can rewrite the expression for power as

$$P = \frac{V_0 I_0}{2} \frac{R}{Z}, \tag{5.51}$$

$$\therefore \quad P = \frac{1}{2} V_0 \frac{V_0}{Z} \frac{R}{Z} = \frac{1}{2} \frac{V_0^2 R}{Z^2}. \tag{5.52}$$

Obviously, power will be maximum when $\cos \phi = 1$ or impedance is equal to resistance, i.e., the circuit is resistive only:

$$P_{max} = \frac{V_0^2}{2R}.$$

Corresponding to half of maximum power, we have

$$\frac{1}{2} \frac{V_0^2}{2R} = \frac{V_0^2 R}{2Z^2} \tag{5.53}$$

$$\therefore \quad 2R^2 = Z^2 = R^2 + X^2$$

so that

$$X^2 = R^2$$

or

$$X = \pm R. \tag{5.54}$$

This result shows that X can have two values. The value corresponding to $X = R$ is

$$\omega L - \frac{1}{\omega C} = R. \tag{5.55}$$

$$\Rightarrow \quad LC\omega^2 - RC\omega - 1 = 0.$$

This is quadratic in ω and the admissible value is

$$\omega = \frac{RC + \sqrt{R^2 C^2 + 4LC}}{2LC}. \tag{5.56a}$$

Similarly, for $X = -R$, we get

$$\omega = \frac{-RC + \sqrt{R^2 C^2 + 4LC}}{2LC}. \tag{5.56b}$$

In Eq. (5.38), $\omega_2 > \omega_1$. Therefore, we choose the greater value of ω as ω_2 and the smaller value as ω_1. Hence, we can write

$$\omega_2 - \omega_1 = \frac{2RC}{2LC} = \frac{R}{L}.$$

$$\therefore \quad Q = \frac{\omega_0 L}{R}. \tag{5.57}$$

REVIEW EXERCISES

5.1 A particle of mass m is acted on by a restoring kx, where x is its distance from a fixed point. An external force of magnitude $F \cos 2\omega t$ acts on it along the line joining it to the fixed point, where $\omega = \sqrt{k/m}$. Describe its subsequent motion if it is initially at rest at the fixed point. Express the limits of motion on either side of the fixed point in terms of m, F and ω.

$$\left[\textbf{Ans.} \quad x = -\frac{3}{8}\frac{F}{m\omega^2} \text{ to } x = \frac{2}{3}\frac{F}{m\omega^2}\right]$$

5.2 (a) A particle hangs at rest at the end of an elastic string whose unstretched length is a. In equilibrium position, the length of the string is b and its time period of oscillation is T. At $t = 0$, the point of suspension begins to move so that its downward displacement at time t is $c \sin \omega t$. Show that length of the string at time t is given by

$$b - \frac{c\omega\omega_0}{\omega_0^2 - \omega^2}\sin \omega_0 t + \frac{c\omega^2}{\omega_0^2 - \omega^2}\sin \omega t,$$

where $\omega_0 = (2\pi/T)$.

(b) What would be the length of the string at time t if in the above case, $\omega = \omega_0$.

$$\left[\textbf{Ans.} \quad b - \frac{c}{2}\sin \omega_0 t - \frac{1}{2}\omega_0 ct \cos \omega_0 t\right]$$

5.3 A body of mass 10 g performs damped oscillations with maximum amplitude 7 cm. The damping coefficient is 1.6 s^{-1} and the epoch is zero. The body is acted upon by an external force which produces forced oscillation whose equation is $x = 5 \sin (10\pi t - 0.75\pi)$, where x is in cm and t is in seconds. Write down (i) the differential equation of motion of natural oscillations and (ii) expression for external periodic force.

[**Ans.** $x = 7e^{-1.6t} \sin 10.5\pi t$, $F = 7.2 \times 10^{-2} \sin 10\pi t$ N]

5.4 A telephone diaphragm of effective mass 1 g is acted on by a restoring force of 100 N per cm of displacement, a retarding force of 0.04 N per unit velocity (in cm s^{-1}) and a driving force of sin 1000 t newton. Calculate its (i) mechanical impedance, (ii) maximum amplitude and (iii) maximum velocity. At what frequencies do (ii) and (iii) occur?

[**Ans.** (i) 9 kgs^{-1}, (ii) 8 cm, (iii) 25ms^{-1}]

5.5 Derive an expression for the acceleration of a damped oscillator driven by a force $F \cos \omega t$. At what frequency will this be maximum? Obtain the limiting value of acceleration amplitude at high frequencies. Under what condition will this value be equal to the acceleration amplitude at velocity resonance?

$$\left[\textbf{Ans.} \quad f = \frac{F\omega^2}{m[(\omega_0^2 - \omega^2)^2 + 4b^2\omega^2]^{1/2}}\cos(\omega t - \phi), \text{ where } \phi = \tan^{-1}\left(\frac{2b\omega}{\omega_0^2 - \omega^2}\right)^{-\pi/2};\right.$$

$$\left. f\Big|_{\text{high}\,\omega} = \frac{F}{m}; \text{ high } \omega_0\right]$$

5.6 Show that for a forced oscillation, the total energy of a system is not constant.

Annexure

Alternative Methods for solving the Differential Equation of a Forced Oscillator

Method 1: Using operator $D \equiv (d/dt)$

We start from Eq. (5.5) and write

$$\frac{d^2 x_2}{dt^2} + 2b\frac{dx_2}{dt} + \omega_0^2 x_2(t) = f_0 \cos \omega t .$$

In operator notation, we can rewrite as

$$(D^2 + 2bD + \omega_0^2)\, x_2(t) = f_0 \cos \omega t .$$

On rearranging terms, we get

$$x_2 = \frac{1}{D^2 + 2b + \omega_0^2} f_0 \cos \omega t$$

$$= f_0 \frac{D^2 + \omega_0^2 - 2bD}{(D^2 + \omega_0^2)^2 - 4b^2 D^2} \cos \omega t$$

$$= \frac{f_0}{(\omega_0^2 - \omega^2)^2 + 4b^2\omega^2}(D^2 - 2bD + \omega_0^2) \cos \omega t$$

$$= F_0 [(-\omega^2 + \omega_0^2) \cos \omega t + 2b\omega \sin \omega t], \quad \text{(i)}$$

where $F_0 = \dfrac{f_0}{(\omega_0^2 - \omega^2)^2 + 4b^2\omega^2}$. For convenience, we put $a_1 = \omega_0^2 - \omega^2$ and $a_2 = 2b\omega$. Then the expression for x_2 takes a compact form:

$$x_2(t) = F_0\, (a_1 \cos \omega t + a_2 \sin \omega t)$$

$$= F_0 \sqrt{a_1^2 + a_2^2} \cos (\omega t - \theta)$$

$$= \frac{f_0 \cos(\omega t - \theta)}{\sqrt{(\omega_0^2 - \omega^2)^2 + 4b^2\omega^2}}$$

and phase θ is given by

$$\theta = \tan^{-1}\left(\frac{2b\omega}{\omega_0^2 - \omega^2}\right).$$

We expect you to use this method to obtain expressions for charge and current starting from Eq. (5.40).

Method 2: Use of Complex numbers

In the text of the chapter, we obtained solutions of Eq. (5.5) by starting from a harmonic function and obtained instantaneous velocity by differentiating the expression for displacement. Here we illustrate a more elegant and compact method using complex numbers for determining instantaneous velocity. To do so, we write the RHS of Eq. (5.5) as

$$\text{RHS} = \text{real part of } f_0 \exp(j\omega t) = \text{Re}\,[f_0 \exp(j\omega t)]. \quad \text{(i)}$$

We assume a solution of the form

$$v(t) = \text{Re}[v_0 \exp(j\omega t)]. \quad \text{(ii)}$$

For brevity, we drop the notation 'Re' and write

$$v = v_0 \exp(j\omega t) = \frac{dx_2}{dt}. \quad \text{(iii)}$$

If we differentiate this expression with t, we obtain

$$\frac{d^2x}{dt^2} = \frac{dv}{dt} = j\omega v_0 \exp(j\omega t). \quad \text{(iv)}$$

And if we integrate (iii) with respect to t, we get expression for displacement:

$$x_2 = \int v\,dt = \frac{v_0}{j\omega}\exp(j\omega t).$$

We rewrite Eq. (5.5) as

$$j\omega v_0 \exp(j\omega t) + 2bv_0 \exp(j\omega t) + \frac{\omega_0^2}{j\omega} v_0 \exp(j\omega t) = f_0 \exp(j\omega t).$$

$$\therefore \quad v_0\left[2b - j\left(-\omega + \frac{\omega_0^2}{\omega}\right)\right] = f_0$$

or

$$v_0 = \frac{f_0}{A - jB}$$

$$= \frac{f_0(A + jB)}{A^2 + B^2} = \frac{f_0 \exp(j\alpha)}{\sqrt{A^2 + B^2}},$$

where $A = 2b$, $B = -\omega + \dfrac{\omega_0^2}{\omega}$ and $\alpha = \tan^{-1}\dfrac{B}{A} = \tan^{-1}\left(\dfrac{\omega_0^2 - \omega^2}{2b\omega}\right)$.

You can express this result in terms of ω, ω_0 and α:

$$v_0 = \frac{f_0 \exp(j\alpha)}{\sqrt{4b^2 + [(\omega_0^2/\omega) - \omega]^2}}$$

$$\therefore \quad v = \frac{f_0 \exp\{j(\omega t + \alpha)\}}{\sqrt{4b^2 + [(\omega_0^2/\omega) - \omega]^2}}.$$

$$\therefore \quad v(t) = \text{Re}\ v = \frac{f_0 \cos(\omega t + \alpha)}{\sqrt{4b^2 + [(\omega_0^2/\omega) - \omega]^2}}.$$

You can now use this method to obtain expressions for $x_2(t)$, q and I.

6

Coupled Oscillations

Expected Learning Outcomes

In this Chapter, you will acquire capability to:

- state the effect of coupling on the oscillations of individual oscillators;
- derive the equations of motion of a coupled system executing longitudinal and transverse oscillations;
- analyse the motion of coupled masses in terms of normal modes;
- write, by analogy, the normal mode frequencies of coupled pendulums;
- derive expressions for energy of different coupled systems;
- obtain normal modes of a forced coupled oscillator; and
- establish wave equation for a large number of coupled systems.

6.1 INTRODUCTION

So far you have learnt how an isolated (single) 1-D free, damped or forced oscillating system such as a spring-mass system or a simple pendulum behaves, once it is disturbed from the mean equilibrium position. But in actual practice, isolated systems are not many. So you may like to ask: Why then we studied such systems at all? The answer to this question is: Such idealised systems are easy to analyse and facilitate our understanding of the behaviour of coupled oscillators, which are, in general, quite complex systems. From your school physics, you may recall that in a nucleus, protons and neutrons cling together via nuclear forces. Similarly, atoms in a molecule are coupled by interatomic forces. For example, in a carbon dioxide molecule, two oxygen atoms are coupled to one carbon atom and in a water molecule, two hydrogen atoms are coupled to one oxygen atom. Atoms in a solid are coupled to their nearest neighbours and the coupling is so strong that these can only vibrate about their respective mean position. So we can say that in all substances, motion of their constituent atoms/molecules is affected by the presence of coupling, which can be mechanical, electrical, magnetic or electromagnetic. In radio and TV transmission; we use electrical circuits with inductive-capacitative couplings.

An important aspect of a coupled system is that even though dynamical behaviour of each individual oscillator in the system is influenced by all other oscillators, they are treated on equal footing. You will discover that the net effect of coupling manifests as exchange of energy between individual oscillators.

In Section 6.2, we begin our study of coupled systems by analysing longitudinal oscillations of two coupled masses. Do you expect the motion of these masses to be simple harmonic? You will learn that though their motion is not simple harmonic, we can analyse it in terms of *normal modes*, each having a definite frequency and is analogous to SHM. The normal modes specify the number of independent ways in which a coupled system can oscillate. You will also learn about transverse oscillations of two coupled masses in this section. In Section 6.3, you will learn to determine, by analogy, normal mode frequencies of two coupled pendulums. The general method for calculating normal modes is presented in Section 6.4. Discussion of energy relations in coupled oscillations forms the subject matter of Section 6.5. The study of the behaviour of forced coupled oscillations is presented in Section 6.6 and N coupled oscillators in Section 6.7.

When the number of oscillators is so large that the medium can be treated as homogeneous and continuous, exchange of energy due to coupling leads to the spectacular phenomena of *wave motion*. Do you know that this phenomenon is responsible for our audio and visual contact with the outside world? We are able to adore the beauty of nature in its various hues and colours due to perception of light waves. Similarly, sound waves carry music to us. These are also used in exploration of oil and prospecting for minerals; the commodities which drive present day world economy. Electromagnetic waves of low wavelength (high frequency) are used in medical diagnostics, care and treatment. In short, without waves, life would not have been as enjoyable or comfortable as it is now. (In Chapter 7, you will learn that waves can cause disaster and bring miseries also.) So you can say that coupling provides the link between oscillations and waves, and it is extremely important to understand the motion of coupled oscillators.

6.2 OSCILLATIONS OF COUPLED MASSES

To understand the effect of coupling on the behaviour of coupled oscillators, refer to Fig. 6.1, which shows two identical spring-mass systems placed on a horizontal, frictionless surface and joined together (coupled) by another spring. We assume that all springs are of negligible mass. In the equilibrium state, masses m_A and m_B (= m_A), placed at A and B respectively and attached to springs '1' and '2', both of force constant k', are coupled by spring '3' of force constant k. In this (equilibrium) state, neither mass will experience any force due to either spring. The motion of this

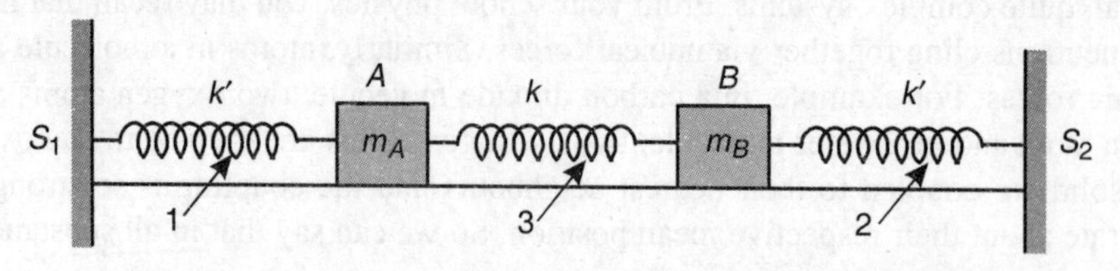

Fig. 6.1 Equilibrium configuration of two coupled masses.

system will, therefore, be determined by the initial conditions. That is, the motion can be transverse or longitudinal depending on how the masses are set in motion. For simplicity, we first consider longitudinal motion of these two coupled masses.

6.2.1 Longitudinal Oscillations

Refer to Fig. 6.2. It shows the instantaneous configuration of two coupled masses executing longitudinal oscillations. (For ease in visualisation, we have also shown the equilibrium configuration of these masses.) Note that the mass m_B has been pulled horizontally, towards the right. The coupling spring '3' pulls the mass m_A. As soon as mass m_B is released, the restoring forces generated in the coupling spring and spring '2' attached to the right of mass m_B tend to bring it back to its equilibrium position. However, mass m_B overshoots its equilibrium mark and the coupling spring begins to push mass m_A. As a result, both masses start oscillating longitudinally. What do you conclude from this? This means that in a coupled system, motion imparted to one component (mass m_B in the instant case) is not confined to it alone; it is transmitted to the other component (mass m_A) as well. You may now like to learn to establish the equations of motion of these masses and solve these to discover changes in the nature of oscillations.

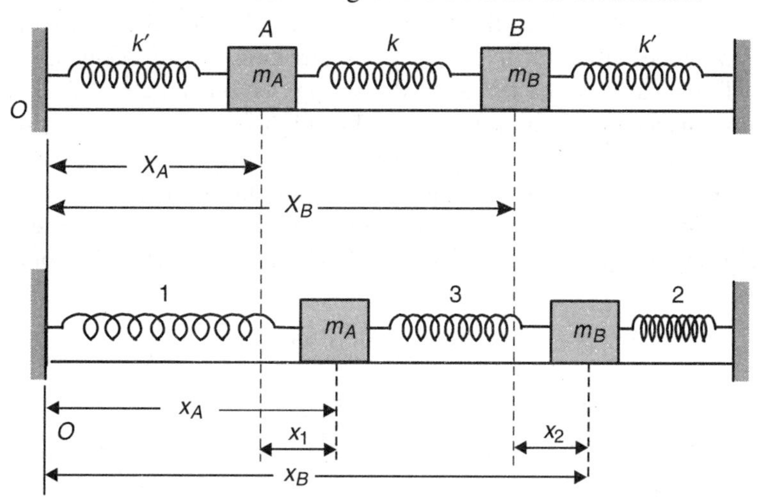

Fig. 6.2 Instantaneous configuration two coupled masses executing longitudinal oscillations.

The differential equation

We choose x-axis along the length of the springs with O as the origin. Let X_A and X_B be the initial co-ordinates of the masses m_A and m_B, respectively. When the system is executing longitudinal oscillations, let us assume that the instantaneous positions of these masses are $x_A(t)$ and $x_B(t)$, respectively. Therefore, the instantaneous displacements of the masses m_A and m_B from their respective equilibrium positions can be expressed as

$$x_1(t) = x_A(t) - X_A$$

and

$$x_2(t) = x_B(t) - X_B.$$

Since these masses are assumed to be moving on a frictionless surface, at any instant of time, mass m_A will be simultaneously subject to forces from springs '1' and '3' attached to it. To calculate the magnitude of these forces, we note that:

(i) The force due to spring '1' (of force constant k' on the left of A) is analogous to the restoring force generated in an isolated spring-mass system (Chapter 2). Thus, the restoring force on mass A due to the spring of force constant k' can be written as

$$F_{k'} = -k'[x_A(t) - X_A] = k'x_1(t).$$

(ii) The force exerted on A by the coupling spring '3' (of spring constant k) is referred to as the *coupling force*. (Since spring '3' joins masses m_A and m_B, it exerts forces on both masses simultaneously.) Note that the coupling force is not necessarily a restoring force always for each mass in the system. That is, the coupling force does not tend to always bring the oscillating mass back to its equilibrium position. The nature of coupling force depends on
- the particular oscillating mass of the coupled system under consideration; and
- relative displacement of the masses with respect to their equilibrium positions.

To write the expression for coupling force acting on the mass m_A, we assume that $x_2(t) > x_1(t)$. It means that at a given instant of time, the coupling spring '3' is stretched more than spring '1' (on the left of mass m_A) and will exert force along the direction of displacement of mass m_A. Since extension is small compared to the length of the spring, the magnitude of coupling force will be proportional to the change in length of the spring. Recall that instantaneous co-ordinates of masses m_A and m_B before and after they have been displaced are X_A and X_B and $x_A(t)$ and $x_B(t)$, respectively. Therefore, the equilibrium length of the coupling spring is equal to $(X_B - X_A)$ and the instantaneous (extended) length of the coupling spring will be $x_B(t) - x_A(t)$. Hence change in length of the coupling spring is $[x_B(t) - x_A(t)] - (X_B - X_A)$. Thus, we can write the coupling force F_C on mass m_A as:

$$\begin{aligned} F_C &= k \times [\text{Change in the length of the coupling spring}] \\ &= k\,\{[x_B(t) - x_A(t)] - (X_B - X_A)\} \\ &= k\,\{[x_B(t) - X_B] - [x_A(t) - X_A]\} \\ &= k\,[x_2(t) - x_1(t)]. \end{aligned}$$

Note that the coupling force on mass m_A is positive. It implies that the force and displacement are in the same direction. This is indeed consistent with the physical situation of mass m_A when $x_2(t) > x_1(t)$; in this condition, the coupling spring pulls mass m_A along the direction of its displacement.

So, dynamics of motion of m_A will be determined by the interplay of two competing forces—the restoring force due to spring '1' (of force constant k') and the coupling force due to spring '3' (of force constant k) acting on it simultaneously. You can write the equation of motion of mass m_A using Newton's second law of motion:

$$m_A \frac{d^2 x_A(t)}{dt^2} = -k'x_1(t) + k\,[x_2(t) - x_1(t)]. \tag{6.1}$$

Since $dX_A/dt = 0$, adding it on both sides will not make any difference. Then we can rewrite Eq. (6.1) as

$$m_A \frac{d^2(x_A - X_A)}{dt^2} = m_A \frac{d^2 x_1(t)}{dt^2} = -k' x_1(t) + k\,[x_2(t) - x_1(t)].$$

On dividing throughout by m_A and rearranging terms, we get

$$\frac{d^2 x_1(t)}{dt^2} + \omega_0^2 x_1(t) - \omega_s^2 [x_2(t) - x_1(t)] = 0, \tag{6.2}$$

where $\omega_0 = (k'/m_A)^{1/2}$ is *natural frequency* of mass m_A and $\omega_s^2 = \sqrt{k/m_A}$ denotes *coupling frequency*.

Similarly, the equation of motion of mass m_B can be written as

$$m_B \frac{d^2 x_2(t)}{dt^2} = -k' x_2(t) - k[(x_2(t) - x_1(t)]. \tag{6.3}$$

Let us pause for a while and reflect on how and why Eq. (6.3) differs from Eq. (6.1). You will note two important differences: (i) In Eq. (6.3), the sign of coupling force is negative. This is because the coupling force as well as the restoring force due to spring '2' tend to bring mass m_B back to its equilibrium position at B. (ii) The coupling force acts differently on the coupled masses; while it is along the direction of displacement in case of mass m_A, it is directed against displacement for mass m_B.

On dividing throughout by m_B and rearranging terms, we can rewrite the equation of motion of mass m_B as

$$\frac{d^2 x_2(t)}{dt^2} + \omega_0^2 x_2(t) + \omega_s^2 [x_2(t) - x_1(t)] = 0. \tag{6.4}$$

Note that the expressions for natural frequency ω_0 and coupling frequency ω_s have not changed. Do you know the reason? It is because the coupled masses are taken to be equal; $m_A = m_B$. For unequal masses, mathematics will be quite involved and we will not consider it here.

Before proceeding further, you may like to guess whether Eqs. (6.2) and (6.4) represent simple harmonic motion? If your guess is no, then you are thinking logically and correctly. But what made you to make this guess? You may recall that SHM is characterised by the fact that acceleration of the oscillating mass is proportional to its displacement. For the coupled masses shown in Fig. 6.2, Eq. (6.2) shows that acceleration of mass m_A depends on its own displacement $[x_1(t)]$ as also on the displacement $[x_2(t)]$ of the mass m_B through the term $[x_2(t) - x_1(t)]$. The same holds for Eq. (6.4) as well. In general, the motion described by Eqs. (6.2) and (6.4) is not simple harmonic because of the presence of the *coupling term*: $\omega_s^2\,[x_2(t) - x_1(t)]$. This means that the analysis of previous units will not apply to these equations as such.

You may now ask: Can we find a way to reduce these equations to a form analogous to the standard equation of SHM? If so, how? To discover answer to this very pertinent query, we have to recast these equations. We first add Eqs. (6.2) and (6.4). This gives

$$\frac{d^2}{dt^2}(x_1 + x_2) + \omega_0^2 (x_1 + x_2) = 0. \tag{6.5}$$

Next we subtract Eq. (6.4) from Eq. (6.2) and rearrange terms to obtain. This gives

$$\frac{d^2}{dt^2}(x_1 - x_2) + (\omega_0^2 + 2\omega_s^2)(x_1 - x_2) = 0. \tag{6.6}$$

We can rewrite Eqs. (6.5) and (6.6) in a form similar to the differential equation for SHM by introducing two new variables ξ_1 and ξ_2 by defining

$$\xi_1 = x_1 + x_2 \tag{6.7a}$$

and

$$\xi_2 = x_1 - x_2. \tag{6.7b}$$

In terms of ξ_1 and ξ_2, Eqs. (6.5) and (6.6) take the form

$$\frac{d^2\xi_1}{dt^2} + \omega_1^2\xi_1 = 0 \tag{6.8}$$

and

$$\frac{d^2\xi_2}{dt^2} + \omega_2^2\xi_2 = 0, \tag{6.9}$$

where we have put

$$\omega_1^2 = \omega_0^2 = \frac{k'}{m} \tag{6.10}$$

$$\omega_2^2 = \omega_0^2 + 2\omega_s^2 = \frac{k' + 2k}{m}, \tag{6.11}$$

and $m_A = m_B = m$.

Note that by introducing the variables ξ_1 and ξ_2, we have decoupled Eqs. (6.5) and (6.6) into two equations, which are analogous to the equation for simple harmonic motion, corresponding to frequencies ω_1 and ω_2 ($> \omega_1$). In other words, the motion of a coupled system can be described in terms of two uncoupled and independent equations [Eqs. (6.8) and (6.9)] with the help of new co-ordinates ξ_1 and ξ_2. These new co-ordinates are known as *normal co-ordinates* and the motion associated with each co-ordinate is referred to as a *normal mode*. It is characterised by its own *normal mode frequency*.

You may now ask: Can the normal mode frequencies be equal? To answer this question, we note from Eqs. (6.10) and (6.11) that $\omega_1 \approx \omega_2$ for $k << k'$. That is, when the coupling is very weak, the normal mode frequencies will be nearly equal; but never the same. On the other hand, if coupling is strong (coupling spring is very stiff), one of the normal mode frequencies will be significantly greater than the frequency of the other mode.

If the force constant of coupling spring equals the force constant of other two springs ($k = k'$), i.e., all three springs are identical, the expressions for ω_1 and ω_2 take simple form:

$$\omega_1 = \left(\frac{k}{m}\right)^{1/2} \tag{6.12}$$

and

$$\omega_2 = \left(\frac{3k}{m}\right)^{1/2}. \tag{6.13}$$

Let us now learn about normal co-ordinates and normal modes in some detail.

Normal co-ordinates and normal modes

You now know that normal co-ordinates ξ_1 and ξ_2 are not a measure of displacement like Cartesian co-ordinates $x_1(t)$ and $x_2(t)$. Yet they specify some configuration of a coupled system at a given instant of time. To discover the significance of normal co-ordinates and normal modes, you may like to discuss the solutions of Eqs. (6.8) and (6.9). As may be noted, these equations are similar to the differential equation describing the SHM of a spring-mass system discussed in Chapter 2. Therefore, we can readily write their general solutions as

$$\xi_1(t) = a_1 \cos(\omega_1 t + \phi_1) \tag{6.14}$$

and

$$\xi_2(t) = a_2 \cos(\omega_2 t + \phi_2), \tag{6.15}$$

where a_1 and a_2 are the amplitudes and ϕ_1 and ϕ_2 are initial phases.
Since $x_1(t) = (\xi_1 + \xi_2)/2$, we can express the displacement of mass m_A as

$$x_1(t) = \frac{1}{2}[a_1 \cos(\omega_1 t + \phi_1) + a_2 \cos(\omega_2 t + \phi_2)]. \tag{6.16}$$

Similarly, we can write the displacement of the mass m_B as

$$x_2(t) = \frac{1}{2}[a_1 \cos(\omega_1 t + \phi_1) - a_2 \cos(\omega_2 t + \phi_2)]. \tag{6.17}$$

The constants a_1, a_2, ϕ_1, and ϕ_2 are determined by the initial conditions. Once these are specified, you can completely specify the motion of coupled masses. To appreciate how this is done, study Example 6.1 carefully.

EXAMPLE 6.1

Solve Eqs. (6.16) and (6.17) subject to the following initial conditions:

$$x_1(0) = a,\ x_2(0) = a,\ \left.\frac{dx_1}{dt}\right|_{t=0} = 0, \text{ and } \left.\frac{dx_2}{dt}\right|_{t=0} = 0.$$

Solution: From Eqs. (6.16) and (6.17), we know that

$$x_1(t) = \frac{1}{2}[a_1 \cos(\omega_1 t + \phi_1) + a_2 \cos(\omega_2 t + \phi_2)] \tag{i}$$

and

$$x_2(t) = \frac{1}{2}[a_1 \cos(\omega_1 t + \phi_1) - a_2 \cos(\omega_2 t + \phi_2)]. \tag{ii}$$

Hence, the first time-derivatives are

$$\frac{dx_1(t)}{dt} = -\frac{1}{2}[a_1\omega_1 \sin(\omega_1 t + \phi_1) + a_2\omega_2 \sin(\omega_2 t + \phi_2)] \tag{iii}$$

and

$$\frac{dx_2(t)}{dt} = \frac{1}{2}[-a_1\omega_1 \sin(\omega_1 t + \phi_1) + a_2\omega_2 \sin(\omega_2 t + \phi_2)]. \tag{iv}$$

Using the first two initial conditions in Eqs. (i) and (ii), we get

$$2a = a_1 \cos\phi_1 + a_2 \cos\phi_2 \tag{v}$$

and

$$2a = a_1 \cos\phi_1 - a_2 \cos\phi_2. \tag{vi}$$

Similarly, using the last two initial conditions in Eqs. (iii) and (iv), we get

$$0 = a_1\omega_1 \sin\phi_1 + a_2\omega_2 \sin\phi_2 \tag{vii}$$

and

$$0 = a_1\omega_1 \sin\phi_1 - a_2\omega_2 \sin\phi_2\,. \tag{viii}$$

On adding Eqs. (v) and (vi), we get

$$a_1 \cos\phi_1 = 2a. \tag{ix}$$

Similarly, on subtracting Eq. (vi) from Eq. (v), we get

$$a_2 \cos\phi_2 = 0. \tag{x}$$

By similar mathematical operations with Eqs. (vii) and (viii), we get

$$a_1\omega_1 \sin\phi_1 = 0, \tag{xi}$$

and

$$a_2\omega_2 \sin\phi_2 = 0. \tag{xii}$$

As a_1, ω_1, and ω_2 are not equal to zero, the conditions given in Eqs. (ix)–(xii) will be satisfied only if $\phi_1 = \phi_2 = 0$, $a_1 = 2a$, and $a_2 = 0$.

Hence, the displacements are given by

$$x_1(t) = a\cos\omega_1 t,$$

and

$$x_2(t) = a\cos\omega_1 t.$$

So we can write

$$\xi_1(t) = x_1(t) + x_2(t) = 2a\cos\omega_1 t \tag{xiii}$$

and

$$\xi_2(t) = x_1(t) - x_2(t) = 0.$$

From the discussion in Example 6.1, you must have noted that *when both masses are initially displaced equally in the same direction before being released*, i.e., $x_1(t) = x_2(t)$, then $\xi_2 = 0$ at all times. The resultant motion represented by (xiii) describes one of the normal modes of oscillations of the coupled masses. In this case, the normal mode frequency is the same as that of either spring-mass system in isolation. This means that *coupling has no influence in this case and both masses oscillate in phase.* In this mode of vibration (Fig. 6.3), the coupling spring is neither stretched nor compressed. That is, it is as if it is not there.

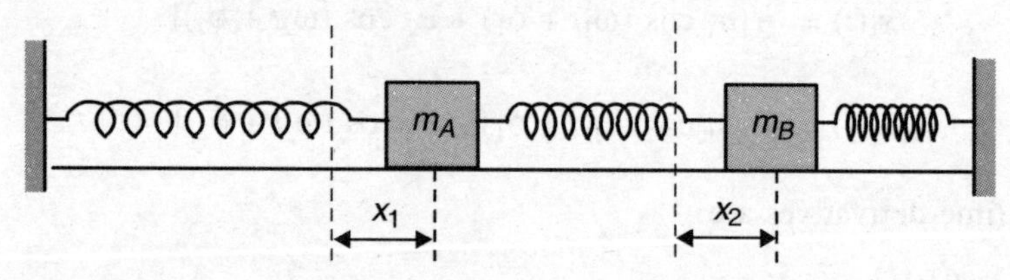

Fig. 6.3 A normal mode of two coupled spring-mass systems when the coupled masses are given equal initial displacement along the same direction.

As you have seen in Example 6.1, the displacements $x_1(t)$ and $x_2(t)$ of the coupled masses are determined by the initial conditions. In other words, initial conditions determine nature of oscillations—the normal mode—of the masses.

Like the initial conditions specified in Example 6.1, another set of initial conditions give rise to a normal mode of coupled spring-mass system. We would like you to analyse the motion of the system under this new set of initial conditions by solving Practice Exercise 6.1.

Practice Exercise 6.1: Consider two spring-mass systems coupled together as shown in Fig. 6.1. These are pulled equally towards each other and their subsequent motion is described by the following initial conditions:

$$x_1(0) = a,\ x_2(0) = -a,\ \left.\frac{dx_1}{dt}\right|_{t=0} = 0, \text{ and } \left.\frac{dx_2}{dt}\right|_{t=0} = 0.$$

Analyse the resultant motion of the coupled oscillator assuming that the masses are equal.

[***Ans.*** $x_1(t) = a \cos \omega_2 t,\ x_2(t) = -a \cos \omega_2 t;\ \xi_1(t) = 0,\ \xi_2(t) = 2a \cos \omega_2 t$]

While answering Practice exercise 6.1, you must have realised that when two coupled masses are initially pulled equally towards each other before being released, their instantaneous displacements are equal but in opposite directions and out of phase by π, i.e., $x_1(t) = -x_2(t)$ or $\xi_1(t) = 0$. The normal mode configuration corresponding to this set of initial conditions is shown in Fig. 6.4. In this case, the normal mode frequency ω_2 is higher than that of the uncoupled masses ($\omega_1 < \omega_2$). Physically it means that the coupling spring is either compressed or stretched, and we can say that coupling is playing an effective role.

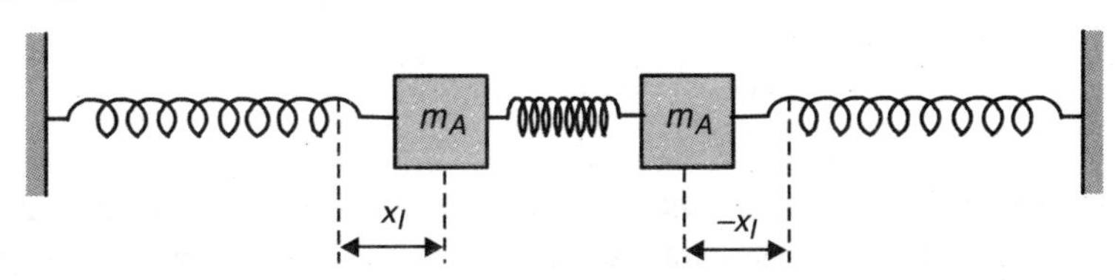

Fig. 6.4 The normal mode corresponding to the initial conditions wherein masses in two coupled spring-mass systems are pulled together equally towards one another.

Can you think of any other physical situation for out-of-phase normal mode? It will correspond to the condition where the coupled masses are pulled in opposite directions away from one another. We expect the motion to be similar to that obtained for initial conditions analysed in Practice Exercise 6.1. Work it out to convince yourself. Does your analysis support our guess?

From the above discussion, we conclude that *normal co-ordinates allow us to decouple the equation of motion of a coupled system into a set of linear differential equations with constant coefficients and each equation so obtained contains only one dependent variable.* You may ask: Does a coupled system oscillate purely in one of its allowed normal modes? It is not so. In addition to normal mode oscillations, the motion of a coupled system may also be regarded as a superposition of its possible normal modes. As you go in higher classes, you will learn that the concept of normal modes of oscillation is fundamental in the study of classical systems.

Modulation of coupled oscillations

In Example 6.1 we considered the cases where coupled masses were pulled equally in the same direction or in opposite directions (Practice Exercise 6.1). You may now like to know: What would happen if only one of the masses was pulled and then released? To understand the nature of resulting oscillations of two coupled spring-mass systems, we solve Eqs. (6.16) and (6.17) subject to the following initial conditions:

$$x_1(0) = 2a,\ x_2(0) = 0,\ \left.\frac{dx_1}{dt}\right|_{t=0} = 0 \text{ and } \left.\frac{dx_2}{dt}\right|_{t=0} = 0. \tag{6.18}$$

Note that this equation represents the initial conditions when one of the masses has been pulled by a distance $2a$ and the other mass is stationary. Applying these initial conditions on Eqs. (6.16) and (6.17) as explained in Example 6.1, you will find that displacements of coupled masses are given respectively by

$$x_1(t) = a\,(\cos \omega_1 t + \cos \omega_2 t) \tag{6.19}$$

and

$$x_2(t) = a\,(\cos \omega_1 t - \cos \omega_2 t). \tag{6.20}$$

Using elementary trigonometric relations for expressing the sum (difference) of two cosine functions in terms of their product (C, D formulae), we can rewrite Eqs. (6.19) and (6.20) in a more familiar and physically informative form as

$$x_1(t) = 2a \cos\left(\frac{\omega_2 - \omega_1}{2}\right) t \cos\left(\frac{\omega_2 + \omega_1}{2}\right) t \tag{6.21}$$

and

$$x_2(t) = 2a \sin\left(\frac{\omega_2 - \omega_1}{2}\right) t \sin\left(\frac{\omega_1 + \omega_2}{2}\right) t. \tag{6.22}$$

Do you recognise Eq. (6.21)? It is essentially the same as Eq. (3.16) obtained for modulated oscillations. It means that for the resultant oscillations of the system, we can define an average angular frequency $\omega_{av} = (\omega_1 + \omega_2)/2$ and a modulated angular frequency $\omega_{mod} = (\omega_2 - \omega_1)/2$. In terms of ω_{av} and ω_{mod}, we can rewrite Eqs. (6.21) and (6.22) as

$$x_1(t) = a_{mod}\,(t) \cos \omega_{av} t \tag{6.23}$$

and

$$x_2(t) = b_{mod}(t) \sin \omega_{av} t, \tag{6.24}$$

where

$$a_{mod}(t) = 2a \cos \omega_{mod} t \tag{6.25}$$

and

$$b_{mod}(t) = 2b \sin \omega_{mod} t \tag{6.26}$$

denote modulated amplitudes of oscillation of the coupled masses.

From these results, we note that Eqs. (6.23) and (6.24) represent modulated oscillations of coupled masses. That is, from these equations, we can not say that the coupled masses execute SHM, since amplitudes of oscillations [$a_{mod}(t)$ and $b_{mod}(t)$] vary with time.

Note that $x_1(t)$, the displacement of mass m_A varies as a cosine function, whereas $x_2(t)$, the displacement of mass m_B varies as a sine function with time. Can you guess its implications for the motion of the coupled system? If you are thinking that the displacements of coupled masses executing longitudinal oscillations are out of phase by $\pi/2$, you are absolutely right. (Trigonometric sine and cosine functions differ in phase by $\pi/2$).

The displacement-time graphs for two coupled masses are shown in Fig. 6.5. We observe that at $t = 0$, the displacement of the mass m_A is $2a$ (maximum) while that of mass m_B is zero. With passage of time, the displacement of m_A decreases and becomes zero at $t = T/4$ while displacement of m_A becomes $2a$. This trend continues interchangeably for every succeeding quarter period and should repeat itself indefinitely, if there were no damping!

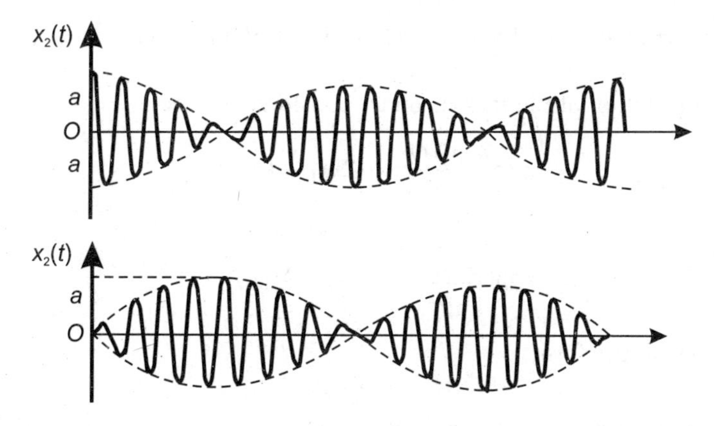

Fig. 6.5 Displacement-time graph for two coupled masses executing longitudinal oscillations. Their displacements change between zero and $2a$ alternately in every quarter period.

If coupling between the masses is weak, ω_2 *will be only slightly greater than* ω_1, *and* ω_{mod} *will be very small.* It implies that a_{mod} and b_{mod} will take long time to show any observable change. We can then assume that a_{mod} and b_{mod} are practically constant over one cycle of angular frequency ω_{av} and Eqs. (6.23) and (6.24) can be regarded as characterising an almost simple harmonic motion.

You should now go through the following examples to fix your ideas.

EXAMPLE 6.2

Two masses m and $2m$ are coupled through a spring of force constant k and attached to two fixed points by two identical springs having same force constant. The whole arrangement rests on a smooth horizontal table. The springs are stretched so that the tension in each spring is T and its length is l (much greater than its relaxed length). Let us calculate the normal mode frequencies for longitudinal oscillations of small amplitude. The equilibrium configuration of the system is shown in Fig. 6.6.

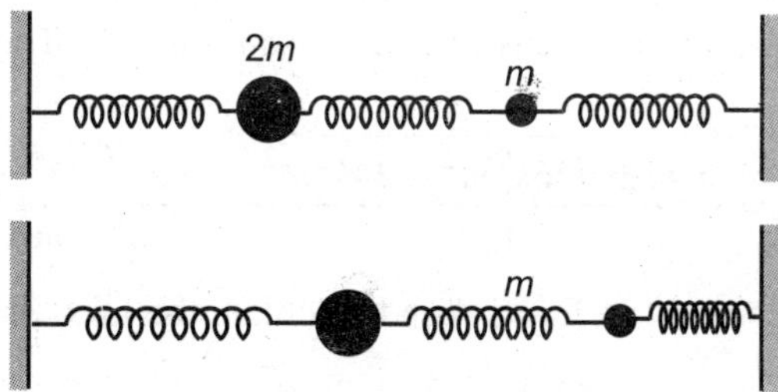

Fig. 6.6 Coupled spring-mass system of unequal masses: (a) equilibrium configuration; and (b) instantaneous configuration.

Solution: Suppose that mass m (let us call this body A) is displaced from its equilibrium position by x_1 and mass $2m$ (body B) is displaced by x_2. The instantaneous configuration of the system is shown in Fig. 6.6b. With respect to equilibrium, the increase in tension in spring 1 will be kx_1. Since spring 2 is extended by $x_2 - x_1$ relative to its length at equilibrium, the increase in its tension will be $k(x_2 - x_1)$. Thus, two forces act on body A: kx_1 due to spring 1 to the left of A and $k(x_2 - x_1)$ due to spring 2 to the right of A. Therefore, the net force on body A in the direction of increasing x_1 is

$$k(x_2 - x_1) - kx_1 = -2kx_1 + kx_2.$$

Similarly, you can write the net force acting on body B in the direction of increasing x_2 as $kx_1 - 2kx_2$.

The equations of motion of the masses under consideration are

$$m\frac{d^2x_1(t)}{dt^2} = -2kx_1 + kx_2 \qquad \text{(i)}$$

and

$$2m\frac{d^2x_1(t)}{dt^2} = kx_1 - 2kx_2. \qquad \text{(ii)}$$

We look for solutions of the form

$$x_1 = A\cos\omega t$$

and

$$x_2 = B\cos\omega t.$$

Substitute these in Eqs. (i) and (ii). On simplification, you will get

$$-m\omega^2 A = -2kA + kB$$

and

$$-2m\omega^2 B = kA - 2kB.$$

On rearranging these equations, you can write

$$(2k - m\omega^2)A = kB \qquad \text{(iii)}$$

and

$$(2k - 2m\omega^2)B = kA. \qquad \text{(iv)}$$

Now, calculate the ratio (A/B) from Eq. (iii) as well as Eq. (iv) and equate the values so obtained. Then cross multiplication will lead to the relation

$$(2k - m\omega^2)\,(2k - 2m\omega^2) = k^2.$$

It can be rearranged to give

$$2m^2\omega^4 - 6km\omega^2 + 3k^2 = 0.$$

This is a quadratic equation in ω^2. Compare it with the standard form $(ax^2 + bx + c = 0)$. You can identify that constants a, b, c are respectively $2m^2$, $-6km$ and $3k^2$. Then you can easily write its roots:

$$\omega^2 = \frac{6km \pm \sqrt{36k^2m^2 - 24k^2m^2}}{4m^2} = \frac{3 \pm \sqrt{3}}{2}\,\frac{k}{m}.$$

Now, substitute these values of ω^2 in Eq. (iv) and calculate the ratio A/B:

$$\frac{A}{B} = \frac{2k - 2m\omega^2}{k}$$

$$= 2 - \frac{2m\omega^2}{k}$$

$$= 2 - (3 \pm \sqrt{3})$$

$$= \pm\sqrt{3} - 1.$$

You will get the same value for the ratio A/B with Eq. (iii) as well.

From this result, we can conclude that the ratio of amplitudes of normal mode corresponding to lower-frequency is $A/B = \sqrt{3} - 1$, which is real and positive, with a numerical value of 0.732. The two bodies thus oscillate in phase, and the amplitude of oscillation of the lighter body is 0.732 times that of the heavier mass.

For higher-frequency solution, $A/B = -1 - \sqrt{3}$, which is real and negative, with a numerical value of –2.732. The negative sign indicates that two masses oscillate out of phase; the lighter mass having an amplitude of oscillation 2.732 times that of the heavier mass. You should depict the normal mode configurations.

EXAMPLE 6.3

An object of mass m is suspended from a rigid support with the help of a spring of force constant k. It has natural frequency ν_0. Next two identical objects P and Q, each of mass m, are joined together by a spring of force constant k'. Then these are connected to rigid supports R and S by two identical springs, each of force constant k. Now, if P is clamped, Q oscillates with frequency 2.5 Hz. Calculate ν_0 if the frequency of the first normal mode is 2 Hz. Also calculate the frequency of the second normal mode.

Solution:

$$f_0 = \frac{1}{2\pi}\sqrt{\frac{k}{m}} = \frac{\omega_0}{2\pi}$$

$$\therefore \qquad \frac{k}{m} = \omega_0^2 = 4\pi^2 f_0^2. \qquad \text{(i)}$$

When P is clamped, the equation of motion of Q will be

$$m\frac{d^2 x_Q}{dt^2} = -(k + k')\, x_Q$$

or

$$\frac{d^2 x_Q}{dt^2} + \left(\frac{k}{m} + \frac{k'}{m}\right) x_Q = 0.$$

This equation represents simple harmonic motion of frequency

$$f_Q = \frac{1}{2\pi}\sqrt{\left(\frac{k}{m} + \frac{k'}{m}\right)} = 2.5\,\text{Hz}$$

$$\therefore \qquad \frac{k}{m} + \frac{k'}{m} = 4\pi^2 f_Q^2 = 4\pi^2\,(2.5\ \text{Hz})^2 = 25\pi^2\ (\text{Hz})^2. \qquad \text{(ii)}$$

Subtracting Eq. (i) from Eq. (ii), we get

$$\frac{k'}{m} = [25\pi^2 - \omega_0^2]\,(\text{Hz})^2.$$

The angular frequencies of normal modes of vibration are

$$\omega_1^2 = 4\pi^2\nu_1^2 = \omega_0^2 = \frac{k}{m}.$$

We are told that $\nu_1 = 2$ Hz.

$$\omega_2^2 = 4\pi^2 2^2 = 4\pi^2 \nu_0^2$$

$$\Rightarrow \quad \nu_0 = 2 \text{ Hz} \quad \text{and} \quad \frac{k}{m} = 4\pi^2 \times 4 = 16\pi^2$$

Hence
$$\frac{k'}{m} = 25\pi^2 - 16\pi^2 = 9\pi^2.$$

and
$$\omega_2^2 = 4\pi^2 \nu_2^2 = \frac{k}{m} + \frac{2k'}{m} = 34\pi^2 \text{ (Hz)}^2$$

$\therefore$
$$\nu_2 = \sqrt{\frac{17}{2}} \text{ Hz} = 2.9 \text{ Hz}.$$

EXAMPLE 6.4

Two equal masses (m) are connected to each other with the help of a spring of force constant k. The upper mass is connected to a rigid support through an identical spring. The system is made to oscillate in the vertical direction. Calculate the normal frequencies.

Solution: Equations of motion of masses A and B are

$$m\frac{d^2x_1}{dt^2} = -k(x_1 - x_2)$$

and

$$m\frac{d^2x_2}{dt^2} = -kx_2 - k(x_2 - x_1).$$

Hence

$$\frac{d^2x_1}{dt^2} + \frac{k}{m}x_1 - \frac{k}{m}x_2 = 0 \quad \text{(i)}$$

and

$$\frac{d^2x_2}{dt^2} - \frac{k}{m}x_1 + \frac{2k}{m}x_2 = 0. \quad \text{(ii)}$$

Let us assume that

$$x_1 = A_1 \cos(\omega t + \phi)$$

and

$$x_2 = A_1 \cos(\omega t + \phi).$$

Then

$$\ddot{x}_1 = -\omega^2 x_1$$

and

$$\ddot{x}_2 = -\omega^2 x_2.$$

Using these results in Eqs. (i) and (ii), we get

$$\left(\frac{k}{m} - \omega^2\right)x_1 - \frac{k}{m}x_2 = 0$$

and

$$\left(\frac{2k}{m} - \omega^2\right)x_2 - \frac{k}{m}x_1 = 0.$$

For non-zero values of x_1 and x_2, this set of simultaneous equations can be solved for normal mode frequencies by equating the following determinant to zero:

$$\begin{vmatrix} \frac{k}{m} - \omega^2 & -\frac{k}{m} \\ -\frac{k}{m} & \frac{2k}{m} - \omega^2 \end{vmatrix}$$

Hence

$$\omega^4 - \frac{3k}{m}\omega^2 + \frac{k^2}{m^2} = 0.$$

$\therefore$

$$\omega^2 = \frac{3k}{2m} \pm \frac{\sqrt{5}k}{2m}.$$

Thus, the normal mode frequencies are given by

$$\omega^2 = (3 \pm \sqrt{5})\frac{k}{2m}.$$

In section 6.2.1, you have learnt to characterise the longitudinal oscillations of two coupled masses. You may recall that depending on the initial conditions, such a system can also execute transverse oscillations. Let us now learn about these.

6.2.2 Transverse Oscillations

Refer to Fig. 6.7a. It shows the instantaneous configuration of the system depicted in Fig. 6.1 when the coupled masses are executing transverse oscillations. Note that the mass m_B has been pulled vertically upwards and the instantaneous displacement is $y_B(t)$. The coupling spring '3' pulls the mass m_A and we denote its instantaneous displacement as $y_A(t)$. As soon as mass m_B is released, the restoring forces generated in the coupling spring and spring '2' attached to the right of mass m_B tends to bring it back to its equilibrium position. However, mass m_B overshoots its equilibrium mark and the coupling spring begins to pull mass m_A. As a result, both masses begin to execute transverse oscillations.

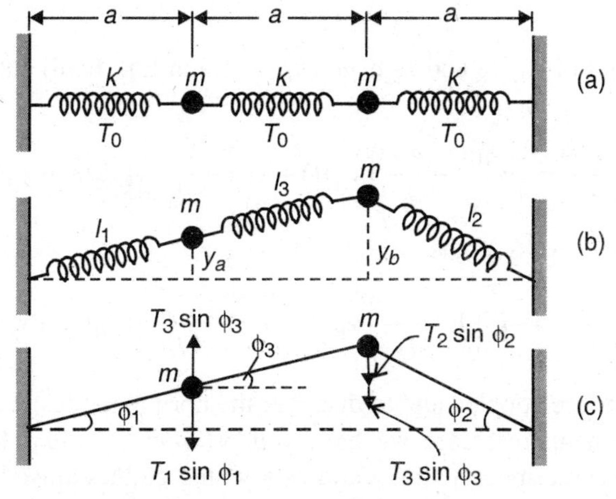

Fig. 6.7 Transverse oscillations of two coupled masses (a) equilibrium configuration, (b) instantaneous configuration and (c) forces acting on the masses.

As before, we note that the motion imparted to one component is transmitted to the other component. You may now like to learn to establish the equations of motion of these masses and solve these to discover the effect of coupling on their oscillations. We choose x-axis along the length of the springs when these are in equilibrium state. Let us suppose that relaxed length of each spring is a_0 and in equilibrium state, it increases to a. But in the instantaneous configuration, the lengths of springs '1' , '2' (of force constant k') and '3' (of force constant k) increase to l_1, l_2 and l_3, respectively. So we can write the expression for tension in the springs as

$$T_1 = k'(l_1 - a_0),$$

$$T_2 = k'(l_2 - a_0),$$

and

$$T_3 = k(l_3 - a_0). \tag{6.27}$$

To establish the equation of motion of mass m_A, we assume that $y_B(t) > y_A(t)$. Note that in the instantaneous configuration, the net force acting on the mass under consideration is determined by the interplay of two competitive forces: spring '1' tries to pull it back, whereas coupling spring '3' pulls it upwards. By resolving these forces into components and referring to Fig. 6.7c, you will note that the net force acting on mass m_A is $(-T_1 \sin \phi_1 + T_3 \sin \phi_3)$, where ϕ_1 and ϕ_3 are the angles at which springs '1' and '3' are inclined to the horizontal. Hence we can write the equation of motion as

$$m_A \frac{d^2 y_A(t)}{dt^2} = -T_1 \sin \phi_1 + T_3 \sin \phi_3$$

$$= -T_1 \frac{y_A(t)}{l_1} + T_3 \frac{y_B(t) - y_A(t)}{l_3}. \tag{6.28}$$

Similarly, for mass m_B, you will note by referring to Fig. 6.7c that the restoring force is $(-T_2 \sin \phi_2 - T_3 \sin \phi_3)$, where ϕ_2 and ϕ_3 are the angles at which springs '2' and '3' are inclined to the horizontal. And the equation of motion of the second mass can be written as

$$m_B \frac{d^2 y_B(t)}{dt^2} = -T_2 \sin \phi_2 - T_3 \sin \phi_3$$

$$= -T_2 \frac{y_B(t)}{l_2} - T_3 \frac{y_B(t) - y_A(t)}{l_3}. \tag{6.29}$$

On substituting the values of T_1, T_2 and T_3 from Eq. (6.27) in Eq. (6.28) and rearranging terms, we get

$$m_A \frac{d^2 y_A(t)}{dt^2} = -k'\left(1 - \frac{a_0}{l_1}\right) y_A(t) + k\left(1 - \frac{a_0}{l_3}\right)[y_B(t) - y_A(t)]. \tag{6.30}$$

Similarly, we can rewrite Eq. (6.28) as

$$m_B \frac{d^2 y_B(t)}{dt^2} = -k'\left(1 - \frac{a_0}{l_2}\right) y_B(t) - k\left(1 - \frac{a_0}{l_3}\right)[y_B(t) - y_A(t)]. \tag{6.31}$$

Note that these equations are coupled and to discover the normal modes, we have to resort to some approximation. For mathematical case we have *Slinky approximation.* In your school Physics classes, you may have learnt about a slinky and as a young child, enjoyed playing with it. Let us learn about it now.

Slinky approximation

In Slinky approximation, we assume that relaxed length of the springs is much less than their extended length in the equilibrium state ($a_0 << a$), which, in turn, is much less than their respective lengths in instantaneous state, i.e., $a << l_1, l_2,$ or l_3. Physically it means that the force constant is very small so that the ratios like $a_0/l_1 = (a_0/a)\,(a/l_1)$ will be two orders of magnitude less than unity and can be ignored in comparison to unity. Under this approximation, Eqs. (6.30) and (6.31) simplify to

$$m_A \frac{d^2 y_A(t)}{dt^2} = -k' y_A(t) + k[y_B(t) - y_A(t)] \tag{6.32}$$

and

$$m_B \frac{d^2 y_B(t)}{dt^2} = -k' y_B(t) - k[y_B(t) - y_A(t)]. \tag{6.33}$$

Note that Eqs. (6.32) and (6.33) are analogous to Eqs. (6.1) and (6.3), respectively. To decouple these, discover normal modes and write their solutions, you can follow the steps followed in section 6.2.1. These are outlined below:

(i) Divide Eq. (6.32) throughout by m_A and Eq. (6.33) by m_B.
(ii) Define angular frequencies ω_0 and ω_s as before.
(iii) Add and subtract the resulting equations and introduce two new variables in terms and difference of $y_A(t)$ and $y_B(t)$. This will lead you to equations identical to Eqs. (6.8) and (6.9) for the normal modes corresponding to frequencies ω_1 and ω_2 ($> \omega_1$), provided coupled masses are taken to be equal.
(iv) Write the expressions for normal modes and corresponding normal mode frequencies by analogy.

We leave this as an exercise for you to fix your concept and gain confidence in working out mathematical steps.

Practice Exercise 6.2: Solve Eqs. (6.32) and (6.33) to obtain normal modes and the corresponding frequencies.

Let us now pause for a while and reflect on transverse oscillations of the coupled masses. From the above analysis, you will discover that in Slinky approximation, the frequencies of the normal modes characterising transverse oscillations are the same as those for longitudinal oscillations. That is, when relaxed length of the spring is much less than its extended length in equilibrium configuration, the initial conditions do not seem to influence the normal mode frequencies. We express this as degeneracy. Did you expect this result physically?

6.2.3 Small Oscillation Approximation

In Chapter 2, you had learnt that single isolated systems execute SHM in small oscillation approximation, where we assume that the oscillating parts do not deviate much from the equilibrium position. In the present case, it means that the equilibrium length of the springs is nearly equal to their lengths in instantaneous configuration, i.e., $a \approx l_1 \approx l_2 \approx l_3$. In this approximation, Eqs. (6.30) and (6.31) respectively simplify to

$$m_A \frac{d^2 y_A(t)}{dt^2} = -k'\left(1 - \frac{a_0}{a}\right) y_A(t) + k\left(1 - \frac{a_0}{a}\right)[y_B(t) - y_A(t)] \tag{6.34}$$

and

$$m_B \frac{d^2 y_B(t)}{dt^2} = -k'\left(1 - \frac{a_0}{a}\right) y_B(t) - k\left(1 - \frac{a_0}{a}\right)[y_B(t) - y_A(t)]. \tag{6.35}$$

If we now introduce two new constants through the relations

$$K_1 = k'\left(1 - \frac{a_0}{a}\right) \tag{6.36a}$$

and

$$K_2 = k\left(1 - \frac{a_0}{a}\right). \tag{6.36b}$$

we can rewrite Eqs. (6.34) and (6.35) as

$$m_A \frac{d^2 y_A(t)}{dt^2} = -K_1 y_A(t) + K_2 [y_B(t) - y_A(t)] \tag{6.37}$$

and

$$m_B \frac{d^2 y_B(t)}{dt^2} = -K_1 y_B(t) - K_2 [y_B(t) - y_A(t)]. \tag{6.38}$$

These equations are analogous to Eqs. (6.32) and (6.33), and you can solve these by following the same steps. However, just for variety, we assume that the normal mode has an angular frequency ω and phase ϕ. So we take

$$y_A(t) = a \cos(\omega t + \phi) \tag{6.39a}$$

and

$$y_B(t) = b \cos(\omega t + \phi). \tag{6.39b}$$

On differentiating these expressions for $y_A(t)$ and $y_B(t)$ twice with respect to t, and substituting in Eqs. (6.37), we get

$$-\omega^2 m_A a = -K_1 a + K_2(b - a)$$

or

$$[-\omega^2 m_A + (K_1 + K_2)]\, a = K_2 b.$$

Hence, the ratio of amplitudes of oscillations of coupled masses is given by

$$\frac{b}{a} = \frac{[(K_1 + K_2)/m_A] - \omega^2}{K_2/m_A}. \tag{6.40a}$$

Similarly, Eq. (6.38) will lead to the relation

$$\frac{b}{a} = \frac{K_2/m_B}{[(K_1 + K_2)/m_B] - \omega^2}. \tag{6.40b}$$

On equating the RHS of these two equations and cross-multiplication, we get

$$\frac{K_2^2}{m_A m_B} = \left(\frac{K_1 + K_2}{m_A} - \omega^2\right)\left(\frac{K_1 + K_2}{m_B} - \omega^2\right).$$

If the coupled masses are equal, i.e., $m_A = m_B = m$, say, then you can solve this equation for normal mode frequencies:

$$\frac{K_1 + K_2}{m} - \omega^2 = \pm \frac{K_2}{m},$$

so that

$$\omega_1 = \sqrt{\frac{K_1}{m}}$$

and

$$\omega_2 = \sqrt{\frac{K_1 + 2K_2}{m}}. \tag{6.41}$$

Note that

(i) one of the normal mode frequencies corresponds to a mode which denotes the configuration as if there were no coupling;

(ii) the values of normal mode frequencies for transverse motion of coupled masses in small oscillation approximation are smaller than those for corresponding frequencies for longitudinal motion.

The in-phase and out-of-phase normal modes of two coupled masses executing transverse oscillations corresponding to frequencies given by Eq. (6.41) are shown in Fig. 6.8.

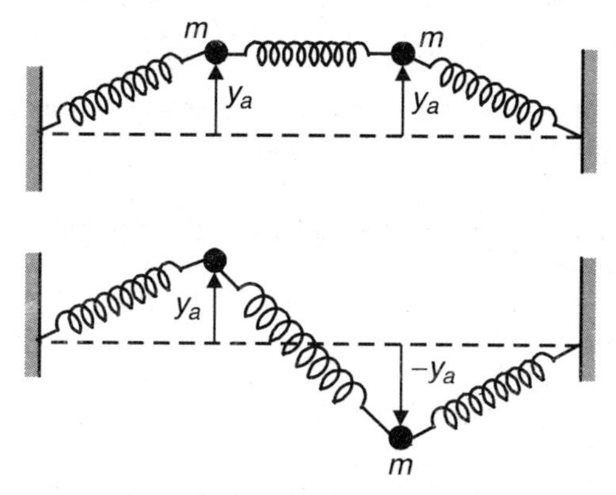

Fig. 6.8 (a) In-phase and (b) out-of-phase normal modes of two coupled masses executing transverse oscillations.

6.3 NORMAL MODE ANALYSIS OF OTHER OSCILLATING SYSTEMS

So far we have analysed the longitudinal and transverse motion of two coupled masses. You may now ask: Can we extend this analysis to other physically different oscillating systems. If so, how? To discover answers to these questions, let us first consider two coupled simple pendulums and obtain normal modes and corresponding normal mode frequencies. To the extent possible, we shall draw analogies from the preceding analysis.

6.3.1 Coupled Simple Pendulums

Let us take two identical simple pendulums A and B having bobs of same mass m and suspended by strings of equal length l, as shown in Fig. 6.9a. Let us assume that the bobs of these pendulums are connected by a weightless spring of force constant k. *In the equilibrium position, the distance between the bobs will be equal to the length of the unstretched spring.*

To begin with, we displace both the bobs to the right of their respective equilibrium positions. If we denote the displacements of these bobs at time (t) by $x_1(t)$ and $x_2(t)$ [$< x_1(t)$], as shown in Fig. 6.9b, you can easily express the (magnitude of) tension in the coupling spring as $k(x_1 - x_2)$. Note that we have taken $x_1(t) > x_2(t)$. It implies that the coupling spring is initially stretched. It means that tension in coupling spring will oppose the acceleration of A but support the acceleration of B.

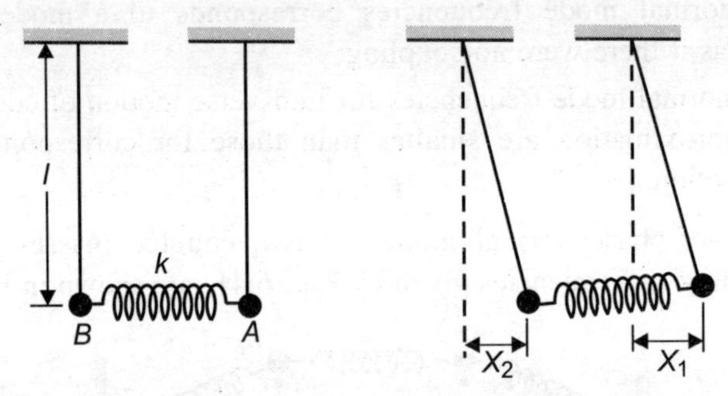

Fig. 6.9 Two identical simple pendulums coupled by a spring of force constant k: (a) equilibrium configuration and (b) instantaneous configuration.

By analogy with the coupled masses, we can write the equations of motion of pendulums A and B in small angle approximation as

$$m\frac{d^2x_1(t)}{dt^2} = -\left(\frac{mg}{l}\right)x_1(t) - k[x_1(t) - x_2(t)] \tag{6.42}$$

and

$$m\frac{d^2x_2(t)}{dt^2} = -\left(\frac{mg}{l}\right)x_2(t) + k[x_1(t) - x_2(t)]. \tag{6.43}$$

Note that negative sign in the second term on the RHS in Eq. (6.42) for the motion of pendulum A indicates that coupling spring opposes its acceleration. On the other hand, the second term $k[x_1(t) - x_2(t)]$ in Eq. (6.43) supports the motion of pendulum B. As such, the term $\pm k(x_1 - x_2)$ signifies the presence of coupling between the two pendulums.

Dividing throughout by m and rearranging terms, we get

$$\frac{d^2x_1(t)}{dt^2} + \omega_0^2 x_1(t) + \omega_s^2[x_1(t) - x_2(t)] = 0 \tag{6.44}$$

and

$$\frac{d^2x_2(t)}{dt^2} + \omega_0^2 x_2 - \omega_s^2[x_1(t) - x_2(t)] = 0, \tag{6.45}$$

where $\omega_0^2 = g/l$ is the natural frequency of individual pendulums and $\omega_s^2 = k/m$ signifies the coupling frequency of the spring.

You will recognise that Eqs. (6.44) and (6.45) are identical to Eqs. (6.2) and (6.4), respectively. It means that we can use the analysis of section 6.2.1 to understand the motion of coupled pendulums by drawing analogies. The normal modes of this system are shown in Fig. 6.10. Note that

(i) In mode 1 ($x_1 = x_2$), the pendulums oscillate in phase with the same frequency $\omega_1 = \omega_0 = \sqrt{g/l}$. *And the spring always retains its natural length, and coupling has no effect on the motion of coupled pendulums.*

(ii) In mode 2 ($x_1 = -x_2$ or $x_2 = -x_1$), the coupled pendulums will oscillate with frequency $\omega_2 = [\omega_0^2 + 2\omega_s^2]^{1/2} = [(g/l) + 2(k/m)]^{1/2}$, which is greater than ω_0 but in opposite phase (i.e., 180° out of phase). *In this mode, the spring is stretched or compressed alternately and the role of coupling is visible.*

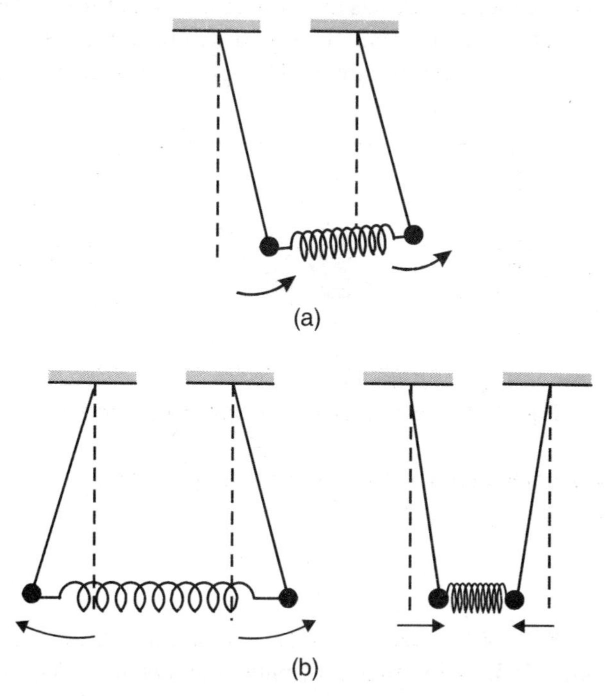

Fig. 6.10 Normal modes of two coupled simple pendulums: (a) in-phase normal mode and (b) out-of-phase normal mode.

You may recall that frequency of oscillations of a simple pendulum does not depend on the mass of the bob. But normal mode frequency ω_2 of the out-of-phase mode of the coupled pendulums exhibits inverse dependence on square root of the mass of the bob. Also note that in our analysis, we have not included the effect of air resistance on the motion of the pendulums. That is, we have considered undamped oscillations of coupled pendulums. The presence of air resistance will lead to gradual dissipation of energy. As a result, the amplitude of oscillations will slowly die out and the pendulums will ultimately come to rest. For sustaining oscillations, energy has to be provided from outside. Such a coupled oscillator is said to be forced coupled oscillator and has been discussed later in this chapter.

So far, we have discussed the salient features of coupled systems which were coupled mechanically. It is worthwhile to mention here that two or more systems can also be coupled magnetically or inductively. In section 6.3.2, you will learn about an inductively coupled *LC* circuit, and discover how magnetic and electrostatic couplings influence its behaviour.

6.3.2 Inductively Coupled *LC* Circuits

In Chapter 2, you learnt that in a *LC* circuit, charge oscillates back and forth between capacitor and inductor and the energy changes repeatedly from electric to magnetic and vice versa. What will happen if two such circuits are coupled? We should expect some energy to be exchanged between them. To understand as to what happens, refer to Fig. 6.11. It shows two *LC* circuits. For simplicity, we assume that neither circuit has any resistance so that no dissipation of energy takes place. When current is passed through circuit '1', say, it will be inductively coupled with circuit 2, since change in magnetic flux linked with circuit '1' begins to induce e.m.f in circuit '2'.

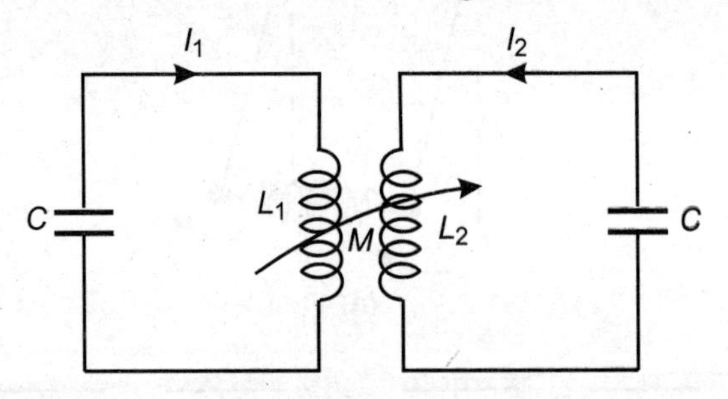

Fig. 6.11 Two inductively coupled *LC* circuits.

Let I_1 and I_2 be the instantaneous values of currents in circuits '1' and '2', respectively. The equation of motion of charge in circuit '1' is given by

$$\frac{q_1}{C_1} = -L_1\frac{dI_1}{dt} + M\frac{dI_2}{dt}, \tag{6.46}$$

where second term on the RHS of this equation $M(dI_2/dt)$ signifies the e.m.f produced in circuit '1' due to current I_2 in circuit '2'; M is coefficient of mutual inductance. (You must have learnt about it in your + 2 classes.) This induced e.m.f tends to increase current I_1 in circuit '1'.

Similarly, for circuit '2', we can write

$$\frac{q_2}{C_2} = -L_2\frac{dI_2}{dt} + M\frac{dI_1}{dt}. \tag{6.47}$$

Since $I = (dq/dt)$ and $(dI/dt) = (d^2q/dt^2)$, we can rewrite these equations as

$$\frac{d^2q_1}{dt^2} + \omega_p^2 q_1 = \frac{M}{L_1}\frac{d^2q_2}{dt^2} \tag{6.48a}$$

and

$$\frac{d^2q_2}{dt^2} + \omega_s^2 q_2 = \frac{M}{L_2}\frac{d^2q_1}{dt^2}, \tag{6.48b}$$

where $\omega_p^2 = 1/L_1C_1$ and $\omega_s^2 = 1/L_2C_2$.

Note that Eqs. (6.48a,b) are coupled equations. As in case of the coupled spring-mass system, we determine the normal modes by assuming a solution of the form $q_1 = q_{01} \cos(\omega t + \phi)$ and $q_2 = q_{02} \cos(\omega t + \phi)$. On inserting these in Eqs. (4.48a,b), we get

$$(\omega_p^2 - \omega^2)q_1 = -\frac{M}{L_1}\omega^2 q_2 \tag{6.49a}$$

and

$$(\omega_s^2 - \omega^2)q_2 = -\frac{M}{L_2}\omega^2 q_1. \tag{6.49b}$$

On equating the values of q_1/q_2 obtained from Eqs. (6.49a,b), we get

$$\frac{M}{L_1}\frac{\omega^2}{\omega_p^2 - \omega^2} = \frac{L_2}{M}\frac{\omega_s^2 - \omega^2}{\omega^2}$$

or

$$\frac{M^2\omega^4}{L_1L_2} = (\omega_p^2 - \omega^2)(\omega_s^2 - \omega^2). \tag{6.50}$$

Note that Eq. (6.50) is quadratic in ω^2 and the roots of this equation will give normal modes. Therefore, to solve it, we define coupling coefficients as $\mu = M/\sqrt{L_1L_2}$ and for simplicity, assume that $\omega_p = \omega_s = \omega_0$. Then we can write

$$(\omega_0^2 - \omega^2)^2 = \mu^2\omega^4$$

or

$$\omega_0^2 - \omega^2 = \pm\mu\omega^2$$

$\therefore$

$$\omega = \pm\frac{\omega_0}{\sqrt{1 \pm \mu}}. \tag{6.51}$$

Since a negative value of ω will not be physically acceptable. Therefore, the admissible values of ω are given by

$$\omega_1 = \frac{\omega_0}{\sqrt{1+\mu}} \tag{6.52a}$$

and

$$\omega_2 = \frac{\omega_0}{\sqrt{1-\mu}} \tag{6.52b}$$

When $u << 1$, i.e., coupling is weak, $\omega_1 \approx \omega_2 \approx \omega_0$, and the two circuits behave as essentially independent. But when coupling is strong, the values of ω_1 and ω_2 will be much different.

In the preceding sections, you must have noted that for systems which can be modelled as a coupled oscillator, we can follow a general method to calculate their normal mode frequencies. You will learn about it now.

6.4 GENERAL PROCEDURE FOR CALCULATING NORMAL MODE FREQUENCIES

To calculate the normal mode frequencies of a physical system, which can be modelled as a coupled system, we follow the following steps:

(i) Write the equations of motion of coupled oscillators;
(ii) Assume a normal mode solution;
(iii) Substitute these solutions in the equations of motion and compare the ratios of normal mode amplitudes; and
(iv) Solve the resultant equation to obtain normal mode frequencies.

We now illustrate this procedure for two pendulums with bobs of unequal masses m_1 and m_2 coupled through a spring of force constant k. Suppose at any instant t, the bob of mass m_1 is displaced by x_1 and bob of mass m_2 is displaced by x_2 such that $x_1 > x_2$. In this condition, we can write the equations of motion of two coupled pendulums as (STEP 1)

$$m_1 \frac{d^2 x_1(t)}{dt^2} = -\left(\frac{m_1 g}{l}\right) x_1 - k(x_1 - x_2) \tag{6.53}$$

and

$$m_2 \frac{d^2 x_2(t)}{dt^2} = -\left(\frac{m_2 g}{l}\right) x_2(t) + k[x_1(t) - x_2(t)]. \tag{6.54}$$

Next, we assume solutions of the form (STEP 2)

$$x_1(t) = a_1 \cos(\omega t + \phi) \tag{6.55a}$$

and

$$x_2(t) = a_2 \cos(\omega t + \phi), \tag{6.55b}$$

where ω is angular frequency and ϕ is initial phase. Note that instantaneous displacements are in phase but their amplitudes are different.

Differentiating these expressions twice with respect to time, we get

$$\frac{d^2 x_1(t)}{dt^2} = -\omega^2 x_1(t) \tag{6.56a}$$

and

$$\frac{d^2 x_2(t)}{dt^2} = -\omega^2 x_2(t). \tag{6.56b}$$

Now, substitute for $d^2x_1(t)/dt^2$ and $x_1(t)$ respectively from Eqs. (6.56a) and (6.55a) in Eq. (6.53) (STEP 3). You will note that the cosine term cancels out as it is common in the resultant expression. Then Eq. (6.53) simplifies to

$$-\omega^2 m_1 a_1 = -\left(\frac{m_1 g}{l}\right) a_1 - k a_1 + k a_2$$

or

$$\left(\frac{m_1 g}{l} + k - m_1\omega^2\right) a_1 = k a_2. \tag{6.57}$$

Similarly, on substituting for $d^2x_2(t)/dt^2$ and $x_2(t)$ in Eq. (6.54) from Eqs. (6.56b) and (6.55b), respectively (STEP 3) and rearranging terms as before, we obtain

$$\left(\frac{m_2 g}{l} + k - m_2\omega^2\right) a_2 = k a_1. \tag{6.58}$$

We can rewrite Eqs. (6.57) and (6.58) as

$$\left(\omega_0^2 + \frac{k}{m_1} - \omega^2\right) a_1 = \frac{k}{m_1} a_2 \tag{6.59}$$

and

$$\left(\omega_0^2 + \frac{k}{m_2} - \omega^2\right) a_2 = \frac{k}{m_2} a_1, \tag{6.60}$$

where $\omega_0^2 = (g/l)$ is *natural frequency* of the pendulums and is independent of the mass the bob. From Eq. (6.59), we can write the ratio of amplitudes as

$$\frac{a_1}{a_2} = \frac{k}{m_1[\omega_0^2 + (k/m_1) - \omega^2]}. \tag{6.61}$$

Similarly, Eq. (6.60) gives

$$\frac{a_1}{a_2} = \frac{m_2[\omega_0^2 + (k/m_2) - \omega^2]}{k}. \tag{6.62}$$

For non-zero values of a_1 and a_2, we can equate the expressions for a_1/a_2 given in Eqs. (6.61) and (6.62) (STEP 4). This leads us to the result

$$\frac{k}{m_1[\omega_0^2 + (k/m_1) - \omega^2]} = \frac{m_2[\omega_0^2 + (k/m_2) - \omega^2]}{k}. \tag{6.63}$$

On cross-multiplication, we get

$$\left(\omega_0^2 + \frac{k}{m_1} - \omega^2\right)\left(\omega_0^2 + \frac{k}{m_2} - \omega^2\right) = \frac{k^2}{m_1 m_2}.$$

In the expanded form, we can rewrite it as

$$\omega_0^4 + \frac{k}{m_1}\omega_0^2 - \omega^2\omega_0^2 + \frac{k}{m_2}\omega_0^2 + \frac{k^2}{m_1 m_2}$$

$$-\frac{k}{m_2}\omega^2 - \omega_0^2\omega^2 - \frac{k}{m_1}\omega^2 + \omega^4 = \frac{k^2}{m_1 m_2}.$$

Note that the term $k^2/m_1 m_2$ occurs on LHS as well as RHS and cancels out. Then, on collecting the coefficients of ω^4, ω^2 and ω_0^2, we can write

$$\omega^4 - \left(2\omega_0^2 + \frac{k}{m_1} + \frac{k}{m_2}\right)\omega^2 + \left(\omega_0^2 + \frac{k}{m_1} + \frac{k}{m_2}\right)\omega_0^2 = 0$$

or

$$\omega^4 - \left(2\omega_0^2 + \frac{k}{\mu}\right)\omega^2 + \left(\omega_0^2 + \frac{k}{\mu}\right)\omega_0^2 = 0, \quad 6.64)$$

where μ is a constant, defined as:

$$\frac{1}{\mu} = \frac{1}{m_1} + \frac{1}{m_2}.$$

It is analogous to reduced mass.

The roots of the equation $ax^2 + bx + c = 0$ are given by

$$x = \frac{-b \pm \sqrt{b^2 - 4ac}}{2a}.$$

In Eq. (6.64), $x = \omega^2$, $a = 1$, $b = -\left(2\omega_0^2 + \frac{k}{\mu}\right)$ and $c = \left(\omega_0^2 + \frac{k}{\mu}\right)\omega_0^2$.

Hence

$$x = \frac{1}{2}\left(2\omega_0^2 + \frac{k}{\mu}\right) \pm \frac{1}{2}\sqrt{\left(2\omega_0^2 + \frac{k}{\mu}\right)^2 - 4\left(\omega_0^2 + \frac{k}{\mu}\right)\omega_0^2}$$

$$= \frac{1}{2}\left(2\omega_0^2 + \frac{k}{\mu}\right) \pm \frac{1}{2}\frac{k}{\mu}$$

$$= \omega_0^2,\ \omega_0^2 + \frac{k}{\mu}.$$

Proceeding further, we note that Eq. (6.64) is quadratic in ω^2 and has roots (see the box)

$$\omega_1^2 = \omega_0^2 \tag{6.65a}$$

and

$$\omega_2^2 = \omega_0^2 + \frac{k}{\mu}$$

$$= \omega_0^2 + k\left(\frac{1}{m_1} + \frac{1}{m_2}\right). \tag{6.65b}$$

For $m_1 = m_2 = m$, these results reduce to those obtained in the preceding section.

Note that the roots ω_1^2 and ω_2^2 obtained here are positive. You may now ask: What would be our conclusion if one of the roots was negative? In such a situation, ω would be imaginary and from what you have learnt in Chapter 4, you can say that it will indicate gradual growth or decay of displacement without oscillation.

6.5 ENERGY OF UNDAMPED COUPLED SYSTEMS

6.5.1 Coupled Masses

From Chapter 2, you may recall that the total energy of an oscillator executing SHM is given by $E = 1/2\ m\omega^2 a^2$. Therefore, energy of mass m_A in the coupled system shown in Fig. 6.1 can be written as

$$E_1 = \frac{1}{2} m_A \omega_{av}^2 a_{mod}^2(t)$$

$$= 2ma^2\omega_{av}^2 \cos^2 \omega_{mod} t. \tag{6.66a}$$

Similarly, for mass m_B, we have

$$E_2 = \frac{1}{2} m\omega_{av}^2 b_{mod}^2(t)$$

$$= 2ma^2\omega_{av}^2 \sin^2 \omega_{mod} t, \tag{6.66b}$$

where we have put $m_A = m_B = m$.

Note that we have used Eqs. (6.25) and (6.26) in writing the expressions for E_1 and E_2. The total energy of these masses assumed to be coupled through a spring, which stores almost no energy, is obtained by adding E_1 and E_2:

$$E = 2ma^2\,\omega_{av}^2 (\cos^2\omega_{mod} t + \sin^2\omega_{mod} t)$$

$$= 2ma^2\,\omega_{av}^2, \tag{6.67}$$

since $\sin^2\theta + \cos^2\theta = 1$. Note that the total energy of coupled masses remains constant with time. You can rewrite Eqs. (6.66a) and (6.66b) in terms of total energy of the coupled masses as

$$E_1 = E\cos^2\omega_{mod}t$$

$$= \frac{E}{2}[1 + \cos(\omega_2 - \omega_1)t] \tag{6.67a}$$

and

$$E_2 = E\sin^2\omega_{mod}t$$

$$= \frac{E}{2}[1 - \cos(\omega_2 - \omega_1)t]. \tag{6.67b}$$

From these equations, we note that

(i) At $t = 0$, $E_1 = E$ and $E_2 = 0$. That is, to begin with, entire energy is stored in mass A.

(ii) As time passes, energy of mass at A starts decreasing and mass at B begins to gain energy, but the total energy of the system remains constant. This energy exchange takes place through the coupling spring.

(iii) When $(\omega_2 - \omega_1)t = \pi/2$, two masses share energy equally; $E_1 = E_2 = E/2$.

(iv) When $(\omega_2 - \omega_1)t = \pi$, $E_1 = 0$ and $E_2 = E$, i.e., mass at B possesses all the energy, though it was at rest initially.

(v) With passage of time, the energy exchange process continues. That is, in one cycle, the total energy flows back and forth twice between the coupled masses.

The time period of coupled oscillations is given by

$$T = \frac{2\pi}{\omega_2 - \omega_1}. \tag{6.68}$$

If the force constant of coupling spring is small compared to the force constant of the spring attached to individual masses, i.e. $k << k'$, the energy transfer is a slow process and the amplitude of oscillation of one mass takes a long time to build up or die down.

In actual physical systems such as H_2O, HCl, CO_2, NH_4, etc., the number of coupled masses is very large. Also, the coupled masses may not necessarily be of equal mass. In such situations, the above analysis will not be applicable directly. However, it provides a useful starting point.

We now discuss the case of coupled pendulums.

6.5.2 Coupled Pendulums

To obtain expressions for the kinetic and potential energies of an undamped coupled pendulum in terms of its normal co-ordinates we start with the general expression for kinetic and potential energies of a coupled pendulums. These are

$$\text{KE} = \frac{m}{2}\left[\left(\frac{dx_1}{dt}\right)^2 + \left(\frac{dx_2}{dt}\right)^2\right] \tag{6.69a}$$

and

$$\text{PE} = \frac{1}{2}\left(\frac{mg}{l}\right)(x_1^2 + x_2^2) + \frac{1}{2}k(x_1 - x_2)^2. \tag{6.69b}$$

From Eqs. (6.7a) and (6.7b), we recall that

$$x_1 = \frac{1}{2}(\xi_1 + \xi_2)$$

and

$$x_2 = \frac{1}{2}(\xi_1 - \xi_2).$$

On differentiating these with respect to time, we get

$$\frac{dx_1}{dt} = \frac{1}{2}\left(\frac{d\xi_1}{dt} + \frac{d\xi_2}{dt}\right)$$

and

$$\frac{dx_2}{dt} = \frac{1}{2}\left(\frac{d\xi_1}{dt} - \frac{d\xi_2}{dt}\right).$$

On combining these results with Eq. (6.69a), we get

$$\text{KE} = \frac{m}{2}\left[\frac{1}{4}\left(\frac{d\xi_1}{dt} + \frac{d\xi_2}{dt}\right)^2 + \frac{1}{4}\left(\frac{d\xi_1}{dt} - \frac{d\xi_2}{dt}\right)^2\right]$$

$$= \frac{m}{4}\left[\left(\frac{d\xi_1}{dt}\right)^2 + \left(\frac{d\xi_2}{dt}\right)^2\right]. \qquad (6.70)$$

You may now like to express PE of a coupled oscillator in terms of normal co-ordinates. This is the subject matter of Practice Exercise 6.3.

***Practice Exercise* 6.3** Express potential energy of a coupled pendulum in terms of normal co-ordinates.

[***Ans.*** $\text{PE} = \frac{m}{4}(\omega_0^2\zeta_1^2 + \omega_2^2\zeta_2^2)$ with $\omega_2^2 = \omega_0^2 + 2\omega_s^2$]

While solving Practice Exercise 6.3, you must have learnt that the potential energy of a coupled pendulum can be written as

$$\text{PE} = \frac{m}{4}(\omega_1^2\zeta_1^2 + \omega_2^2\zeta_2^2). \qquad (6.71)$$

We can rewrite Eqs. (6.70) and (6.71) in a more elegant form by writing normal co-ordinates in terms of displacements as

$$\varsigma_1 = \sqrt{\frac{m}{2}}(x_1 + x_2) \text{ and } \varsigma_2 = \sqrt{\frac{m}{2}}(x_1 - x_2). \qquad (6.72)$$

To get an idea about the magnitudes of energies possessed by such system, we advise you to answer Practice Exercise 6.4.

***Practice Exercise* 6.4** Using the definition given in Eq. (6.72), show that the total energy of a system of two coupled pendulums in terms of normal co-ordinates is given by

$$E = \frac{1}{2}[(\dot{\varsigma}_1)^2 + (\dot{\varsigma}_2)^2 + \omega_1^2\varsigma_1^2 + \omega_2^2\varsigma_2^2].$$

At a given instant of time, $\varsigma_1 = 1.6 \times 10^{-3}\,\text{m}^{1/2}$ and $\varsigma_2 = 0.4 \times 10^{-3}\,\text{m}^{1/2}$. Calculate $x_1(t)$ and $x_2(t)$ at the same instant. Take $m = 0.1$ kg.

So far we have considered coupled systems (spring-mass and pendulums), which execute oscillations determined by the initial conditions. That is, there was no external force whatsoever and the modifications in the nature of oscillations of individual oscillators were introduced by coupling only. From Chapter 5, you may recall that when we apply an external periodic force on a (weakly) damped system, we observe the phenomenon of resonance. You may now logically ask: Will resonance occur even if a period force is applied on a coupled system? Will there be one resonant frequency or more (two) corresponding to two coupled masses or pendulums. To discover answer to such questions, let us discuss the behaviour of a forced coupled oscillator.

6.6 NORMAL MODE ANALYSIS OF A FORCED COUPLED OSCILLATOR

Let us consider a system of two coupled identical masses such that a periodic force $F \cos \omega t$ is applied on one of the masses, say mass at A in Fig. 6.1. The equations of motion of masses m_A and m_B can be written as

$$m_A \frac{d^2x_1(t)}{dt^2} = -k'x_1 + k(x_2 - x_1) + F \cos \omega t \tag{6.73a}$$

and

$$m_B \frac{d^2x_2(t)}{dt^2} = k'x_2 - k(x_2 - x_1). \tag{6.73b}$$

Since we are forcing the system at frequency ω, we assume a solution of the form:

$$x_1(t) = a_1 \cos \omega t$$

and

$$x_2(t) = a_2 \cos \omega t.$$

Differentiating these expressions for x_1(t) and x_2(t) twice with respect to time, substituting in Eqs. (6.73a) and (6.73b) and cancelling $\cos \omega t$ from both sides, we get

$$-\omega^2 m_A a_1 = -k'a_1 + k(a_2 - a_1) + F$$

and

$$-\omega^2 m_B a_2 = -k'a_2 - k(a_2 - a_1).$$

For simplicity, we choose $m_A = m_B = m$. Then on dividing throughout by m and recognising that $k'/m = \omega_0^2$ and $k/m = \omega_s^2$, we can rewrite these equations as

$$(\omega_0^2 + \omega_s^2 - \omega^2)a_1 - \omega_s^2 a_2 = \frac{F}{m} \tag{6.74a}$$

and

$$(\omega_0^2 + \omega_s^2 - \omega^2)a_2 = \omega_s^2 a_1. \tag{6.74b}$$

We now use Eq. (6.74b) to express a_2 in terms of a_1:

$$a_2 = \frac{\omega_s^2}{\omega_1^2 - \omega^2} a_1, \tag{6.75}$$

where $\omega_1^2 = \omega_0^2 + \omega_s^2$.

On substituting this result in Eq. (6.74a), we can eliminate a_2 and obtain

$$\left[(\omega_1^2 - \omega^2) - \frac{\omega_s^4}{\omega_1^2 - \omega^2} \right] a_1 = \frac{F}{m}.$$

Hence

$$a_1 = \frac{(\omega_1^2 - \omega^2)F}{m[(\omega_1^2 - \omega^2)^2 - \omega_s^4]}$$

$$= \frac{F}{m} \frac{(\omega_1^2 - \omega^2)}{(\omega_1^2 - \omega^2 + \omega_s^2)(\omega_1^2 - \omega^2 - \omega_s^2)}$$

$$= \frac{F}{m} \frac{(\omega_0^2 + \omega_s^2 - \omega^2)}{(\omega_0^2 + 2\omega_s^2 - \omega^2)(\omega_0^2 - \omega^2)}. \tag{6.76}$$

Using this result in Eq. (6.75), you can readily show that

$$a_2 = \frac{F}{m} \frac{\omega_s^2(\omega_0^2 + \omega_s^2 - \omega^2)}{(\omega_0^2 - \omega^2)(\omega_0^2 + 2\omega_s^2 - \omega^2)(\omega_0^2 + \omega_s^2 - \omega^2)}.$$

The term $\omega_0^2 + \omega_s^2 - \omega^2$ cancels out from the numerator as well as the denominator. Then the expression for a_2 simplifies to

$$a_2 = \frac{F}{m} \frac{\omega_s^2}{(\omega_0^2 - \omega^2)(\omega_0^2 + 2\omega_s^2 - \omega^2)}. \tag{6.77}$$

Note that a_1 and a_2 become infinite for $\omega = \omega_0$ and $\omega = \sqrt{\omega_0^2 + 2\omega_s^2}$. It means that there are two resonance conditions in a forced coupled oscillator.

The variations in amplitudes a_1 and a_2 for two identical coupled, undamped forced masses with driving frequency ω are shown in Figs. 6.12a and b respectively. Note that for frequencies up to lower resonance ($\omega = \omega_0$), a_1 and a_2 have the same sign, indicating that coupled masses oscillate in-phase. However, for frequencies beyond the higher resonance ($\omega = \sqrt{\omega_0^2 + 2\omega_s^2}$), a_1 and a_2 have opposite sign, i.e., the masses oscillate 180° out of phase. Further, for $\omega^2 = \omega_0^2 + \omega_s^2$, the numerator in Eq. (6.76) vanishes giving $a_1 = 0$. But the denominator of Eq. (6.77) simplifies to $-\omega_s^4$ giving $a^2 = F/m\omega_s^2$. This frequency corresponds to the natural frequency of a single mass with coupling spring attached to a fixed mass.

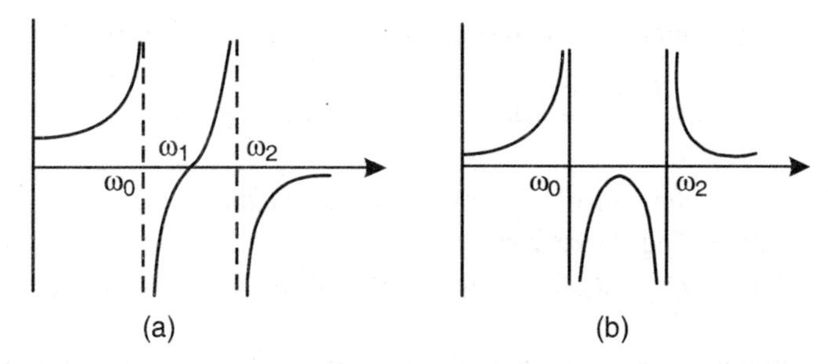

Fig. 6.12 Variation of forced amplitudes of two coupled, undamped, identical masses with driving frequency ω.

You may now like to consider two identical coupled forced pendulums and convince yourself that Eqs. (6.77) and (6.77) will hold in this case also.

6.7 LONGITUDINAL OSCILLATIONS OF *N* COUPLED MASSES: WAVE EQUATION

You now know that two coupled spring-mass systems exchange energy and how their oscillations grow or die down. In practice, every fluid or solid contains very large number of molecules coupled through inter-molecular forces. To know how such a system behaves, we have to extend the analysis of Section 6.2. For simplicity, we first assume that the medium consists of a system of N identical masses. The interatomic forces can be modelled as equivalent to $(N + 1)$ identical springs, each of force constant k, as shown in Fig. 6.13.

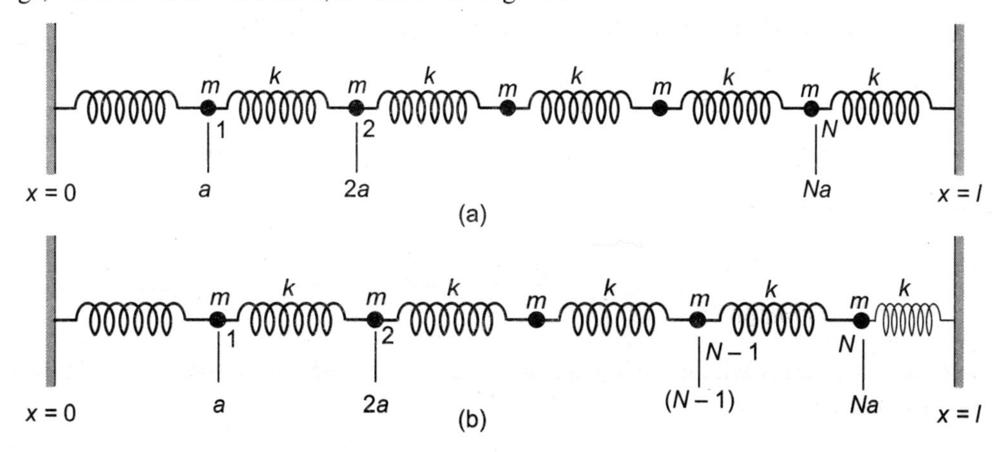

Fig. 6.13 N identical coupled masses: (a) Equilibrium configuration and (b) Instantaneous configuration.

The free ends of the systems are rigidly fixed at $x = 0$ and $x = l$. In the equilibrium state, the masses are situated at $x = a, 2a \dots, Na$ so that the length of the chain comprising coupled masses is given by $l = (N + 1)\,d$. To begin with, we consider three consecutive masses. Let ψ_{n-1}, ψ_n and ψ_{n+1} be the displacements of $(n - 1)$th, nth and $(n + 1)$th masses from their respective mean positions. We assume that motion of the nth mass is influenced by coupling with the nearest neighbours.

Therefore, its equation of motion can be written as

$$m\frac{d^2\psi_n}{dt^2} = k(\psi_{n+1} - \psi_n) - k(\psi_n - \psi_{n-1}), \tag{6.78}$$

where $(\psi_n - \psi_{n-1})$ is the extension of the nth spring from the left and $(\psi_{n+1} - \psi_n)$ denotes the extension of the $(n + 1)$th spring. Here $k(\psi_{n+1} - \psi_n)$ signifies the force pulling the nth mass in the positive x-direction and $k(\psi_n - \psi_{n-1})$ is the force pulling it backwards, i.e., in the negative x-direction.

Now, let us assume that $N \to \infty$ and separation a between any two consecutive masses becomes very small. We then use middle point approximation, where the value of a function over an interval is expressed as a product of the spatial rate of change of the function at the middle point and the range of the interval:

$$\psi_{n+1} - \psi_n = \Delta\psi\Big|_{x=[n+(1/2)]a} = \left(\frac{\partial\psi}{\partial x}\right)_{x=[n+(1/2)]\Delta x} \Delta x$$

and

$$\psi_n - \psi_{n-1} = \Delta\psi\Big|_{x=[n-(1/2)]a} = \left(\frac{\partial\psi}{\partial x}\right)_{x=[n-(1/2)]\Delta x} \Delta x, \tag{6.79}$$

where we have replaced a by Δx.

On making these substitutions in Eq. (6.78) and dividing throughout by m, we obtain

$$\frac{d^2\psi}{dt^2} = \frac{k}{m}\left[\left(\frac{\partial\psi}{\partial x}\right)_{x=[n+(1/2)]\Delta x} - \left(\frac{\partial\psi}{\partial x}\right)_{x=[n-(1/2)]\Delta x}\right]\Delta x. \tag{6.80}$$

If σ denotes mass per unit length, of the medium you can replace m by

$$m = \Delta x\sigma.$$

We now insert this relation in Eq. (6.80) and use Taylor series expansion for $\left(\frac{\partial\psi}{\partial x}\right)_{x=[n\pm(1/2)]\Delta x}$. If we retain terms only upto first order in Δx, we get

$$\frac{d^2\psi}{dt^2} = \frac{k}{\sigma\Delta x}\left[\left(\frac{\partial\psi}{\partial x}\right)_{x=n\Delta x} + \left(\frac{\partial^2\psi}{\partial x^2}\right)\frac{\Delta x}{2} - \left(\frac{\partial\psi}{\partial x}\right)_{x=n\Delta x} + \left(\frac{\partial^2\psi}{\partial x^2}\right)\frac{\Delta x}{2}\right]\Delta x.$$

The first and third terms within the square brackets on the RHS cancel out. Then the resultant expression simplifies to

$$\frac{d^2\psi}{dt^2} = \frac{k}{\sigma}\left(\frac{\partial^2\psi}{\partial x^2}\right)\cdot\Delta x$$

or

$$\frac{d^2\psi}{dt^2} = \frac{F}{\sigma}\left(\frac{\partial^2\psi}{\partial x^2}\right), \tag{6.81}$$

where F is the force $k\Delta x$ caused by extension of a spring through a distance Δx. Note that the quantity F/σ has dimensions of square of velocity. For this reason, Eq. (6.81) is referred to as the *wave equation*.

Note that if F is large and σ is small, the velocity of the wave will be large. To accomplish this, the material medium must have low density and high modulus of rigidity. We thus find that longitudinal motion of a large number of coupled masses results in wave motion.

You can obtain a similar equation for a system of large number of coupled masses executing transverse oscillations. Study Example 6.5 carefully.

EXAMPLE 6.5

Several hundred beads, each of mass 1 g are connected in series with springs 8 mm long, each weighing 1.5 g. The beads occupy 2 mm each. A force of 0.5 N is required to compress or extend a spring by 1 mm. Calculate the velocity of propagation of longitudinal elastic wave along the spring-mass system.

Solution: Here unit length has mass $\sigma = \dfrac{(1+1.5)\,\text{g}}{(8+2)\,\text{mm}} = 2.5\ \text{g cm}^{-1}.$

$$= 0.25\ \text{kg m}^{-1}.$$

Since $F = 0.5\ \text{N (mm)}^{-1} = 500\ \text{Nm}^{-1}$, we have

$$v = \sqrt{\frac{F}{\sigma}} = \sqrt{\frac{500}{0.25}} = 44.7\ \text{ms}^{-1}.$$

REVIEW EXERCISES

6.1 A thin ring R of mass m and radius a lies flat on a frictionless table (see Fig. 6.14). It is held by two stretched springs, each having relaxed length l_0 such that $a << l_0$. The spring constant of each spring is k.

(a) Determine the normal modes and their corresponding frequencies.

(b) Obtain the differential equations of motion of the system if the relaxed lengths were increased to $2l_0$.

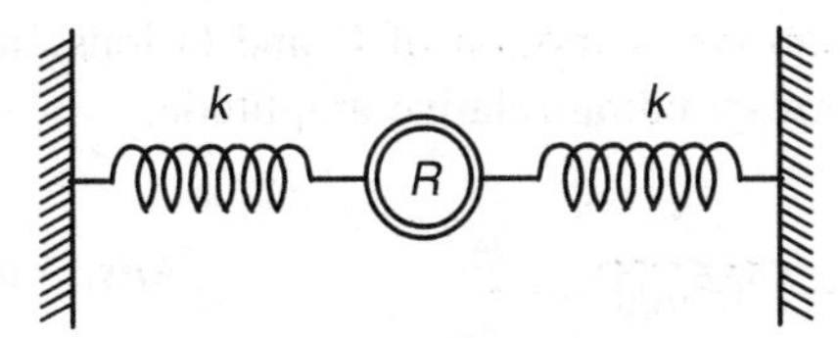

Fig. 6.14

$$\left[\textbf{\textit{Ans.}}\quad \text{(a)}\ \omega_1 = \sqrt{\frac{2k}{m}},\ \ \omega_2 = \sqrt{\frac{k}{m}},\ \text{(b)}\ \ddot{x} + \frac{2k}{M}x = 0,\ \ \ddot{y} + \frac{kxy}{Ml_0} = 0\right]$$

6.2 A uniform thin cylindrical rod of length l and mass m is supported at its ends by two light springs of spring constants k_1 and k_2. In equilibrium state, the rod is held horizontally (see Fig. 6.15). The springs can execute only vertical oscillations. Under small oscillation approximation, calculate the frequencies of normal modes for (a) $k_1 > k_2$, and (b) $k_1 = k_2$.

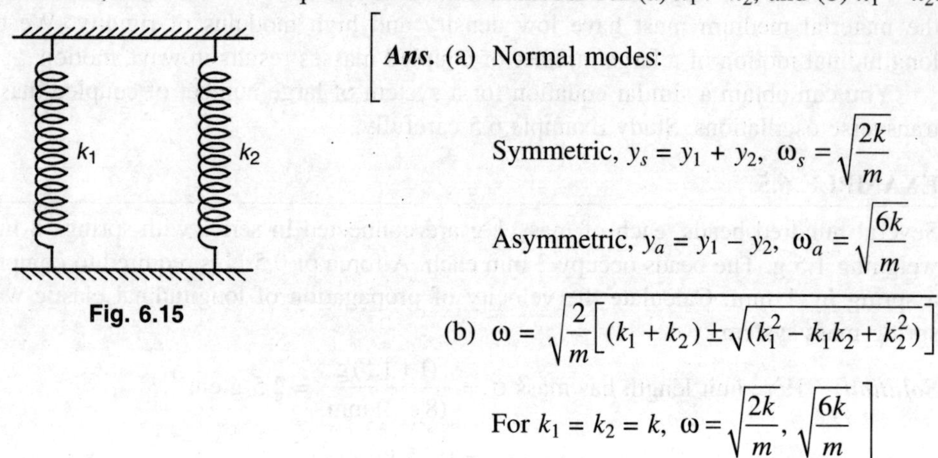

Fig. 6.15

[***Ans.*** (a) Normal modes:

Symmetric, $y_s = y_1 + y_2$, $\omega_s = \sqrt{\dfrac{2k}{m}}$

Asymmetric, $y_a = y_1 - y_2$, $\omega_a = \sqrt{\dfrac{6k}{m}}$

(b) $\omega = \sqrt{\dfrac{2}{m}\left[(k_1 + k_2) \pm \sqrt{(k_1^2 - k_1k_2 + k_2^2)}\right]}$

For $k_1 = k_2 = k$, $\omega = \sqrt{\dfrac{2k}{m}}, \sqrt{\dfrac{6k}{m}}$]

6.3 A thin disc of mass m and radius a is connected by two springs of spring constant k to two fixed points on a frictionless table (see Fig. 6.16). The disc is free to rotate but it is constrained to move in a plane. Each spring has an unstretched length l_0. Initially both springs are stretched to l and then released. Calculate the normal mode frequencies.

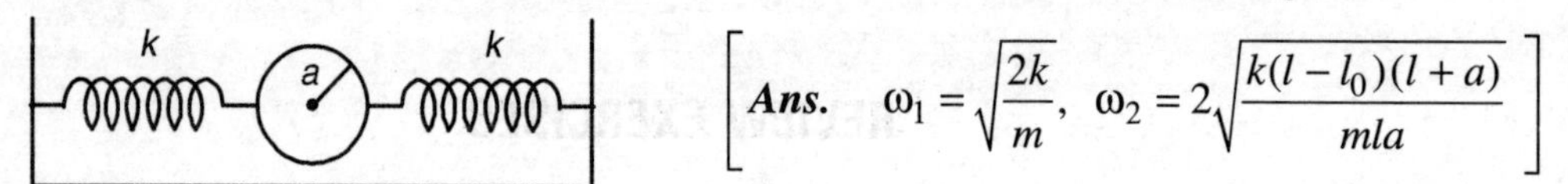

[***Ans.*** $\omega_1 = \sqrt{\dfrac{2k}{m}}$, $\omega_2 = 2\sqrt{\dfrac{k(l - l_0)(l + a)}{mla}}$]

Fig. 6.16

6.4 Suppose that the electrical forces between the ions in the linear structure of CO_2 molecule are represented by two identical springs of equilibrium length l and force constant k (see Fig. 6.17). If oscillations are possible about the equilibrium positions of atoms with no rotation.

Write down the equations of motion of C and O ions, and calculate the normal mode frequencies and the corresponding relative amplitude.

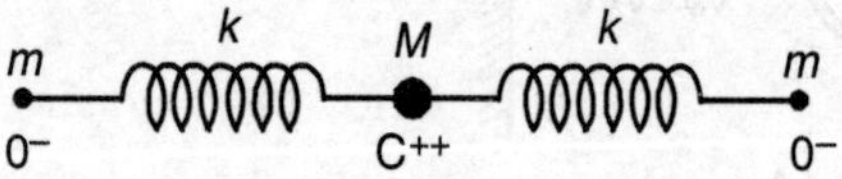

m k M k m
O^- C^{++} O^-

Fig. 6.17

[***Ans.*** $\omega_1 = \sqrt{\dfrac{k}{m}}$, $\omega_2 = \sqrt{\dfrac{(2m + M)k}{mM}}$

Relative amplitudes, $\begin{pmatrix} 1 \\ 0 \\ -1 \end{pmatrix}$ $\begin{pmatrix} 1 \\ -2m/M \\ 1 \end{pmatrix}$]

6.5 Consider a long chain of beads connected by springs of force constant k as shown in Fig. 6.18. Each bead can oscillate along x-axis. All are beads, except one, and have same mass m. The mass of the different body is m_0 ($< m$). If the springs can be assumed to be massless, calculate the relation between the wave vector k and the frequency ω of the oscillations.

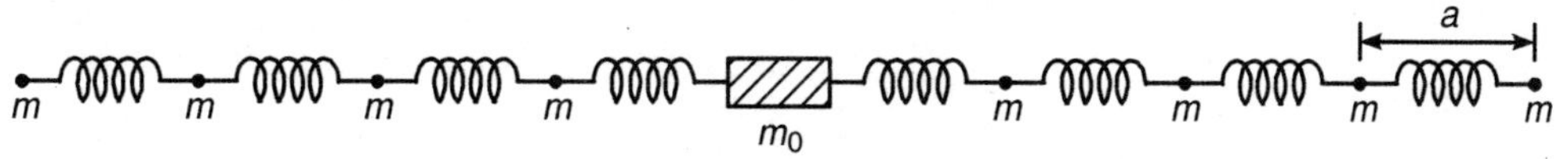

Fig. 6.18

$$\left[\textit{Ans.} \quad \omega^2 = \frac{2k}{m}(1 - \cos ka) \right]$$

Part II

Waves

- Wave Motion
- Reflection and Refraction of Waves
- Vibrations of Strings
- Vibrations of Bars
- Vibrations of Air Columns
- Large Amplitude Oscillations

7

Wave Motion

Expected Learning Outcomes

In this Chapter, you will acquire capability to:

- explain formation and propagation of waves in a medium;
- represent a wave at a fixed position and at a fixed time graphically;
- write mathematical expression of a progressive wave corresponding to a given set of wave parameters and travelling along $+x$ and $-x$ directions;
- explain the concept of phase and phase difference in relation to waves;
- derive expressions for (i) the energy carried by a wave and (ii) its intensity at a point in space;
- derive 1-D wave equation for (i) transverse waves on a stretched string, and (ii) longitudinal waves in a solid rod as well as a gaseous medium;
- write wave equation for two and three dimensional systems; and
- explain Doppler effect and obtain expressions for apparent frequency of sound when the source or listener or both are in motion.

7.1 INTRODUCTION

You have studied about wave motion in your school physics course. You will agree that the study of wave motion is important because waves are present all around us and are responsible for all phenomena associated with our communication—seeing, speaking and hearing—with the outside world. When we *speak*, our *vocal cords* inside our throat *vibrate*. These vibrations cause the surrounding air molecules to vibrate and the effect—speech—manifests as *sound*. When this sound reaches the ears of another person, it sets her/his ear drums into vibrations making it heard by her/him. Sound is a form of energy and is carried by waves. The chirping of birds, humming of bees, whisper of lovers, the fascinating music as well the irritating noise is carried by sound waves. Can you list other situations in which waves carry energy from one place to another? If you have ever stood at the sea shore or dropped pebbles in a still pool of water, you must have observed ripples which propagate in the form of *water waves*. Apart from sound and water waves, *electromagnetic waves* are other more familiar and useful waves. These include visible light, radio waves, microwaves, X-rays and gamma rays, among others. You know that our visual contact with the world around us depends on light waves. We are able to appreciate different hues and colours that Mother Nature has created because light makes us to see.

Sound waves find applications in SONAR (*So*und *Na*vigation and *R*anging) and prospecting for underground mineral deposits and oil; the commodities driving the world economy in present times. The *ultra-sound waves* (of frequency greater than 20 kHz) are used to obtain images of soft tissues in the interior of human bodies and locate abnormalities such as kidney and gall bladder stones, growth of foetus or fibroids in the uterus, etc. The modern communication tools such as radio, television, telephone (land line as well as mobile), fax, internet, etc. are based on transmission and receipt of signals in the form of electromagnetic waves. The use of X-rays in medical diagnosis, for taking images of bones to diagnose fractures, say and use of gamma rays in radiation therapy for treatment of malignant tumours is well known.

Shock waves and *seismic waves* are other less familiar but equally important type of waves. The havoc created by the earthquakes in January, 2001 in the state of Gujarat, in October, 2005 in J&K and Tsunami in Tamil Nadu in December, 2004 are vivid evidences of the huge energy carried by seismic and tidal waves. If harnessed properly, these waves can emerge suitable alternatives of renewable and green energy and help us to off-set disadvantages of our dependence on oil. To understand the nature of atoms, molecules and nuclei, we use the concept of *matter waves*.

We hope now you agree that understanding wave phenomena is of fundamental importance to us. The waves mentioned above can be broadly categorised into two main types: mechanical waves and electromagnetic waves. In this book, we will confine ourselves to the study of mechanical waves with particular emphasis on sound waves.

From Chapter 2, you may recall that our discussion of oscillations was simplified because we could draw analogies for different physical systems. We can make similar simplification in the study of waves. The basic description of a wave and the parameters required to quantify this description remain the same when we deal with waves in one, two and three dimensions such as waves travelling along a string, on the surface of water and sound waves in air, respectively. You have read about these in your school Physics classes and we would like you to refresh your knowledge. Yet for completeness, we begin our study of wave motion by revisiting/revising basic concepts of wave motion.

In Section 7.2, we begin by discussing how waves are formed and how these propagate. We take the examples of waves on a string and sound waves. In Section 7.3, we depict wave motion graphically and describe it mathematically. The wave parameters such as amplitude, frequency, time period, and wave number are introduced here. We explain the concept of the *phase of a wave* and *phase difference* in Section 7.4. You will learn to derive expression for energy transported by progressive waves in Section 7.5. The concept of *intensity* and the inverse square law for a wave form the subject matter of discussion in Section 7.6. The vocabulary, language and ideas developed here will be used to derive expressions for velocities of 1-D transverse waves on strings and longitudinal waves in gases in Section 7.7. The effect of temperature, pressure, humidity, etc. on these waves is also discussed here. Extension of these concepts to two and three dimensional systems is presented in Section 7.8.

You may recall that frequency of a wave is a fundamental physical parameter. That is, frequency of a wave is not affected by the properties of the medium. But it may seem to change when there is a relative motion between the listener and the source or the medium. You must have experienced that the pitch (or frequency) of the whistle of a train moving towards or going away from you is different from the original frequency emitted by the train. This *apparent change in frequency* due to the relative motion between the source and the listener (or detector) is known as *Doppler effect.* This effect is observed for sound (mechanical waves) as well as light (electromagnetic waves) and has found wide applications. In Section 7.9, we have discussed Doppler effect for sound.

7.2 WAVE FORMATION AND PROPAGATION

You will perhaps agree that water waves are the most familiar type of waves as we can generate and observe these easily. If you drop small pebbles in still water, say, in a pond or a water tub, you will observe that circular ripples spread out from the point where pebbles strike the water surface, as shown in Fig. 7.1. When you look casually at these ripples, you may get an impression that water moves with them. However, if you observe the ripples carefully, you will note that this is not true: *water does not move along the ripples.* You can easily verify this by keeping a paper boat or a dry leaf on the surface of water and observing its motion. What do you observe? You will note that the paper boat (or the leaf) bounces up and down at the same place on the water surface; it does not move with the ripples. It means that at no point in space, water has any translational motion.

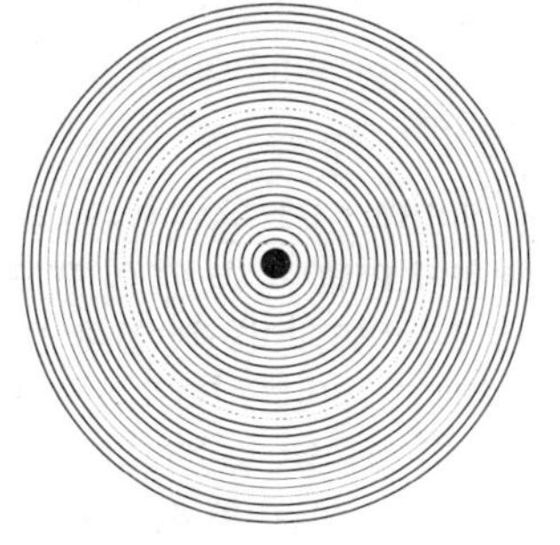

Fig. 7.1 Waves generated on water surface.

Next, you may like to know: Does water undergo any motion at all and if so, what is its nature. Alternatively, if water does not move along the surface, what is it that moves in a water wave? The answer to the first question is: Water molecules execute oscillatory motion and the answer to the second question is: The *disturbance* caused in still water by dropping pebbles in it moves as a wave. This disturbance is transferred by energetic molecules progressively to adjacent water molecules.

> The word **disturbance** has been used here as a general term which refers to the deformation in the shape of the water surface (or any other medium such as a string) with respect to its undisturbed horizontal surface.

To understand how it happens, refer to Fig. 7.2, which depicts N masses, each of mass m, joined to each other by massless, identical springs. You may recall that when we continuously disturb (displace) the first mass from its equilibrium position, subsequent individual masses gradually begin to oscillate about their respective equilibrium positions successively. That is, neither the masses (or connecting springs) nor the system as a whole move from their respective positions; what moves instead is a disturbance. Similarly, ripples on the water surface spread out due to transfer of deformation to successive molecules along the 2-D sheet of water molecules.

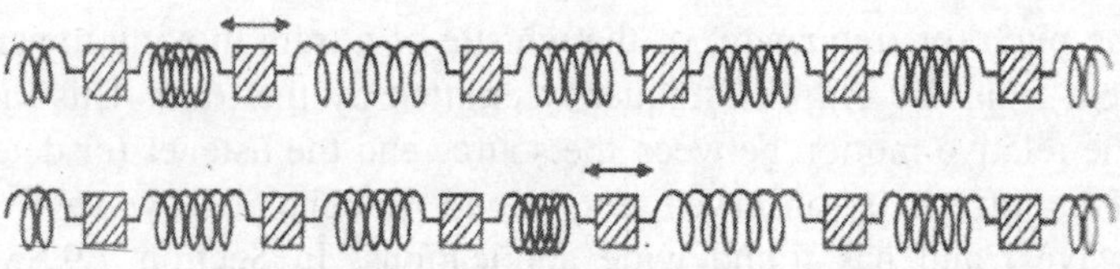

Fig. 7.2 The oscillatory motion of a system of *N*-coupled masses. The disturbance in the form of compression and elongation of the springs, say at mass 1, is transferred successively to adjacent masses.

You can produce a mechanical wave by performing a simple activity. Take a long but thin elastic string and fix its one end to a rigid support such as a wall. Hold the other end of the string in your hand so that the string is stretched and taut. Now, quickly move your hand up and down once. You will see a disturbance travelling along the length of the string. (An isolated disturbance is called a *pulse*.) If you keep your hand moving up and down, you should see a series of pulses moving along the string giving rise to a *wave*.

From this description of waves, we may conclude that:

(i) A wave is generated due to two simultaneous but distinct motions: (i) oscillatory motion of the particles of the medium; and (ii) linear motion of the disturbance.

(ii) In wave motion, propagation of a disturbance in a medium does not take place due to physical movement of the particles of the medium. The disturbance propagates due to transfer of energy from one particle to another particle progressively. This fact manifests in the form of compressions and elongations/rarefactions. Thus, we can conclude that *waves transport energy not matter.*

While considering formation and propagation of 1-D waves on a string, you must have noticed that oscillations of the string held by you and propagation of wave (along the string) are intimately connected. To appreciate the nature of relationship (between the oscillations of the particles of a medium and propagation of a wave in the medium) further, refer to Fig. 7.3. It shows a string tied to a spring-mass system, which can execute vertical oscillations. The other end of the string is tied to a rigid support like a wall. We assume that motion of the spring–mass system is frictionless and the mass executes vertical oscillations without any lateral motion.

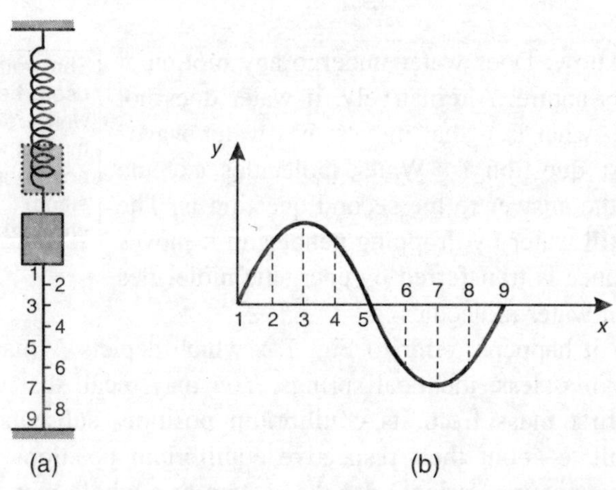

Fig. 7.3 (a) A string fastened to a vertically oscillating spring-mass system, and (b) snapshot of the motion of string.

Now, refer to Figs. 7.4a to 7.4i, which depict the snapshots of the waveform on the string taken at intervals of $T/8$. These snapshots indicate the positions of nine particles at intervals of $T/8$ during the period from $t = 0$ to $t = T$. The arrows attached to each particle indicate the directions along which these particles are about to move at the given instant. At the instant $t = 0$, all particles are at their mean position (Fig. 7.4a).

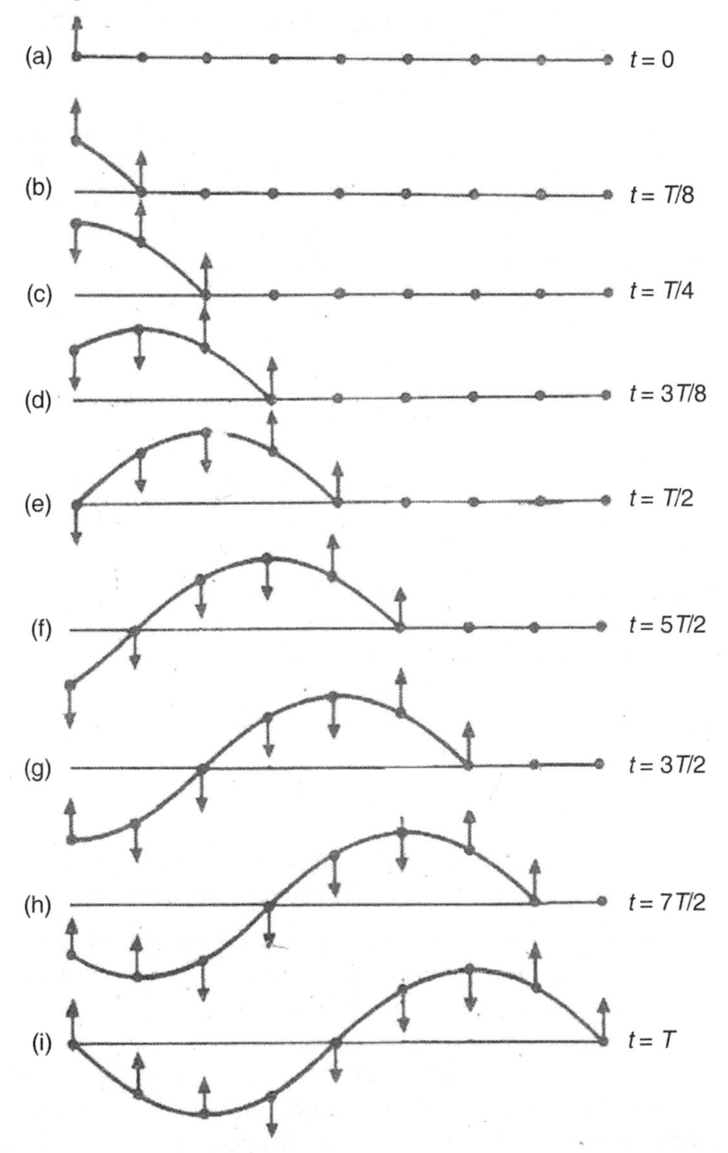

Fig. 7.4 Snapshots of the motion of the particles 1 to 9 of the string beginning at the instant $t = 0$ and up to the instant $t = T$ at intervals of $T/8$.

You may now like to analyse this activity to discover answers to some of the questions that were raised in the beginning of this chapter. We first address the question: How do particles in the string begin to oscillate? This is due to transfer of mechanical energy and momentum from the spring-mass system to particle 1 of the string and the sustenance of motion of the particle due to elasticity of the medium. As particle 1 oscillates, it transfers energy and momentum to particle 2. In the process particle 2 accepts the energy and momentum and it is set into oscillation due to the elasticity of the medium. As particle 2 starts oscillating, this, in turn, transfers energy and momentum to particle 3. This process continues till energy that activated particle 1 reaches particle 9. And we say that a wave has been generated in the string. This process continues as long as the spring-mass system keeps oscillating. So we may conclude that *a wave motion is a disturbance which travels from one region to another region in a medium due to transfer of energy and momentum from one particle to another successively.*

(i) Let us now discover answer to the question: How do particles of the medium move? To do so we study the motion of each of the nine particles shown in Fig. 7.4a to 7.4i at intervals of $T/8$ from $t = 0$ to $t = T$. First consider particle 1. Suppose that we see its motion through a slit placed at $x = 0$. How does it move? It goes up (Fig. 7.4b) to a maximum displacement from its mean position (Fig. 7.4c), moves to its mean position (Fig. 7.4e), goes further down to a maximum displacement from its mean position on the opposite side (Fig. 7.4f and g) and then comes up to the mean position once again (Fig. 7.4i). By imagining slits at the positions of other particles, you can see from Fig. 7.4a to 7.4i that they too oscillate about their mean positions. The curve joining the positions of all the particles at $t = T$ in Fig. 7.4i) represents a *wave*.

You must note that (i) *All particles of the string oscillate up and down about their respective mean positions with time period T.* (ii) *The wave moves along the string with the same time period.* From the example and the activities discussed above, we conclude that

(i) A disturbance is generated in a medium by the oscillatory motion of particles. The disturbance may take any shape—from a finite width pulse to an infinitely long sine wave or any other shape, depending on the driving force.
(ii) The particles oscillate about their mean equilibrium position; they do not travel with the disturbance nor do they show any translational motion.
(iii) The disturbance propagating in the medium *transfers energy* and *momentum from one particle to another in the medium.* Thus, we can conclude that *waves transport energy and momentum; not matter.*
(iv) Every particle on receiving energy and momentium by way of transfer from the previous particle is *set into oscillations due to the elasticity of the medium.*

> The mechanism of wave propagation described here is peculiar to mechanical waves. The scenario is different in case of electromagnetic waves.

So we can now say that wave *is a disturbance that transfers energy and momentum progressively from one particle to another particle in a medium.*

Note that in our discussion so far we have considered the propagation of mechanical waves on strings and springs for introducing wave motion. You have also learnt that tidal waves, seismic waves, water waves and sound waves are mechanical waves. *Mechanical waves require material medium such as water, air, rocks, etc. to transfer mechanical energy and momentum from one point to another.* The constituents of such waves obey Newton's laws of motion as well as Hooke's law.

Note that sound waves travelling in air columns as well as on a string are mechanical waves. But you may recall an important difference between them. While the former are longitudinal, the latter are transverse waves. In your school physics course, you have learnt the basic difference between these waves but we discuss it briefly for completeness.

7.2.1 Transverse and Longitudinal Waves

In the case of waves propagating on a string, you must have noted that the medium particles *oscillate in a direction perpendicular to the direction of propagation of the wave.* Such a wave is said to be *transverse wave.* Waves propagating on the strings of musical instruments such as sitar ektara, sarangi, veena and violin are transverse waves. Electromagnetic waves are also transverse in nature.

In longitudinal waves such as the sound waves, *particles oscillate along the direction of propagation of the wave.* Waves in musical instruments such as the flute, organ pipes, harmonium, etc. in which an air column vibrates to produce sound are *longitudinal waves.* Mechanical waves in gaseous media are longitudinal, whereas liquids and solids support transverse waves as well. Longitudinal waves are accompanied by alternate regions of *compression* (or the regions of high density or high pressure) and *rarefaction* (or the regions of low density or low pressure).

Now, that you understand the role of the motion of the particles of a medium in generating wave motion, you may like to know: How to describe wave motion? In this context, it is important to note that we shall confine our discussion in this chapter to progressive (or travelling) waves. For simplicity, we first consider waves moving along a straight line, i.e., one-dimensional waves.

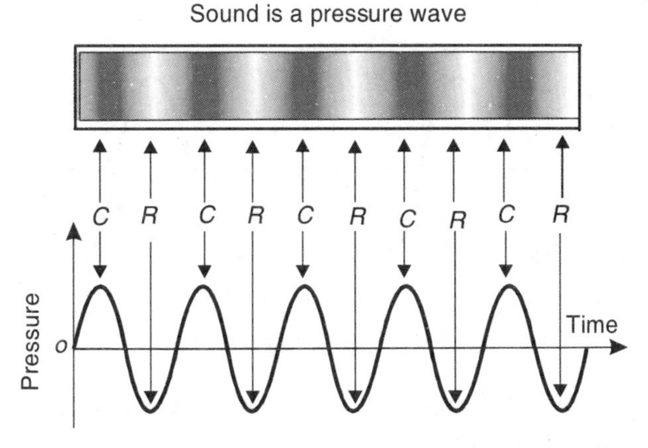

Fig. 7.5 Sound waves in air show alternate regions of compression (C) and rarefaction (R).

7.3 DESCRIBING WAVE MOTION

In the earlier discussion, you must have noted that when a wave moves, the displacements of the particles change with time as well as with position. And *in one complete period of oscillation, the particles of the medium are displaced in one direction from their mean position to a position of maximum displacement (corresponding to the crest in the wave), come back to their mean position and move in the opposite direction to the position of maximum displacement (corresponding to the trough in the wave). Finally, they come back to their mean position.*

The wave generated under these assumptions propagates with the same time period as the period of vibration of each particle of the medium (which is equal to the period of vibration of the generator of disturbance). The amplitude of the wave equals the maximum positive displacement of the particles from their mean position.

Since frequency is reciprocal of time period, the frequency of a wave is equal to the frequency of vibration of the medium particles. The distance between any two points in the same state of motion defines the wavelength of a wave. Physically, it means that wavelength is equal to the distance between two consecutive crests or two consecutive troughs. Thus, the wavelength and the period respectively signify the *spatial* and the *temporal properties* of a wave. We can now say that when a wave propagates in a medium, it travels with the same amplitude, time period, frequency and wavelength as those of the particles oscillating in the medium. Therefore, we can say that in a wave, variation with position and time follow the same pattern as that of the oscillating particles. It means that we can represent wave motion graphically as well as mathematically.

In graphical representation, we can display information by

- keeping the position x fixed and changing time t; or
- keeping the time t fixed and changing position x.

Some authors prefer to refer to the first type of graph as *vibration graph.* Note that *a vibration graph shows the wave behaviour at one position* in the path of a wave with time. You can obtain it by fixing a slit at a point in space and observing wave motion at different times. It is a sinusoidal graph, as shown in Fig. 7.6.

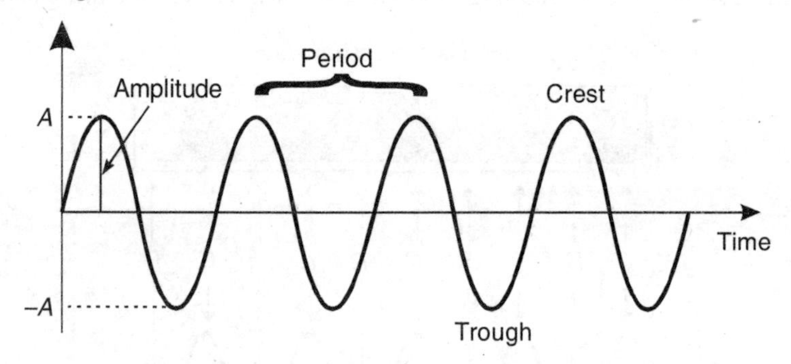

Fig. 7.6 The vibration graph of a wave.

> Note that even though we are using the sine function to represent the vibration graph, we can as well use the cosine function. Then Eq. (7.1) will be given by
>
> $y(t) = a \cos \omega t.$

We know that such a graph is mathematically represented by the equation

$$y(t) = a \sin \omega t, \tag{7.1}$$

where a is amplitude of the wave (and is equal to the amplitude of oscillating particles) and ω is its angular frequency. It is related to the frequency f by the relation $\omega = 2\pi f$ and to the time period T by $\omega = 2\pi/T$. Thus, Eq. (7.1) can be rewritten as

$$y(t) = a \sin 2\pi\left(\frac{t}{T}\right). \tag{7.2}$$

When time is kept fixed and position is changed, the graph is called a *waveform graph.* It is the same as the snapshot shown in Fig. 7.4i but taken at a later instant of time. *A waveform graph displays the wave behaviour simultaneously at different locations.* It is depicted in Fig. 7.7.

Note the similarities in the shapes of vibration and waveform graphs. But you should not confuse. While vibration graph gives us information about the *shape* of the wave, its *amplitude*, and *time period,* waveform graph tells us about wave's *shape*, *amplitude*, and *wavelength.* This basic difference is reflected in the labelling along the x-axis: For the *vibration graph*, we plot *time* t along the x-axis, whereas for the *waveform graph*, we have *position* x along the x-axis.

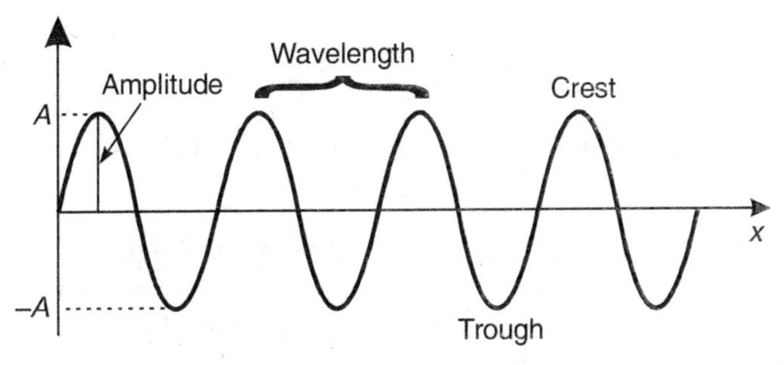

Fig. 7.7 The waveform graph of a wave. It is analogous to the snapshot at any instant of time, say at $t = T$.

We can mathematically represent the waveform graph as

$$y(x) = a \sin kx, \tag{7.3}$$

where k is a constant. You will learn that k is related to the wavelength.

Do not confuse this k with spring constant.

Recall that wavelength λ denotes the distance between any two points on the wave in the same state of motion, as shown in Fig. 7.7. This helps us to express the constant k in terms of λ by noting that the displacement y is the same at both ends of this wavelength, i.e., the values of y are equal at $x = x_1$ and at $x = x_1 + \lambda$. Thus, from Eq. (7.3), we can write

$$y(x) = a \sin kx_1 = a \sin k(x_1 + \lambda) = a \sin (kx_1 + k\lambda).$$

From Chapter 1, you may recall that repetition period of a sine function is 2π. So we can write $\sin (t + 2\pi) = \sin t$, where t is any variable. This condition is satisfied only when

$$k = \frac{2\pi}{\lambda}. \tag{7.4}$$

k is called the *wave number*. It signifies *how quickly the wave oscillates in space*, whereas *angular frequency* ω *tells us how quickly the wave oscillates in time.*
On combining Eqs. (7.3) and (7.4), we can write

$$y(x) = a \sin 2\pi\left(\frac{x}{\lambda}\right). \tag{7.5}$$

You now know that a wave is either periodic or a pulse. You have also learnt that wave motion requires a source which moves or vibrates with a particular frequency. It means that *the frequency of a wave is a property of the source, not of the medium, through which it propagates.*

You now know how to represent a longitudinal and a transverse wave. You have also learnt about parameters which characterise spatial and temporal properties—wavelength and time period—associated with a wave. Let us now find out how these parameters are related to the velocity of propagation of a wave in a medium.

7.3.1 Relation between Wave Velocity, Frequency and Wavelength

We can establish the relationship between the velocity of a wave, its frequency and its wavelength using their definitions. Note that a wave moves a distance equal to one full wavelength in one period T. Therefore, the wave speed is given by

$$v = \frac{\text{Wavelength}}{\text{Period}} = \frac{\lambda}{T}. \tag{7.6}$$

Since frequency f of a wave is reciprocal of its period T, we can also write

$$v = f\lambda. \tag{7.7}$$

That is, the velocity of a wave is equal to the product of its frequency and wavelength. This equation predicts that in a given medium, the velocity of a wave of given frequency is constant.

Note that Eq. (7.7) holds for a transverse as well as a longitudinal wave. You may recall that sound travels the fastest in solids and the slowest in gases. At STP, the speeds of sound waves in air, water and steel are 332 ms^{-1}, 1500 ms^{-1} and 5100 ms^{-1} respectively. (This explains why the whistle of an approaching train may be heard twice: first as the sound travels through the railroad track and again as it travels through the air.) Ripples on the surface of a pond move with a speed of about 0.2 ms^{-1}. The seismic waves move with speed of the order of 6×10^3 ms^{-1} in Earth's outer crust and light moves with a speed of 3×10^8 ms^{-1}. That is why light that originates on or near the Earth reaches us almost instantaneously. For the same reason, we are able to see lightening produced by clouds before we hear the thunder.

You may now like to answer a Practice Exercise.

***Practice Exercise* 7.1**

(a) Light moves with a speed of 3×10^8 ms^{-1}. The visible region begins from 400 nm. Calculate the corresponding frequency.

(b) Sound travels in air with a speed of 332 ms^{-1}. The upper limit of audible range is 20,000 Hz. Calculate the corresponding wavelength.

[***Ans.*** (a) 7.5×10^{14} Hz; (b) 1.66×10^{-2} m]

You now know that time variation of a wave at a given position is described by the vibration graph which can be modelled by Eqs. (7.1) and (7.2). You also know that the waveform graph depicts how a wave changes at different points in space at a given instant of time and Eq. (7.5) describes this graph. However, as a wave travels, it changes both with time and position. This information is contained neither in Eq. (7.2) nor in Eq. (7.5). That is, neither of these equations describes a wave completely. You may, therefore, like to know: How to obtain an equation that gives us complete mathematical description of a wave as it travels in space and in time? To do so, we have to combine the aforesaid equations. Let us learn to do so now.

7.3.2 Mathematical Description of Wave Motion

Refer to Figs. 7.8a and b, which show the snapshots of a wave travelling along the positive x-axis at the instant $t = 0$ and at a later time t. Suppose that the wave travels with velocity v. Let us consider the particle at point C on the wave at some position x' in Fig. 7.8a at the instant $t = 0$. The wave travels a distance vt in time t from point C to the point D displacing the particle at that point. Thus, at time t, the displacement of the particle at point D (Fig. 7.8b) is the same as that of the particle at C at the instant $t = 0$. Suppose that the position of D is given by x. Then you can see from Fig. 7.8b that $x = x' + vt$ or $x' = x - vt$.

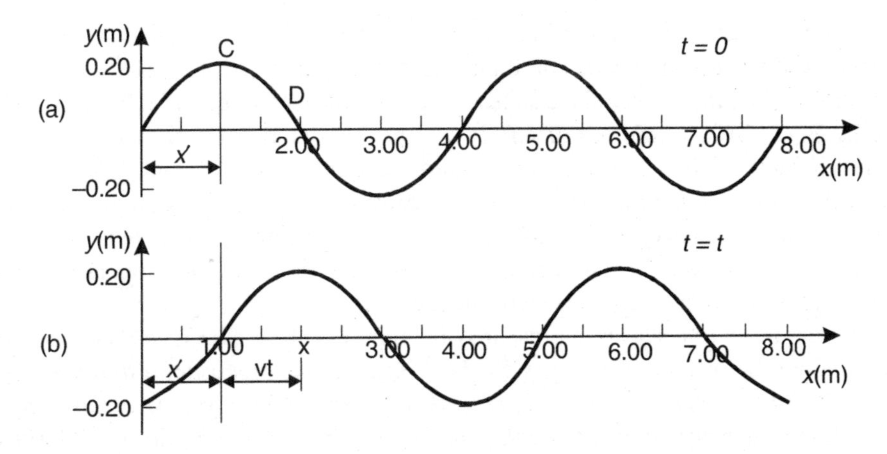

Fig. 7.8 Snapshot of a wave travelling along positive x-axis at (a) $t = 0$ and (b) $t = t$.

Let us denote the displacement of the particle at point D at time t by $y(x, t)$. It is equal to the displacement of the particle at point C at the instant $t = 0$, which is given by $y(x', t) = y(x - vt, 0)$. Thus, we have

$$y(x, t) = y(x', t) = y(x - vt, 0). \tag{7.8}$$

To determine $y(x', t = 0)$, we replace x by x' in Eq. (7.5) and write the expression for the displacement $y(x', t = 0)$ as

$$y(x', t = 0) = a \sin 2\pi \left(\frac{x'}{\lambda}\right).$$

On substituting $x' = x - vt$ in this expression, you will get

$$y(x', t=0) = y(x - vt, 0) = a \sin\left[\frac{2\pi}{\lambda}(x - vt)\right]. \tag{7.9}$$

On inserting this result in Eq. (7.8), we get

$$y(x, t) = a \sin\left[\frac{2\pi}{\lambda}(x - vt)\right]. \tag{7.10}$$

Since $v = \lambda/T$, we can write this expression as

$$y(x, t) = a \sin\left[\frac{2\pi}{T}\left(\frac{x}{v} - t\right)\right]. \tag{7.11}$$

We can also write this equation as:

$$y(x, t) = a \sin\left[2\pi\left(\frac{x}{\lambda} - \frac{t}{T}\right)\right]. \tag{7.12}$$

Equations (7.10), (7.11) and (7.12) give equivalent mathematical description of a one-dimensional (1-D) wave travelling along the positive x-direction. However, the use/choice of a particular expression depends on the specific situation. In terms of angular frequency ω (= $2\pi/T$) and wave number k (= $2\pi/\lambda$), you can write Eq. (7.12) as

$$y(x, t) = a \sin (kx - \omega t). \tag{7.13}$$

The simple way in which ω and k enter in this description of a wave explains why these quantities are so often useful in studying wave motion. Equation (7.13) (or its other equivalent forms) describes a *monochromatic* wave, since it has a single constant frequency. Note that the waves represented by Eqs. (7.10)–(7.13) are of *infinite extent*. That is, there is no mathematical limit on the value of x; it can vary from $-\infty$ to ∞ for a fixed value of t.

Note that these equations describe 1-D waves (transverse or as well as longitudinal sinusoidal waves travelling in the positive x-direction. However, $y(x, t)$ represents displacement in case of waves on strings, air pressure or air density in case of sound waves, temperature, electric or magnetic field intensity in case of 1-D electromagnetic waves.

The basic feature of the waves represented by Eqs. (7.10) to (7.13) is that the entire wave pattern moves along the x-axis as time passes. This leads to *one important difference between the displacement of the particles of the medium and the displacement $y(x, t)$ of any point on the waveform: while the former changes periodically, the latter remains constant.* As the wave travels, *the entire waveform shifts*. Hence, the displacement of a point on the waveform remains the same and this holds for all points on the waveform. This point is illustrated in Fig. 7.9 which shows the variation of $y(x, t)$ with both x and t for a wave represented by Eq. (7.10) propagating along the positive x-direction.

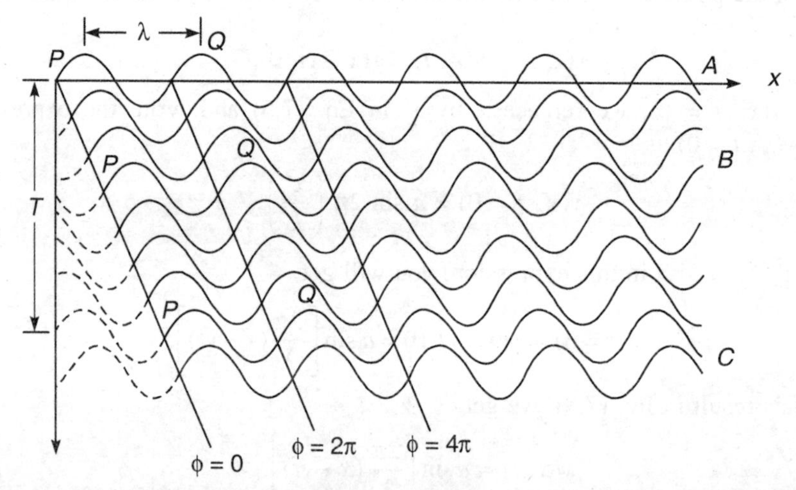

Fig. 7.9 Plot of a sinusoidal wave travelling along the positive x-direction in the x-t plane. The ϕ-values indicate the phases.

It may be pointed out here that several authors prefer to use an alternative equivalent expression for a progressive wave travelling in positive x-direction:

$$y(x, t) = a \sin\left[\frac{2\pi}{\lambda}(vt - x)\right]. \tag{7.10$'$}$$

Let us determine the displacements $y(x, t)$ at different values of x and t as the wave travels. Refer to Fig. 7.9 again and focus on the curves A, B and C, which show the wave at $t = 0$, $t = T/2$ and $t = T$, respectively. Note that the entire waveform has shifted to the right in this time. Now, consider the vertical displacement of the points $P(x = 0, t = 0)$ and $Q(x = 5\lambda/4, t = 0)$ in these three curves. You can see that the displacements of the points P and Q are the same at all times. This is true for all points on the wave for all values of t. Thus, as the wave travels, the displacements of all points on the waveform remain constant with time and equal to their respective values at the reference position and time (in this case $t = 0$ and $x = 0$).

You may now like to know as to how we describe a wave travelling in the negative x-direction. If you think that we should replace x by $-x$ in Eqs. (7.10)–(7.13), you are not so right. To understand the reason, refer to Fig. 7.10. In this case, the displacement of the particle at point D at time t, (Fig. 7.10b) is the same as that of particle at C at the instant $t = 0$. Suppose that the position of C is given by x' and that of D is given by x. Then you can see from Fig. 7.10 that $x' = x + vt$ and we can write

$$y(x, t) = y(x', t = 0) = y(x + vt, 0).$$

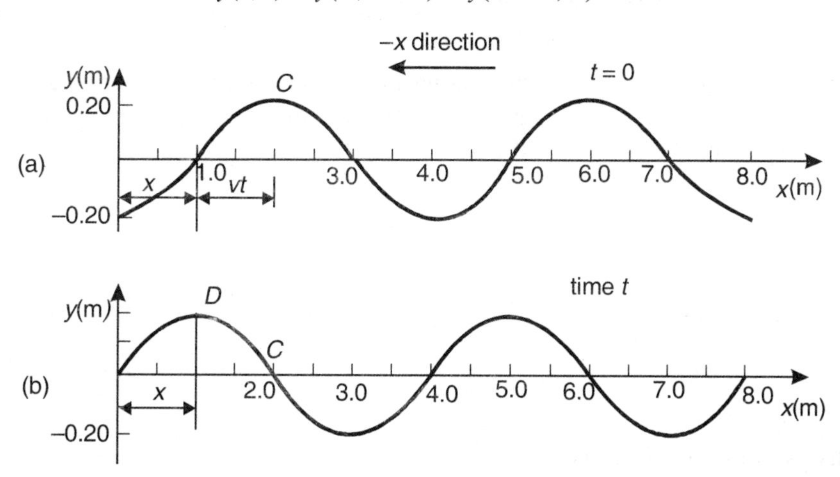

Fig. 7.10 Snapshot of a wave travelling in the negative x-direction at (a) $t = 0$ and (b) time t. The wave travels a distance vt to the left in this time. The particle at point D at the instant t [in part (b)] has the same displacement as the particle at point C at $t = 0$ [in part (a)].

You could repeat the steps outlined in deriving Eqs. (7.8) to (7.13) and convince yourself that the following results hold for a wave travelling in the negative x-direction:

$$y(x, t) = a \sin\left[\frac{2\pi}{\lambda}(x + vt)\right] \tag{7.14}$$

$$y(x, t) = a \sin\left[\frac{2\pi}{T}\left(\frac{x}{v} + t\right)\right] \tag{7.15}$$

$$y(x, t) = a \sin\left[2\pi\left(\frac{x}{\lambda} + \frac{t}{T}\right)\right] \tag{7.16}$$

and

$$y(x, t) = a \sin (kx + \omega t). \tag{7.17}$$

On comparing Eqs. (7.11) and (7.13) or Eqs. (7.15) and (7.17), you can easily convince yourself that wave velocity is related to ω and k as

$$v = \frac{\omega}{k}. \tag{7.18}$$

Now, you should go through Example 7.1 and answer Practice Exercise 7.2 to learn how to find wave parameters from a wave equation and how to write a wave equation for given parameters. This will also give you a feel for the typical values of the parameters associated with a wave.

EXAMPLE 7.1

A wave is represented by

$$y(x, t) = 8\ (\text{cm}) \sin [(10 \text{ rad cm}^{-1})\, x - (10 \text{ rad s}^{-1})\, t]. \qquad \text{(i)}$$

Determine the amplitude, wavelength, angular frequency, wave number and velocity of the wave.

Solution: Compare the above expression [Eq. (i)] for the wave with Eq. (7.13):

$$y(x, t) = a \sin (\omega t - kx). \qquad \text{(ii)}$$

You will find that the given wave is propagating in positive x-direction. Further, comparing the corresponding terms in Eq. (i) and (ii) we obtain amplitude a = 8 cm; angular frequency $\omega = 10 \text{ rad s}^{-1}$, wave number, $k = 10 \text{ rad cm}^{-1}$.

Further, using Eq. (7.18), we get velocity v of the wave

$$v = \frac{\omega}{k} = 1.0 \text{ cms}^{-1}.$$

And using the relation $\lambda = \frac{v}{f} = \frac{2\pi}{k}$ for wavelength, we get

$$\lambda = \frac{2\pi}{10} = 0.63 \text{ cm}.$$

We would like you to solve Practice Exercise 7.2.

Practice Exercise 7.2 (a) A sound wave of frequency 170 Hz travels with speed 340 ms^{-1} along positive x-axis. Each point of the medium moves up and down through 5.0 mm. Calculate (i) wavelength of the wave, and (ii) write down the expression for the displacement of particles of the medium for this wave.

(b) A sinusoidal wave having a maximum height of 6.0 cm above the equilibrium level is propagating in the negative x-direction with a speed of 186 cms^{-1}. The distance between two successive crests is 62.8 cm. Write the expression for the particle displacement for the wave in terms of angular frequency and wave number. Also calculate particle velocity.

Ans. (a) (i) 2 m; (ii) $y(x, t) = 0.0025 \sin (3.14x - 1067.6t)$, (b) $y(x, t) = 6.0 \sin (0.1x + 18.6t)$.

In the preceding paragraphs, you have learnt about wave parameters such as the amplitude, time period, frequency and wavelength which characterise a wave. To complete the mathematical description of a wave, we also need to know the *phase* of a wave and the concept of *phase difference*. This forms the subject matter of discussion of Section 7.4.

7.4 PHASE OF A WAVE

Recall a wave propagating on a string and the profile of the motion of a given particle over one time period. Suppose that the particle under consideration is at rest in its mean equilibrium position at the start of the wave cycle. In one period, it will move upwards from its equilibrium position to the crest, return to the mean position momentarily, then move in the opposite direction to the trough and finally move back to the initial position. (If the external driving force is sinusoidal, the particle will move along a sinusoidal curve.) Let us now understand how we can use this information to arrive at the concept of the *phase* of a wave.

Refer to Fig. 7.11 and examine it carefully. It shows the snapshot of the motion of particles on a vibrating string at the instant $t = T$.

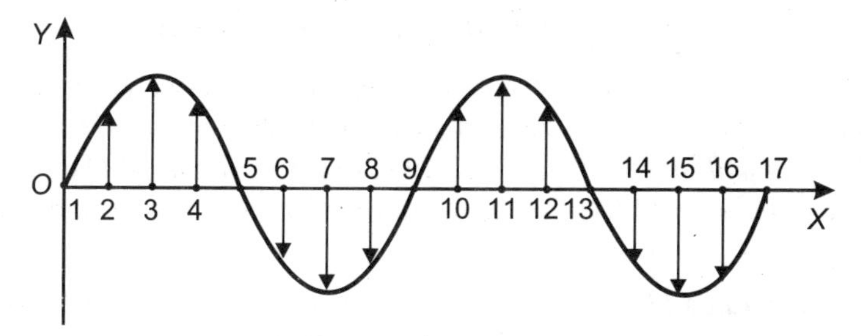

Fig. 7.11 Particles 1 and 2 are at different stages of motion even though both execute simple harmonic motion. But particles 1, 9 and 17 go up and down together around their mean positions.

> At a given instant t,
> $\sin(kx - \omega t + 2\pi) = \sin(kx - \omega t)$.
> Now for a point $t' = t + T$, we have (using the relation $\omega = 2\pi/T$),
>
> $$\begin{aligned}\sin(kx - \omega t') &= \sin[kx - \omega(t + T)] \\ &= \sin(kx - \omega t + \omega T) \\ &= \sin(kx - \omega t + 2\pi) \\ &= \sin(kx - \omega t).\end{aligned}$$
>
> Thus, for a fixed position x, the phase at time $t = T$ on the wave is the same as the phase at the instant $t = 0$ (since the points are separated from each other by one time period).

Now, compare the displacements of any two particles, say, particle 1 and particle 2 in Fig. 7.11. Notice that the motion of both these particles is simple harmonic and corresponds to same amplitude and frequency. But particle 2, starts to oscillate $T/8$ seconds after particle 1. We describe this situation by saying that particle 2 *lags behind* particle 1. Note that particles 1, 9 and 17 in Fig. 7.11 are about to go up and *these are said to be in the same state of motion*. However, particle 1 is at a different state of motion from particles 2 to 8 and 10 to 16. Thus, we may conclude that even when all particles execute SHM and their displacements exhibit the same sinusoidal variation in time, they could be at *different states of motion* at any given instant. We express this *difference in the states of motion of particles of the medium* in terms of the *phase angle* or simply *phase*. We say that particles 1, 9 and 17 vibrate *in phase,* but particle 1 is *out of phase* with particles 2, 3 ,...8 and 10, 11,...16.

Since wave motion arises due to periodic motion of particles around their mean position, we can extend the concept of phase of an SHM, which you have studied in Chapter 2, to define the phase of a wave: The *argument of the periodic function representing a periodic travelling wave* is called the *phase* of the wave. We denote it by the symbol $\phi(x, t)$. It describes the state of motion of a particle. Thus, the phase of a sinusoidal wave at x and t represented by Eq. (7.13) is the argument of the sine function and is given by

$$\phi(x, t) = kx - \omega t. \tag{7.19}$$

Since phase of a wave $\phi(x, t)$ is an angle, it is measured in degrees (or radians); 360° (or 2π) is equivalent phase difference corresponding to one wavelength. Moreover, it is a function of position as well as time and changes both with x and t.

From the definition of phase, it follows that *all points on the wave separated by one wavelength or its integral multiples are in the same phase*. To appreciate this statement, we recall that for a sinusoidal function at a given instant of time t

$$\sin (kx - \omega t + 2\pi) = \sin (kx - \omega t).$$

And for a point $x' = x + \lambda$, using the relation $k = 2\pi/\lambda$, we can write

$$\sin (kx' - \omega t) = \sin [k(x + \lambda) - \omega t)] = \sin (kx + k\lambda - \omega t)$$

$$= \sin (kx - \omega t + 2\pi) = \sin (kx - \omega t).$$

This result shows that, at a given instant t, the phase of particles on the wave at a point $x = \lambda$ is the same as the phase at point $x = 0$. In fact, we can show that *all other particles on the wave separated by an integral multiple of the wavelength have the same phase*. (To prove this result, you have to substitute $x' = x \pm n\lambda$ and use the result $\sin (\theta \pm 2n\pi) = \sin \theta$ for $n = 0, 1, 2, 3,\ldots$)
We can generalise this result as follows:

Particles on a wave separated by λ or its integral multiples are in-phase. However, the particle at a point x has a *finite phase difference* with all other particles at points $x' \neq x$, for which $x' \neq (x \pm n\lambda)$ where $n = 0, 1, 2, \ldots$

We say that two points vibrate *in-phase* if they are separated by an integral multiple of the wavelength. In the same way, we can say that for a fixed position x, *particles on a wave separated by T or its integral multiple are in the same phase*. There is a *finite non-zero phase difference* between the particles in motion at a point at the instant t and all other instants $t' \neq t$, for which $t' \neq (t \pm nT)$, where $n = 0, 1, 2, \ldots$

In terms of angles, in-phase points are separated by $n \times 360°$, where $n = 0, 1, 2, \ldots$, whereas out-of-phase points can be any number of degrees other than $n \times 360°$.

Physically, *the phase of a wave indicates the instantaneous position of the wave relative to a reference position*. To elaborate this statement, let us consider how $y(x, t)$ changes with (i) x for a fixed value of t, and (ii) t for a fixed value of x.

Change in y(x, t) with x for a fixed value of t

Refer to Fig. 7.12 (a-c), which depict the snapshots of a wave travelling in the positive x-direction at three different instants of time: at the reference time ($t = 0$, say) and at two later times t_1 and t_2. Now, let us consider the movement of a particle at point P, say, on the wave at these times simultaneously.

Note that at times t_1 and t_2, the position of P has shifted relative to the (vertical) reference line corresponding to $t = 0$. You can easily verify that all points on the wave shift by the same angle in these time intervals.

From Fig. 7.12 (b and c), you can see that position of the entire wave at t_1 and t_2 with respect to the vertical reference line shifts by $\lambda/4$ and $\lambda/2$, respectively. At $t = t_1$, the *phase angle* of the wave is 90°, whereas at $t = t_2$, the phase angle is 180°. The change in the position of a wave relative to the reference position is called *phase shift*. Thus, the phase shift of the wave under consideration in the time interval $t = 0$ to $t = t_1$ is 90° (Fig. 7.12b). Similarly, the phase shift in the time interval t_1 to t_2 is 90°. However, the phase shift in the time interval $t = 0$ to $t = t_2$ is 180°.

If we consider the wave at a given instant of time, the phase difference between two points on the wave or the phase shift is given by

$$\Delta\phi(x, t) = k\Delta x \qquad \text{for fixed } t. \tag{7.20}$$

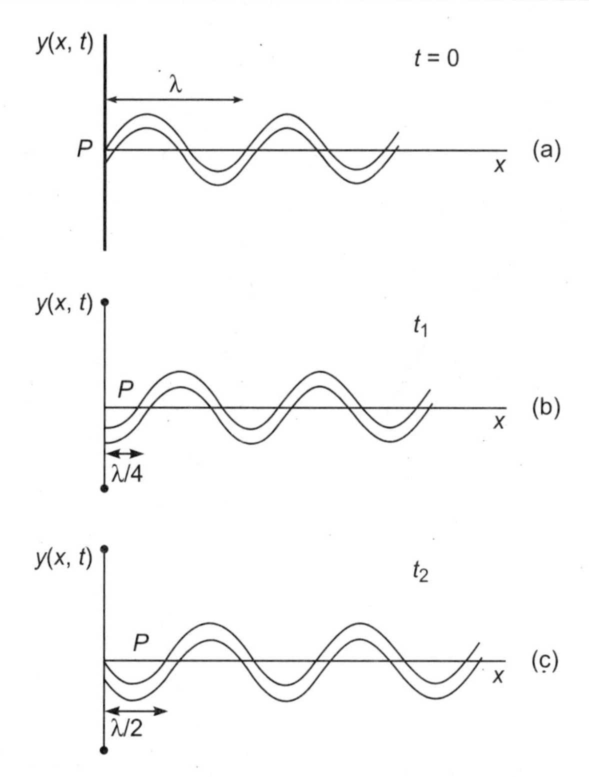

Fig. 7.12 Snapshots of a wave travelling in the positive x-direction at three different instants of time. The wave has shifted to the right relative to the reference position shown by the vertical line.

Change in y(x, t) with t for fixed values of x

Figure 7.12(a–c) shows the displacement $y(x, t)$ of a wave as a function of time for four points: $x_1 = 0$ (reference position), $x_2 = \lambda/4$ and $x_3 = \lambda/2$. Note that for all values of x, $y(x, t)$ is sinusoidal in time. By comparing the waveforms in Fig. 7.13a and b, you will note that these are shifted relative to one another at the points x_1 and x_2. It means that phase of the wave at x_2 is *different* from its phase at x_1.You can see that phase of the wave at x_2 is 90° relative to the reference position shown in Fig. 7.13a. That is, there exists a finite phase difference between these two positions of the wave. You may also like to know the phase difference between the positions shown in Fig. 7.13a and 7.13c. The phase of the wave at these two positions differs by 180°.

If we consider the wave at any fixed position [Fig. 7.13(b or c)], the phase difference of the wave at these two times or the phase shift is given by

$$\Delta\phi(x, t) = -\omega\Delta t \qquad \text{for fixed } x. \tag{7.21}$$

You can extend this concept to a situation where more than one wave are travelling in space and time. Two waves are said to be *in-phase* when the corresponding points on each wave reach their respective maximum or minimum displacements simultaneously. Thus, if the crests and troughs of the two waves coincide, they are said to be in-phase. If the crest of one wave coincides with the trough of the other wave, their phases differ by 180° and the waves are said to be in *opposite phase*. The phase difference between two waves can vary from 0° to 360°.

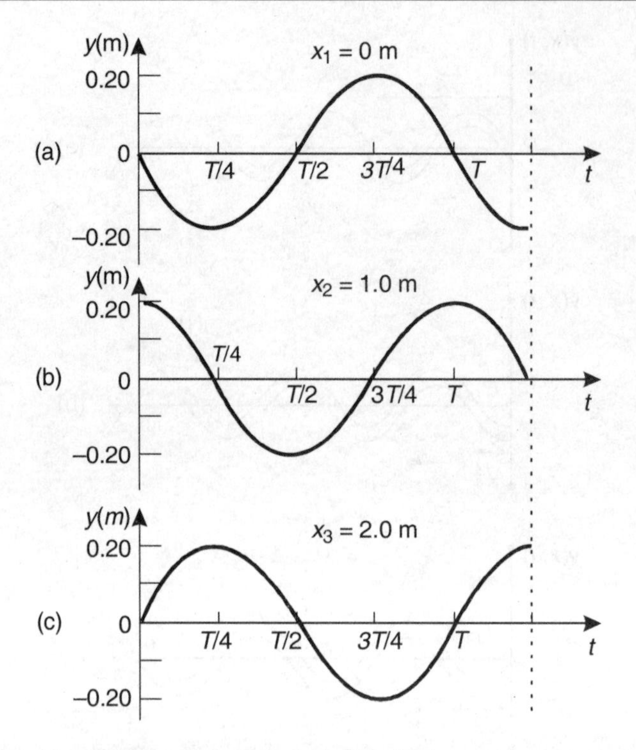

Fig. 7.13 Variation of amplitude of a wave travelling in the +x-direction as a function of time.

For waves travelling in the positive or negative x-directions, the arguments of the most general equation contain an additional factor due to initial phase. If we take it as θ, we can write

$$y(x, t) = a \sin (kx - \omega t + \theta) \tag{7.22a}$$

and

$$y(x, t) = a \sin (kx + \omega t + \theta). \tag{7.22b}$$

Note that these equations represent *waves shifted by an angle* θ from the waves given by Eqs. (7.13) and (7.17), respectively.

You now know that for a travelling wave, the entire waveform shifts with time. If the medium is isotropic and its characteristics do not change with time, the wave propagation takes place at constant velocity. For harmonic travelling waves, this velocity is called the *phase velocity*. For example, ripples on the water surface will travel with a constant velocity if the depth of water remains the same. Let us now derive the expression of the phase velocity of the wave.

7.4.1 Phase Velocity

Let us consider a sinusoidal wave travelling in the positive x-direction and look at its variation with x at time t. Imagine a wave pattern physically moving in the positive x-direction with speed v, as shown in Fig. 7.14. The phase velocity of the wave defines the speed with which the wave pattern travels.

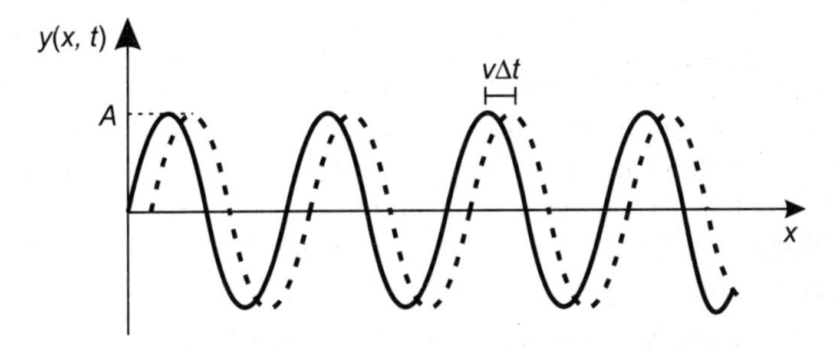

Fig. 7.14 Wave profile of a sinusoidal wave travelling along positive x-axis.

The phase velocity of a wave is defined as the velocity with which a point of constant phase on the wave travels. In terms of the wave parameters, it is given by

$$v_p = \frac{\omega}{k} = \lambda f = \frac{\lambda}{T}. \qquad (7.23)$$

You can derive this expression using Eq. (7.13) by noting that for $y(x, t)$ to be constant, the argument of the sine term should be constant. For a point of constant phase, we can write

$$\phi(x, t) = (kx - \omega t) = \text{constant}. \qquad (7.24)$$

For a function of two variables, a change in dependent variable is expressed in terms of partial derivatives. For example, any change in $y(x, t)$ can be expressed as

$$dy(x, t) = \left(\frac{\partial y}{\partial x}\right)_t dx + \left(\frac{\partial y}{\partial t}\right)_x dt.$$

From Eq. (7.24), you can write

$$\left(\frac{\partial \phi}{\partial t}\right)_x = -\omega$$

and

$$\left(\frac{\partial \phi}{\partial x}\right)_t = k.$$

To obtain an expression for the phase velocity, we use calculus. We first express the infinitesimal change in $\phi(x, t)$ in terms of changes in x and t as

$$d\phi = \left(\frac{\partial \phi}{\partial t}\right)_x dt + \left(\frac{\partial \phi}{\partial x}\right)_t dx.$$

On substituting $\left(\frac{\partial \phi}{\partial t}\right)_x = -\omega$ and $\left(\frac{\partial \phi}{\partial x}\right)_t = k$ from Eq. (7.24), we get

$$d\phi = -\omega\, dt + k\, dx.$$

Since phase velocity is the velocity of a point of constant phase, we must set $d\phi = 0$. This gives

$$\left(\frac{dx}{dt}\right)_\phi = \frac{\omega}{k} = v_p. \qquad (7.25)$$

This relation gives the phase velocity v_p of the wave. On comparing Eqs. (7.18) and (7.25), we find that the wave velocity is the same as the *phase velocity* of the wave.
You may like to read Example 7.2.

EXAMPLE 7.2

A 1-D plane progressive wave of amplitude 1 cm is generated at one end ($x = 0$) of a long string by a tuning fork. At some instant of time, the displacements of the particles at $x = 10$ cm and at $x = 20$ cm are –0.5 cm and 0.5 cm, respectively. The speed of the wave is 100 ms^{-1}. Calculate the frequency of the tuning fork. If the wave is travelling along the positive x-direction and the end $x = 0$ is at the equilibrium position at $t = 0$, write the displacement of wave in terms of amplitude, time period and wavelength.

Solution: We know that the equation of a plane progressive harmonic wave in 1-D is described by

$$y(x, t) = a \sin\left[\frac{2\pi}{\lambda}(vt - x)\right],$$

where $a = 1$ cm. From the first condition, we note that at $x = 10$ cm, $y(x, t) = -0.5$ cm. Therefore,

$$-0.5 \text{ cm} = (1 \text{ cm}) \sin\left[\frac{2\pi}{\lambda}(vt - 10)\right]$$

or

$$\sin\left[\frac{2\pi}{\lambda}(vt - 10)\right] = -0.5 = -\frac{1}{2} = -\sin\frac{\pi}{6}$$

$$\sin\left[\frac{2\pi}{\lambda}(vt - 10)\right] = \sin\left(\pi + \frac{\pi}{6}\right).$$

This equality implies that

$$\frac{2\pi}{\lambda}(vt - 10) = \frac{7\pi}{6}$$

or

$$vt - 10 = \frac{7}{12}\lambda. \qquad \text{(i)}$$

Let us now consider the second condition:
At $x = 20$ cm, $y(x, t) = +\ 0.5$ cm.
Therefore, we can write

$$0.5 \text{ cm} = (1 \text{ cm}) \sin\left[\frac{2\pi}{\lambda}(vt - 20)\right]$$

or

$$\sin\left[\frac{2\pi}{\lambda}(vt - 20)\right] = 0.5 = \sin\frac{\pi}{6}.$$

$$\therefore \qquad \frac{2\pi}{\lambda}(vt - 20) = \frac{\pi}{6}$$

so that

$$vt - 20 = \frac{\lambda}{12}. \qquad \text{(ii)}$$

From Eqs. (i) and (ii), we get

$$(vt - 10) - (vt - 20) = \frac{7\lambda}{12} - \frac{\lambda}{12}$$

$$\therefore \qquad \frac{\lambda}{2} = 10.$$

Hence the wavelength of the wave is 20 cm = 0.2 m.
Since $v = f\lambda$, we find that the frequency of the tuning fork

$$f = \frac{v}{\lambda} = \frac{100}{0.2} = 500 \text{ Hz}.$$

The wave equation in the desired form is

$$y(x, t) = (0.01 \text{ m}) \sin [2\pi (500t - 5x)] = (0.01 \text{ m}) \sin 10\pi (100t - x).$$

The mathematical description of wave motion can be used for deriving an expression for energy transported by a progressive wave. You will learn it now.

7.5 ENERGY TRANSPORTED BY PROGRESSIVE WAVES

Let us consider a one-dimensional mechanical wave travelling along the positive x-direction in air or a string. You may recall that such a wave is described by the equation

$$y(x, t) = a \sin (kx - \omega t).$$

We know that when a sinusoidal wave moves in a medium, it transports energy which is characterised by wave velocity. Since the wave travels a distance equal to one wavelength in one period, the energy is transported one wavelength per period of the wave. For simplicity, we consider a thin segment of the medium and calculate the rate at which total energy is transferred. For mechanical waves, the mechanical energy transported by these can be expressed as the sum of the kinetic energy and the potential energy of the segment under consideration.

Suppose that the thin segment under consideration has thickness dx, cross-sectional area A and is situated at a distance x from the source generating the progressive waves. If density of the medium is ρ, the mass of the layer dm will be equal to $\rho A \Delta x$.

The kinetic energy imparted by the wave propagating with velocity v to the thin layer of mass dm of the medium is given by

$$d\text{KE} = \frac{1}{2} dm v^2, \tag{7.26}$$

We calculate v *by* differentiating $y(x, t)$ with respect to t. But $y(x, t)$ is a function of two variables: position x and time t. Therefore, strictly speaking, to determine v, we have to consider partial derivative of $y(x, t)$ with respect to t. This means that we differentiate v with respect to t treating x as a constant. That is, we calculate

$$v = \frac{\partial y}{\partial t},$$

where $\partial/\partial t$ denotes the partial derivative with respect to t. Thus, we can write

$$v(x, t) = a\omega \cos (kx - \omega t). \tag{7.27}$$

Substituting this expression for $v(x, t)$ in Eq. (7.26), we get the expression for kinetic energy transferred to the segment of mass dm of the medium:

$$d\text{KE}(x, t) = \frac{1}{2} dmv^2 = \frac{1}{2} dm\omega^2 a^2 \cos^2 (kx - \omega t).$$

To calculate the average kinetic energy transported by the wave in one period or one wavelength, we integrate this expression over an interval and divide the resultant expression by the length of interval. For convenience, we set $t = 0$, which corresponds to $x = 0$ for this integration. By definition, the average value of kinetic energy transported over one wavelength is given by

$$\langle \text{KE} \rangle = \frac{1}{\lambda} \left[\frac{1}{2} dm\omega^2 a^2 \int_0^{\lambda} \cos^2 (kx)\, dx \right]. \tag{7.28}$$

To solve this integral, we introduce a change of variable by substituting $kx = \phi$ so that $kdx = d\phi$ or $dx = d\phi/k$. The limits of integration change to 0 and $k\lambda(= 2\pi)$. It means that in effect, we have to integrate a function of the form of $\cos^2 \phi$ over one cycle from 0 to 2π. From elementary calculus discussed in Chapter 1, you may recall that the average value of the square of a cosine function over one full cycle is one-half. On inserting this value in Eq. (7.28), we get the expression for *average kinetic energy* over one wavelength:

$$\langle \text{KE} \rangle = \frac{1}{4} dm\omega^2 a^2. \tag{7.29}$$

Let us now calculate the potential energy of the layer under consideration. Using Newton's second law, we can write the expression for the force acting on the layer of mass dm as

$$F = \text{Mass} \times \text{Acceleration} = dm \frac{\partial v(x,t)}{\partial t}.$$

Differentiating $v(x, t)$ given by Eq. (7.27) with respect to t while keeping x fixed and inserting the resultant expression in the relation of force, we get

$$\begin{aligned} F &= -dm(\omega^2 a) \sin (kx - \omega t) \\ &= -dm\, \omega^2 y\, (x, t), \end{aligned}$$

since $y(x, t) = a \sin (kx - \omega t)$. We know that work done by this force in displacing the layer through a distance Δy from its equilibrium position will be stored in it as its potential energy. You may recall from your course on Elementary Mechanics that we can express potential energy as integral of force. The potential energy (PE) stored in the mass dm when it is displaced from equilibrium through a distance equal to the maximum displacement y is given by

$$\begin{aligned} \text{PE} &= -\int_0^y F dy \\ &= dm\omega^2 \int_0^y y\, dy \\ &= \frac{1}{2} dm\omega^2 y^2 \\ &= \frac{1}{2} dm\omega^2 a^2 \sin^2 (kx - \omega t). \end{aligned} \tag{7.30}$$

You can calculate average potential energy over one wavelength following exactly the same steps as used to calculate average kinetic energy. As before, you may like to set $t = 0$ for convenience. The average potential energy per unit wavelength is, therefore, given by

$$\langle PE \rangle = \frac{1}{\lambda}\left[\frac{1}{2}\, dm\omega^2 a^2 \int_0^{\lambda} \sin^2(kx)\, dx\right].$$

The value of the integral in this function is $\pi/2$ so that expression for the average potential energy takes the form

$$\langle PE \rangle = \frac{1}{4}\, dm\omega^2 a^2. \tag{7.31}$$

On comparing Eqs. (7.29) and (7.31), we find that the average potential energy of the mass dm of a layer in the medium is equal to its average kinetic energy. In Chapter 2, you learnt that the time averages kinetic energy and potential energy are equal for simple harmonic motion. We have obtained similar relation for the energy transported by a travelling harmonic wave.

The total average energy of the segment of the medium under consideration at any instant of time is the sum of its average kinetic and potential energies:

$$E = \langle KE \rangle + \langle PE \rangle = \frac{1}{2} dm\omega^2 a^2. \tag{7.32}$$

This equation gives the total energy carried by a progressive wave and transported per cycle through a thin layer of mass dm of the medium. This result shows that half of the average energy transported by the wave is kinetic and the other half is potential. This energy is transferred to successive layers of the medium and in this process, energy is transported in the medium by a progressive wave.

You may now like to know: What is the average rate of energy flow in the medium per cycle? That is, how much power is transmitted by the wave? Since $dm = \rho A\Delta x$, we can write the expression for power P of the wave as

$$P = \frac{E}{\Delta t} = \frac{2\pi^2 a^2 f^2 \rho A \Delta x}{\Delta x / v}$$

$$= 2\pi^2 (\rho A) a^2 f^2 v, \tag{7.33}$$

where we have used the expression $\omega = 2\pi f$ for the frequency and $\Delta t = \Delta x/v$ for the time taken by the wave to cross the layer of thickness Δx by the wave travelling with velocity v.

This result shows that the rate at which energy is transported by a wave varies linearly with wave velocity and as the square of its amplitude and frequency. It is important to mention here that Eq. (7.33) is valid for all types of sinusoidal waves—mechanical as well as electromagnetic. You should now read Example 7.3 carefully.

EXAMPLE 7.3

A plane progressive wave of amplitude 2×10^{-4} m is generated when a musical instrument is played. If a note of frequency 300 Hz is produced, calculate the rate at which energy is generated per unit volume, if the density of air is 1.29 kg m^{-3}.

Solution: From Eq. (7.33), we know that energy per unit volume, i.e., energy density is given by

$$\frac{E}{V} = 2\pi^2 a^2 f^2 \rho$$

$$= 2 \times (3.1416)^2 \times (2 \times 10^{-4} \text{ m})^2 \times (300 \text{ Hz})^2 \times (1.29 \text{ kg m}^{-3})$$

$$= 9.2 \times 10^{-2} \text{ Jm}^{-3}.$$

We know that energy and power are useful parameters, but these do not account for observations related to the variations in the strength of a progressive wave with distance from the source. Consider progressive waves generated by a stationary source and spreading out in the surrounding medium. If there were no loss of energy of the wave, its strength should remain the same everywhere away from the source. This is, however, not true in practice. It is our common experience that chirping of birds, screaming of a person, vehicular noise, sound of crackers or light from a lamp fade out beyond a certain distance. (If it were not true, noise pollution would have made our life really difficult.) Similarly, the surface temperatures of planets in the solar system depend on their distance from the Sun. (This is one of the reasons why life does not exist on planets far away from the sun.) It may, therefore, make more sense to describe the strength of a wave at a given point in space by specifying its *intensity*. We discuss it now.

7.6 INTENSITY OF A WAVE

Intensity of a wave is defined as *the rate of energy transfer by the wave per unit area A normal to the direction of propagation*. So, by definition, we can write the intensity I of the wave as

$$I = \frac{P}{A}, \tag{7.34}$$

where P is the power (energy transfer rate per unit time) and A is the area of the surface intercepting the wave.

In the context of sound waves, intensity refers to loudness (strength of sound). So, to understand how loudness of sound decreases as we move away from its source, refer to Fig. 7.15. It depicts sound waves emitted by a point source S. Note that the sound waves will spread out in 3-D space over a spherical surface. However, we have depicted these spreading sound waves in 2-D.

Fig. 7.15 Sound waves emitted by a point source *S*.

Note that as we move away from S, sound energy transported by sound waves will be distributed over a greater surface area. For example, at a distance r_1, energy is distributed over a spherical surface of area $4\pi r_1^2$, whereas at a distance r_2 ($> r_1$), the same amount of energy will be distributed over a larger spherical surface (having area $4\pi r_2^2$). So, if we assume that there is no energy loss due to any dissipative mechanism, the energy available per unit cross-sectional area will be more at r_1 than at r_2. That is, energy received per unit cross-sectional area will decrease as we move away from the source. In other words, the intensity of sound decreases as distance from the source increases. For an arbitrary point located at a distance r from the source, the energy carried by sound waves will be distributed over a spherical area, $A = 4\pi r^2$. Thus, from Eq. (7.34), we can write the intensity of sound at a distance r from the source as

$$I = \frac{P}{4\pi r^2}$$

or

$$I \propto \frac{1}{r^2}. \tag{7.35}$$

This result, known as the *inverse square law*, implies that intensity of a wave varies inversely with square of distance from the source.

From Eq. (7.33), we note that

$$P \propto a^2,$$

where a is the amplitude of the wave. So, we can say that

$$I \propto a^2. \tag{7.36}$$

For sound waves, amplitude corresponds to the maximum change in pressure in the medium. So, we can write

$$I \propto (\Delta p)^2$$

On comparing Eqs. (7.35) and (7.36), we get

$$a \propto \frac{1}{r}. \tag{7.37}$$

This result shows that *amplitude of a sound wave is inversely proportional to the distance of observation point from the source*. This explains why we can be heard only up to a certain distance.

Les us pause for a while and ask: How is our perception of hearing related to the intensity of sound waves? As such, the auditory system of a normal person can detect a very wide range of intensities. A sound wave of intensity as low as 1×10^{-12} Wm^{-2} can be detected by us. This faintest sound is said to be the *Threshold of Hearing* (ToH) for human beings. On the other hand, we can hear sound of intensity as high as 10 Wm^{-2} without any damage to our ear. It means that the range of intensities which a human ear can detect is very large (~ 10^{13} Wm^{-2}). It was, therefore, considered pragmatic to use a logarithmic scale, called *decibel* (dB) *scale* to measure intensity of sound. The decibel scale is based on the multiples of 10. The threshold of hearing is taken as sound level of zero decibel. That is, 0 dB corresponds to an intensity of 10^{-12} Wm^{-2}. And a sound which is 10 times more intense (i.e., corresponds to 10^{-11} Wm^{-2}) than the threshold of hearing has a sound level of 10 dB. Similarly, if a sound is 1000 times more intense (10^{-9} Wm^{-2}), it is assigned

The logarithmic scale used to express intensity of sound is bel, named after Alexander Graham Bell. For finer gradation, it is customary to express intensity of sound in decibel, which is one-tenth of a bel (1 bel = 10 decibels).

To understand how the logarithmic scale works, let us consider the relation

$$a = \log b$$

where a and b are variables. If we multiply b by 10, we get

$$\begin{aligned} a &= \log (10\, b) \\ &= \log 10 + \log b \\ &= 1 + a. \end{aligned}$$

That is, when we multiply b by 10, the variable a increases by 1. Similarly, if we multiply b by 10^6, we get

$$\begin{aligned} a &= \log (10^6\, b) \\ &= 6 \log 10 + \log b \\ &= 6 + a. \end{aligned}$$

That is, when b is multiplied by 10^6, a increases by 6, the value in the exponent.

a sound level of 30 dB. Thus, in the decibel scale, we express intensity of sound level of a given sound wave with respect to the threshold of hearing. So, the intensity level β corresponding to intensity I is given in decibel unit by

$$\beta = (10 \text{ dB}) \log \left(\frac{I}{I_0}\right), \tag{7.38}$$

where I_0 is the intensity corresponding to ToH (1×10^{12} Wm^{-2}). The decibel scale is used only for mechanical waves in the auditory range.

In Table 7.1, we have listed intensities of waves generated by different sources of sound. The corresponding intensity levels are also given.

Note that the concepts of wavelength, frequency, power, and intensity discussed in this chapter with reference to mechanical waves are also valid for other types of waves, including electromagnetic waves. You may, therefore, like to know the intensity of electromagnetic waves (radiations) emitted by some familiar sources. Electromagnetic waves from a radio signal, 50 kW transmitter TV signal, typical camera flash and in the interior of a microwave oven are 10×10^{-8}, 5×10^4, 4×10^3 and 6×10^3, respectively.

Now you should read the following examples carefully.

Table 7.1 Intensities of sound at a distance of 1 m from the source

Source of sound	*Intensity* (Wm^{-2})	*Intensity* (dB)
Threshold of hearing (ToH)	10^{-12}	0
Rustling of leaves	10^{-11}	10
Whisper	10^{-10}	20
Normal conversation	10^{-6}	60
Street traffic	10^{-5}	70
Threshold of pain	10	130
Jet take off	10^2	140

EXAMPLE 7.4

In normal conversation, the intensity of sound is 5×10^{-6} Wm^{-2}. The amplitude and velocity of the sine wave are respectively 2.4×10^{-8} m and 332 ms^{-1}. If the density of air at STP is 1.29 kgm^{-3}, calculate the frequency of normal human voice.

Solution: The expression for the intensity is

$$I = 2\pi^2 a^2 f_0^2 \rho v$$

so that

$$f_0 = \frac{1}{\pi a}\sqrt{\frac{I}{2\rho v}}.$$

Here $v = 332\ \text{ms}^{-1}$, $\rho = 1.29\ \text{kgm}^{-3}$, $a = 2.4 \times 10^{-8}$ m and $I = 5 \times 10^{-6}\ \text{Wm}^{-2}$. On inserting these values in the above expression, we get

$$f_0 = \frac{1}{3.14 \times (2.4 \times 10^{-8}\text{m})} \sqrt{\frac{5 \times 10^{-6}\ \text{Wm}^{-2}}{2 \times (1.29\ \text{kg m}^{-3}) \times (332\ \text{ms}^{-1})}}$$

$$\approx 1000\ \text{Hz}.$$

EXAMPLE 7.5

A transverse wave travelling in positive x-direction is represented as

$$y(x, t) = 5 \sin (4.0t - 0.02x).$$

Calculate velocity of the wave, maximum particle velocity, acceleration and intensity. Note that y is measured in cm, t in seconds and density of media is $1.25\ \text{gcm}^{-3}$.

Solution: The equation of transverse wave travelling in + x-direction is given as

$$y(x, t) = 5 \sin (4.0t - 0.02x).$$

Comparing it with standard equation

$$y(x, t) = a \sin (\omega t - kx),$$

we find that $\omega = 4.0\ \text{rads}^{-1}$, $k = 0.02\ \text{radcm}^{-1}$ and $a = 5$ cm. Hence, velocity of the wave is given by

$$v = \frac{\omega}{k} = \frac{4.0\ \text{rads}^{-1}}{0.02\ \text{radcm}^{-1}} = 200\ \text{cms}^{-1}.$$

Instantaneous velocity of the particle

$$\frac{dy(x,t)}{dt} = 5 \times 4 \cos (4.0t - 0.02x)$$

$$= 20 \cos (4.0t - 0.02x).$$

$\therefore$ Maximum velocity of the particle = $20\ \text{cms}^{-1}$.

Particle acceleration

$$\frac{d^2y(x,t)}{dt^2} = -20 \times 4 \sin (4.0t - 0.02x)$$

$$= -80 \sin (4.0t - 0.02x).$$

$\therefore$ Maximum particle acceleration = $80\ \text{cms}^{-2}$.

The intensity of the wave is given by

$$I = 2\pi^2 a^2 f_0^2 \rho v$$

$$= \frac{1}{2} a^2 \omega^2 \rho v$$

$$= \frac{1}{2} \times 25\ \text{cm}^2 \times 16\ \text{rad}^2\text{s}^{-2} \times 1.25\ \text{gcm}^{-3} \times 200\ \text{cms}^{-1}$$

$$= 5 \times 10^4\ \text{erg cm}^{-2}\text{s}^{-1}.$$

7.7 WAVES IN STRINGS AND GASES: 1-D WAVE EQUATION

As a child, the music produced by *ektara*, a single stringed musical instrument, must have attracted you. The performances of sitar genius Bharat Ratna Pt. Ravi Shankar, late shahnai maestro Ustad Bismilla Khan or music queen Bharat Ratna Lata Mangeshkar must have delighted you. Do you know how this music is produced before it reaches you? What determines whether or not waves can propagate in a medium and when these move, how fast these do so? Experimental investigations show that speed of a wave does not depend on its wavelength or period. This means that answers to such questions should lie in the physical properties of the medium in which a wave propagates. To discover this, we consider some typical physical systems. For simplicity, we first study waves on stretched strings.

7.7.1 Waves on Stretched Strings

Consider a uniform stretched string, having mass per unit length m. Under equilibrium conditions, it can be considered to be straight. Let us choose x-axis along the length of the string in its equilibrium state. Suppose that we displace the string perpendicular to its length by a small amount so that a small section of length Δx is displaced through a distance y from its mean equilibrium position, as shown in Fig. 7.16. What happens when the string is released? It results in wave motion. Let us understand how.

We wish to know the speed of this wave. We have studied that the wave disturbance travels from one particle to another due to their masses, i.e., inertia and the factor responsible for the periodic motion of the particle is the elasticity of the medium. So we can legitimately expect that the interplay of inertia and elasticity of the medium will determine the wave speed. For a stretched string, the elasticity is measured by the tension F in it and inertia is measured by mass per unit length or linear density m. You may like to answer Practice Exercise 7.3.

***Practice Exercise* 7.3** Using dimensional analysis, show that $v = K\sqrt{F/m}$, where K is a dimensionless constant.

On solving this exercise, you will find that speed of a wave is directly proportional to square root of tension in the string and inversely proportional to half-power of mass per unit length. Let us now analyse the problem by considering a small element along the string.

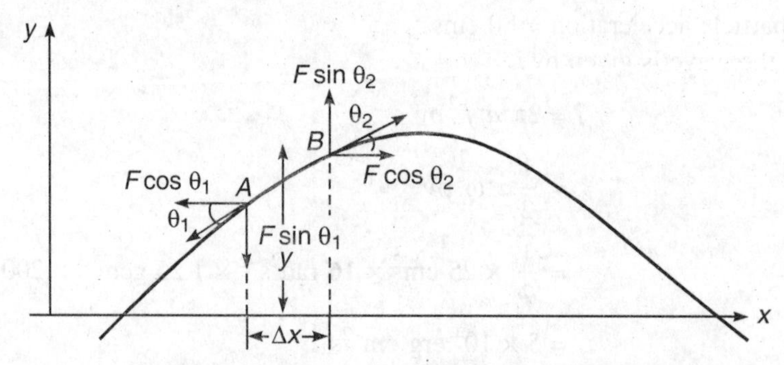

Fig. 7.16 Forces acting on a small element of a transversely displaced string. The net force acting on it is non-zero. (The distortion has been exaggerated for clarity).

Suppose that the tangential force at each end of element AB is $\mathbf{F}$; the force at B is produced by the pull of the string to the right and the one at A is due to the pull of the string to the left. Note that due to the curvature of the element under consideration, the two forces are not directly opposite to each other. Instead, they make angles θ_1 and θ_2 with x-axis. This means that the forces pulling the element at opposite ends, though of equal magnitude, do not exactly cancel out. To calculate the net force along x- and y-axes, we resolve $\mathbf{F}$ into rectangular components. The difference in the x- and y-components of tension between the right and the left ends of the element is respectively given by

$$F_x = F \cos \theta_2 - F \cos \theta_1$$

and

$$F_y = F \sin \theta_2 - F \sin \theta_1. \tag{7. 39}$$

You may recall that in small angle approximation, $\cos \theta_1 \approx \cos \theta_2 \approx 1$. It implies that there is no net force in the x-direction; $F_x = 0$ and the element AB of the string is subject to a net upward force F_y. Under the action of this force, the string element will move up and down. If the curvature of the string is not very large, the angles θ_1 and θ_2 will be small and the sine of the angles is very nearly the same as their tangents, i.e.,

$$\sin \theta_1 \approx \tan \theta_1$$

and

$$\sin \theta_2 \approx \tan \theta_2.$$

Using these results, we can write the y-component of the force on the element AB as

$$F_y = F (\tan \theta_2 - \tan \theta_1).$$

You may recall that tangent of an angle defines the slope at that point. That is, $\tan \theta_2$ and $\tan \theta_1$ define derivatives (dy/dx) at the ends A and B, respectively of the element under consideration. Thus, the y-component of force on the element is approximately

$$F_y = F\left(\left.\frac{dy(x,t)}{dx}\right|_{x+\Delta x} - \left.\frac{dy(x,t)}{dx}\right|_{x}\right). \tag{7.40}$$

Note that displacement of the string is a function of position as well as time. So, the displacement changes if either of these variables changes. However, Eq. (7.40) is valid for the configuration of the string at a particular instant of time. Therefore, the derivative in this expression should be taken keeping time fixed. Let us put $f(x) = \left.\frac{\partial y}{\partial x}\right|_x$ and $f(x + \Delta x) = \left.\frac{\partial y}{\partial x}\right|_{x+\Delta x}$. Then we can rewrite Eq. (7.40) as

$$F_y = F[f(x + \Delta x) - f(x)].$$

To simplify this, we use Taylor series expansion of the function $f(x + \Delta x)$ about the point x:

$$f(x + \Delta x) = f(x) + \left.\frac{\partial f}{\partial x}\right|_x \Delta x + \frac{1}{2}\left.\frac{\partial^2 f}{\partial x^2}\right|_x (\Delta x^2) + \cdots$$

Since Δx is very small, we can ignore the second and higher orders terms in Δx. Then we obtain

$$f(x + \Delta x) - f(x) = \left.\frac{\partial f}{\partial x}\right|_x (\Delta x)$$

$$= \frac{\partial}{\partial x}\left(\frac{\partial y}{\partial x}\right)\Delta x$$

$$= \frac{\partial^2 y}{\partial x^2} \cdot \Delta x.$$

On inserting this result in Eq. (7.40), we get

$$F_y = F\frac{\partial^2 y(x,t)}{\partial x^2}\Delta x.$$

This equation gives net force on the element Δx. To obtain the equation of motion of this element, we use Newton's second law of motion and equate this force to the product of mass and acceleration of the element *AB*. The mass of the segment of length Δx is $m\Delta x$. Then, we can write

$$m\Delta x \frac{\partial^2 y(x,t)}{\partial t^2} = F\frac{\partial^2 y(x,t)}{\partial x^2}\Delta x.$$

Cancelling out Δx from both sides and rearranging terms, we obtain

$$\frac{\partial^2 y(x,t)}{\partial x^2} = \frac{m}{F}\frac{\partial^2 y(x,t)}{\partial t^2}. \tag{7.41}$$

You have already seen through Practice Exercise 7.2 that F/m has the dimensions of the square of velocity.

Note that Eq. (7.41) has been obtained by applying Newton's second law to a small element of a stretched string. Since there is nothing special about this particular element on the string, we can say that this equation applies to the entire string.

Let us now pause for sometime and ask: What goals we set for ourselves and how Eq. (7.41) helps us in attaining them? We set out to know as to what determines the speed of a wave. To know this, let us assume that a sinusoidal wave propagating on the string is described by the equation

$$y(x, t) = a \sin(\omega_0 t - kx).$$

If this mathematical form is consistent with Eq. (7.41), you can be sure that such waves can move on the string. To check this, you should calculate second spatial ($\partial^2 y/\partial x^2$) and temporal ($\partial^2 y/\partial t^2$) partial derivatives of particle displacement:

$$\frac{\partial^2 y(x,t)}{\partial x^2} = -k^2 a \sin(\omega_0 t - kx)$$

and

$$\frac{\partial^2 y(x,t)}{\partial t^2} = -\omega_0^2 a \sin(\omega_0 t - kx).$$

Substituting these derivatives in Eq. (7.41), we obtain

$$-k^2 a \sin(\omega_0 t - kx) = -\frac{m}{F} \cdot a\omega_0^2 \sin(\omega_0 t - kx).$$

On cancelling $a \sin(\omega_0 t - kx)$ from both sides, we get

$$\left(\frac{\omega_0}{k}\right)^2 = \frac{F}{m}. \tag{7.42}$$

What is implied by this equality? We know that it has followed from Newton's law of motion applied to a stretched string when a harmonic wave is travelling along it. So, the above relation tells us that only those waves can propagate on the string for which wave properties ω_0 and k are related to F and m through the relation

$$\frac{\omega_0}{k} = \sqrt{\frac{F}{m}}.$$

But ω_0/k is the wave speed so that

$$v = \frac{\omega_0}{k} = \sqrt{\frac{F}{m}}. \tag{7.43}$$

This relation tells us that velocity of a transverse wave on a stretched string depends on tension and mass per unit length of the string, not on wavelength or time period. This means that v is not a property only of the material of the string. It involves an external factor—tension—which can be adjusted for fine tuning. This explains why musicians tighten/loosen their stringed instruments. However, no such thing is done in case of a flute or a harmonium.
Using Eq. (7.43), we can write Eq. (7.41) as

$$\frac{\partial^2 y(x,t)}{\partial x^2} = \frac{1}{v^2}\frac{\partial^2 y(x,t)}{\partial t^2}. \tag{7.44}$$

Do you recognise this equation? You learnt to obtain it in Chapter 6 in the so-called continuous approximation, i.e., in the limit where number of oscillators increased to infinity and the separation between them tended to be zero.

This result expresses one-dimensional *wave equation*. It holds so long as we deal with small amplitude waves, i.e., the oscillations of the string have small amplitude. You may now ask: Will Eq. (7.44) hold for large amplitude waves as well? Such waves result in a more complicated equation and the wave speed tends to depend on wavelength. You will learn about it in Chapter 12. Before you proceed further, you may like to answer Practice Exercise 7.4.

***Practice Exercise* 7.4** A one metre long string weighing one gram is stretched with a force of 10 N. Calculate the speed of transverse waves.

[***Ans.*** 100 ms^{-1}]

We now know that speed of a wave is determined by the interplay of elasticity and inertia of the medium. Elasticity gives rise to restoring force and inertia determines the response of the medium. Since a gas lacks rigidity, transverse waves cannot propagate in a gaseous medium; only solids can sustain transverse waves. However, longitudinal waves can propagate in all media—gaseous, liquid and solid—in the form of compressions and rarefactions. We now discuss longitudinal waves in solids.

7.7.2 Longitudinal Waves in Solids

To consider 1-D longitudinal waves in solids, we consider a cylindrical metal rod of uniform cross-sectional area. When the rod is struck with a hammer at one end, the disturbance will propagate along it with a speed determined by its physical properties. For simplicity, we assume that the rod is fixed at the left end.

We choose x-axis along the length of the rod with origin at O. We divide the rod in a large number of small elements, each of length Δx. Let us consider one such element PQ, as shown in Fig. 7.17a. Since the rod has been struck at end O lengthwise, the section at P, which is at a distance x_1 from O, will be displaced along x-axis. Since the force experienced by different sections of the rod is a function of distance, the displacements of particles in different sections will also be a function of position. Let us denote it by $\xi(x)$.

Figure 7.17b shows the deformed state of the rod and displaced position of the element under consideration. Let us denote the x-co-ordinate of the element in the displaced position by $x_1 + \xi(x_1)$ so that $\xi(x_1)$ represents the displacement of the particles in the section at P. Similarly, let $x_2 + \xi(x_2)$ be the new x-co-ordinate of the particles initially located in the section at Q ($x = x_2$) in the equilibrium state so that $\xi(x_2)$ signifies the displacement of the particles in section at Q. Hence change in length of the element is $\xi(x_2) - \xi(x_1)$. Using Taylor series expansion of $\xi(x_2)$ around x_1 and retaining first order terms, as in the case of string, we can write

$$\xi(x_2) - \xi(x_1) = \left(\frac{\partial \xi}{\partial x}\right)_{x=x_1} \Delta x.$$

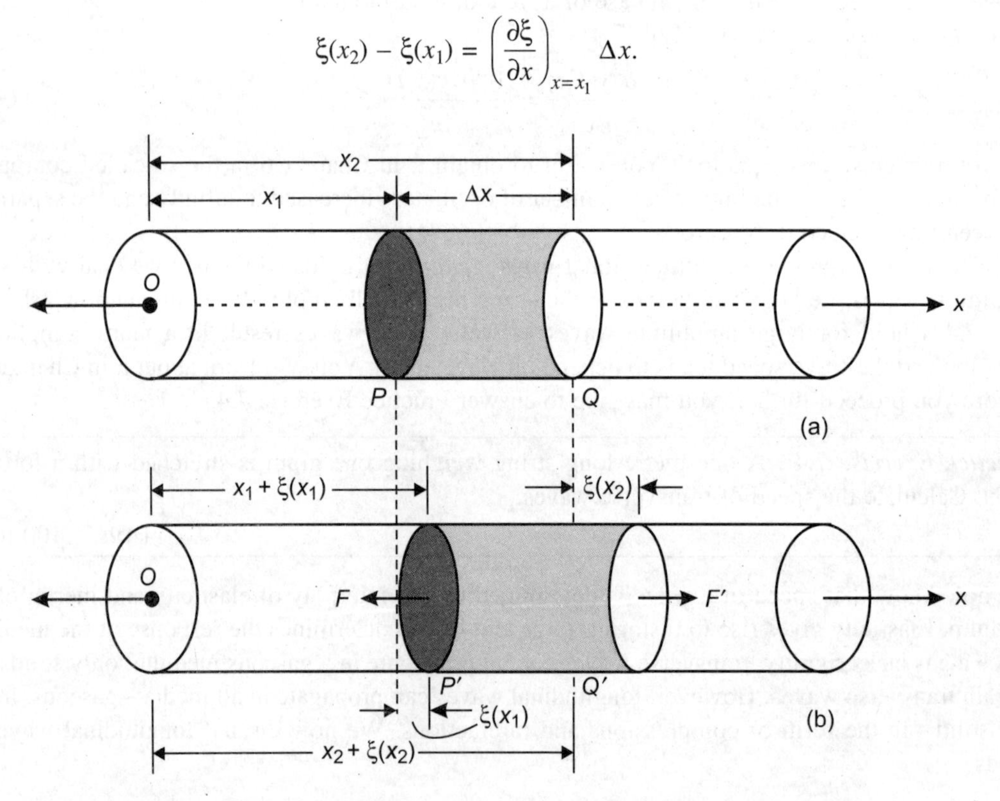

Fig. 7.17 Longitudinal wave propagation in a rod. Element PQ in (a) equilibrium state, and (b) deformed state.

The linear strain produced in the element PQ can be expressed as

$$\varepsilon(x_2) = \frac{\text{Change in length}}{\text{Original length}} = \frac{(\partial\xi/\partial x)_{x=x_1}\,\Delta x}{\Delta x} = \left(\frac{\partial \xi}{\partial x}\right)_{x=x_1}. \tag{7.45}$$

Physically, the net force $F' - F$ on the element $P'Q'$ points to the right. Due to this, the element under consideration will experience stress, which is restoring force per unit area. From your school physics, you may recall that the ratio of stress to longitudinal strain defines Young's modulus Y:

$$Y = \frac{\text{Stress}}{\text{Longitudinal strain}}$$

so that

$$\text{Stress} = Y \times \text{Longitudinal strain.}$$

In view of spatial variation of force, we can say that sections P and Q of the element under consideration will develop different stresses. Therefore, we can write

$$\sigma(x_1) = Y\left(\frac{\partial \xi}{\partial x}\right)_{x_1}$$

and

$$\sigma(x_2) = Y\left(\frac{\partial \xi}{\partial x}\right)_{x_2}.$$

The net stress on the element PQ is

$$\sigma(x_2) - \sigma(x_1) = Y\left[\left(\frac{\partial \xi}{\partial x}\right)_{x_2} - \left(\frac{\partial \xi}{\partial x}\right)_{x_1}\right]$$

$$= Y\,[f(x_2) - f(x_1)],$$

where we have put $f(x) = (\partial\xi/\partial x)$. As before, using Taylor series expansion for $f(x_2)$ about x_1, you can easily show that

$$\sigma(x_2) - \sigma(x_1) = Y\left(\frac{\partial f}{\partial x}\right)\Delta x$$

$$= Y\frac{\partial}{\partial x}\left(\frac{\partial \xi}{\partial x}\right)\Delta x$$

$$= Y\left(\frac{\partial^2 \xi}{\partial x^2}\right)\Delta x. \tag{7.46}$$

If the area of cross-section of the rod is A, the net force on the elements in the x-direction is given by

$$F(x_2) - F(x_1) = A\,[\sigma(x_2) - \sigma(x_1)]$$

$$= YA\left(\frac{\partial^2 \xi}{\partial x^2}\right)\Delta x. \tag{7.47}$$

Under dynamic equilibrium, equation of motion of the element PQ, using Newton's second law of motion, can be written as

$$YA\left(\frac{\partial^2 \xi}{\partial x^2}\right)\Delta x = \rho A \Delta x \left(\frac{\partial^2 \xi}{\partial t^2}\right), \qquad (7.48)$$

where ρ is density of the material of the rod and $\rho A \Delta x$ signifies the mass of the element PQ. On simplification, we find that displacement $\xi(x, t)$ satisfies the equation

$$\frac{\partial^2 \xi(x,t)}{\partial t^2} = \frac{Y}{\rho}\frac{\partial^2 \xi(x,t)}{\partial x^2}, \qquad (7.49)$$

which is of the form of wave equation [Eq. (7.44)] with

$$v = \sqrt{\frac{Y}{\rho}}. \qquad (7.50)$$

This result shows that deformation $\xi(x, t)$ propagates along the rod as a wave, and velocity of longitudinal waves is independent of the cross-sectional area of the rod. You may now like to answer Practice Exercise 7.5.

***Practice Exercise* 7.5** For a steel rod, $Y = 2 \times 10^{11}$ Nm^{-2} and $\rho = 7800$ kgm^{-3}. Calculate the speed of longitudinal waves.

[*Ans.* 5.06×10^3 ms^{-1}]

On working out Practice Exercise 7.5, you will find that speed of longitudinal waves in steel comes out to 5.06×10^3 ms^{-1}. The experimental value is 5.10×10^3 ms^{-1} at 0 °C. If you compare this value with measured value of velocity of sound in air, which is about 332 ms^{-1}, you will conclude that speed of longitudinal waves is an order of magnitude higher in metals than in air.

7.7.3 Longitudinal Waves in Gases

Sound waves in air columns exemplify the most familiar 1-D longitudinal waves in a gas. These can be easily excited by placing a vibrating turning fork at the open end of an air column. You may recall that these waves consist of a series of compressions and rarefactions, which set up pressure fluctuations in the gas. Can you highlight the basic difference between longitudinal waves in a solid rod and a gas column? We know that gases are compressible and pressure variations in a gas are accompanied by fluctuations in the density. However, in a solid rod, its density remains constant.

To understand propagation of 1-D longitudinal waves in a gas, let us consider a gas column in a long pipe or cylindrical tube of cross-sectional area A. As before, we choose x-axis along the length of the tube and divide the column of the gas in small volume elements. One such volume element $PQRSP$ extending between planes at x and $x + \Delta x$ is shown in Fig. 7.18a. The mass of this element is $A\rho\Delta x$. Under equilibrium conditions, pressure and density of the gas remain same throughout the volume of the gas, independent of x. Let us denote the equilibrium pressure as p_0. If the pressure of the gas in the tube is changed, the volume elements $PQRSP$ will be set in motion giving rise to a net force. Let us analyse its motion when it is displaced along the tube.

Let us choose the origin of the co-ordinate system so that the particles in plane *PQ* are as a distance x_1 and those in plane *SR* are at a distance x_2 from it. Figure 7.18b shows the displaced position of the volume element when *PQ* has shifted to *P′Q′* and *RS* has shifted to *R′S′*. Let the new co-ordinates be $x_1 + \psi(x_1)$ and $x_2 + \psi(x_2)$, respectively. It means that $\psi(x_1)$ and $\psi(x_2)$ respectively denote the displacements of the particles originally at x_1 and x_2. Therefore, change in thickness Δl is given by

$$\Delta l = \psi(x_2) - \psi(x_1).$$

If Δl is positive, there is increase in the length, and hence volume of the element and vice versa. Using Taylor series expansion for $\psi(x_2)$ about $\psi(x_1)$, we can write

$$\Delta l = \psi(x_2) - \psi(x_1) = \left(\frac{\partial \psi}{\partial x}\right)\Delta x.$$

This means that the change in volume ΔV is

$$\Delta V = A\Delta l = A\,\Delta x\frac{\partial \psi}{\partial x}.$$

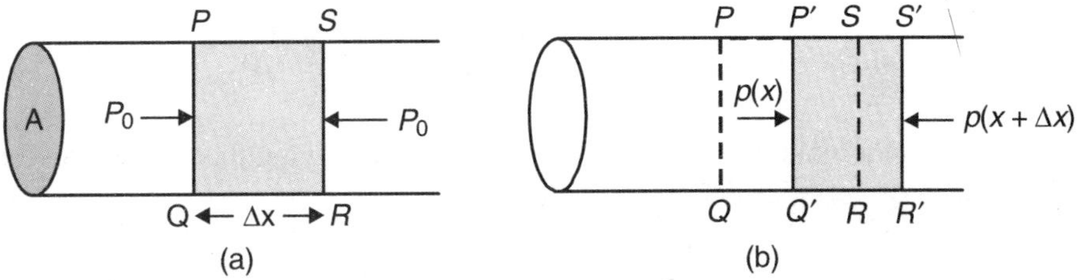

Fig. 7.18 (a) Equilibrium state of the column *PQRSP* of a fluid contained in a long tube of cross sectional area *A*; and (b) displaced position of column under pressure difference.

The *volume strain*, which is defined as the change in volume per unit volume, is given by

$$\frac{\Delta V}{V} = \frac{A\Delta x}{A\Delta x}\left(\frac{\partial \psi}{\partial x}\right) = \frac{\partial \psi}{\partial x}. \tag{7.51}$$

The increase in volume of the element is due to decrease in pressure and vice versa.

Note that so far, all steps that we have followed are identical to the case of the solid rod. However, as mentioned earlier, due to comparatively large compressibility of the gas, change in volume is accompanied by changes in density. This implies that pressure in the compressed/rarefied gas varies with distance.

To proceed further, let us suppose that pressure at *P′Q′* is $p_0 + p(x_1')$ while pressure at *R′S′* is $p_0 + p(x'_2)$. Hence, pressure difference across the ends of the element *P′Q′R′S′P′* can be expressed in terms of pressure gradient:

$$\begin{aligned} p(x_2') - p(x_1') &= \left(\frac{\partial p(x)}{\partial x}\right)_{x=x_1'} \Delta x \\ &= \frac{\partial(p_0 - \Delta p)}{\partial x}\Delta x \\ &= -\frac{\partial(\Delta p)}{\partial x}\Delta x, \end{aligned} \tag{7.52}$$

since p_0 is constant.

To express this result in a familiar form, we note that Δp is connected to the *bulk modulus of elasticity* by the relation

$$E = \frac{\text{Stress}}{\text{Volume strain}} = -\frac{\Delta p}{\Delta V/V}$$

The negative sign is included to account for the fact that when pressure increases, volume decreases. This ensures that E is positive.
We can rewrite the above relation as

$$\Delta p = -E\left(\frac{\Delta V}{V}\right)$$

On substituting for $\Delta V/V$ from Eq. (7.51), we get

$$\Delta p = -E\frac{\partial \psi}{\partial x}.$$

Using this result in Eq. (7.52), we find that pressure difference at the ends of the displaced column is given by

$$p(x_2') - p(x_1') = -\frac{\partial}{\partial x}\left(-E\frac{\partial \psi}{\partial x}\right)\Delta x = E\left(\frac{\partial^2 \psi}{\partial x^2}\right)\Delta x.$$

The net force acting on the volume element is obtained by multiplying this expression for pressure difference by the area of cross-section of the column:

$$F = A[p(x_2') - p(x_1')]$$

$$= EA\Delta x\left(\frac{\partial^2 \psi}{\partial x^2}\right).$$

Under the action of this force, the volume element under consideration shall be set in motion. Using Newton's second law of motion, we find that the equation of motion of the element under consideration can be expressed as

$$\rho \Delta x A\frac{\partial^2 \psi}{\partial t^2} = EA\Delta x\left(\frac{\partial^2 \psi}{\partial x^2}\right).$$

Note that $A\Delta x$ cancels from both sides. On dividing by ρ, we get

$$\frac{\partial^2 \psi}{\partial t^2} = \frac{E}{\rho}\frac{\partial^2 \psi}{\partial x^2}. \tag{7.53}$$

If we identify

$$v = \sqrt{\frac{E}{\rho}} \tag{7.54}$$

as the speed of longitudinal waves, Eq. (7.53) becomes identical to Eq. (7.44). Note that the wave speed is determined only by the bulk modulus of elasticity and density; two properties of the medium through which the wave is propagating.
Let us now consider propagation of sound waves in a gas.

Sound waves in air

When a longitudinal wave propagates through a gaseous medium such as air, the volume elasticity is influenced by the thermodynamic changes that take place in it. These changes can be *isothermal* or *adiabatic*. (In an isothermal process, temperature remains constant. In an adiabatic process, the total thermal energy of the system remains constant.) For sound waves, Newton assumed that isothermal changes take place in the medium. For an isothermal change, volume elasticity equals atmospheric pressure:

$$E = E_T = p.$$

Then, we can write

$$v = \sqrt{\frac{p}{\rho}}. \tag{7.55}$$

> For an isothermal process, using Boyle's law, we can write
> $$pV = \text{constant}.$$
> At constant temperature, small changes in volume and pressure are connected through the relation
> $$pdV + Vdp = 0$$
> Hence, we can write
> $$\left(\frac{\partial p}{\partial V}\right)_T = -\frac{p}{V}$$
> $$E_T = -V\left(\frac{\partial p}{\partial V}\right)_T = p.$$
> i.e., pressure equals isothermal elasticity.

This is *Newton's formula* for velocity of sound.

For air at STP, $\rho = 1.29$ kgm^{-3} and $p = 1.01 \times 10^5$ Nm^{-2}. Hence, velocity of sound in air, using Newton's formula, comes out to be

$$v = \sqrt{\frac{1.01 \times 10^5 \text{ Nm}^{-2}}{1.29 \text{ kgm}^{-3}}} = 280 \text{ ms}^{-1}.$$

But experiments show that velocity of sound in air at STP is 332 ms^{-1}, which is about 15% higher than the value predicted by Newton's formula. It implies that something was wrong with the assumption of isothermal change. You may now like to know as to how was this discrepancy resolved. The problem was with the assumption of isothermal changes and hence use of Boyle's law. The discrepancy was resolved by Laplace when he pointed out that sound waves produced adiabatic changes; the regions of compression were hotter while the regions of rarefaction were cooler. That is, local changes in temperature occur when sound propagates in air. However, the total energy of the system is conserved. This means that adiabatic changes occur in air when sound propagates.

> For *an adiabatic change,* the equation of state is
> $$pV^\gamma = \text{constant}.$$
> The changes in p and V are connected through the relation
> $$V^\gamma dp + p\gamma V^{\gamma-1} dV = 0$$
> or
> $$-V\left(\frac{\partial p}{\partial V}\right)_s \equiv E_S = \gamma p.$$

For an adiabatic change, E_s is γ times the pressure, where γ is the ratio of specific heat capacities of a gas at constant pressure and at constant volume, i.e., $E_s = \gamma p$. Then Eq. (7.55) modifies to

$$v = \sqrt{\frac{\gamma p}{\rho}}. \tag{7.56}$$

This is known as *Laplace's formula.*

For air, $\gamma = 1.4$ and the velocity of sound in air at STP based on Eq. (7.56) works out to be 331 ms^{-1}, which is in excellent agreement with the measured value. This showed that Laplace's argument was indeed correct.

At a given temperature, p/ρ is constant for a gas. So Eq. (7.56) shows that the velocity of a longitudinal wave is independent of pressure.

The equation of state of an ideal gas is

$$pV = nRT$$
$$= Nk_BT$$
$$p = \frac{Nk_BT}{V}.$$

From Eq. (7.56), we know that

$$\therefore \quad v_{sound} = \sqrt{\frac{\gamma p}{\rho}}$$
$$= \sqrt{\frac{\gamma k_BT}{\rho(V/N)}}$$
$$= \sqrt{\frac{\gamma k_BT}{m}}.$$

You may now like to know as to why thermal energy is unable to flow from a compression to a rarefaction and equalize the temperature creating isothermal conditions? To discover answer to this question, we note that to attain this condition, thermal energy must flow through a distance of one-half wavelength in a time much shorter than one-half of the period of oscillation of the particles. Thermodynamically this means that we would need

$$v_{thermal} >> v_{sound}.$$

Since flow of thermal energy in a fluid is mostly due to convection, you may recall from kinetic theory of gases that the root mean square speed of air molecules is given by

$$v_{rms} = \sqrt{\frac{2k_BT}{m}}, \tag{7.57}$$

where m is mass of air molecules and T is absolute temperature. We can similarly write the expression for speed of sound:

$$v_{sound} = \sqrt{\frac{\gamma k_BT}{m}}. \tag{7.58}$$

It has been experimentally observed that at STP, air molecules move randomly in zigzag paths, and the distance between two successive collisions is of the order of 10^{-5} cm. In Section 7.6, you have learnt that the shortest wavelength for audible sound corresponding to the highest frequency (20 kHz) is 1.66 cm. Since this value is many orders of magnitude higher than the distance travelled by a gas molecule between two successive collisions, the adiabatic flow constitutes a very good approximation. Indeed, this result was an important indirect evidence in favour of kinetic theory of gases.

The ability to measure the speed of sound has been put to many uses in defence. During World War I, a technique called *sound ranging* was developed to locate the position of enemy guns by using the sound of cannons in action. Now, it is being used for exploration of under sea minerals and off-shore oil explorations routinely. In recent years, the rush for the under sea wealth of the North pole is driving several European countries to make scientific expeditions in Arctic waters.

7.7.4 Sound Waves in Liquids

Liquids are, in general, almost incompressible. This means that speed of sound in liquids must be significantly higher than that in gases. For water, $\rho = 10^3$ kgm^{-3} and $E = 2.22 \times 10^9$ Nm^{-2}. This gives a wave speed of about 1500 ms^{-1}. Compare it with the speed of sound in air at STP. Though air is about 10^{-3} times dense than water, sound propagates faster in water than air. This means that we can send audio messages from one ship to another faster via water than via air.

7.8 WAVES IN TWO AND THREE DIMENSIONS

So far we have confined our discussion to waves propagating along 1-D as in a stretched string. But many musical instruments are not stringed. You may have enjoyed *tabla* performances of

Ustad Zakir Hussain or Allah Rakha Khan. What happens when a *tabla*, *drum* or *dholak* membrane is suddenly disturbed in a direction normal to the plane of the membrane? Particles of the membrane vibrate along the direction of applied force. But tension in the membrane makes the disturbance to spread over the surface. That is, waves on stretched membranes are two-dimensional (2-D). Similarly, surface waves caused by dropping a pebble into a quite pond are 2-D. In such cases, the displacement is a function of x, y and t, and we write $\psi = \psi(x, y, t)$. You may now ask: What is the equation of a 2-D wave? Will the preceding analysis apply as such in this case?

We will not go into mathematical details to answer these questions. However, for an isotropic medium, extension of Eq. (7.44) for 2-D wave is a straight forward exercise. Since forces along x- and y-axes can be taken to act independently, each one will contribute analogous term to the wave equation so that we can write

$$\frac{\partial^2 \psi(x, y, t)}{\partial t^2} = \frac{F}{\rho}\left(\frac{\partial^2}{dx^2} + \frac{\partial^2}{dy^2}\right)\psi(x, y, t). \qquad (7.59)$$

The solution of this equation has the form

$$\psi(x, y, t) = a \sin(\omega_0 t - \mathbf{k}.\mathbf{r}). \qquad (7.60)$$

Let us pause for a minute and ask: Do sound and light waves emanate radially from a small two-dimensional source? How can we describe seismic waves? These are three-dimensional waves. To analyse 3-D waves, we can extend the preceding arguments and write:

$$\frac{\partial^2 \psi(\mathbf{r}, t)}{\partial t^2} = v^2\left(\frac{\partial^2}{\partial x^2} + \frac{\partial^2}{\partial y^2} + \frac{\partial^2}{\partial z^2}\right)\psi(\mathbf{r}, t). \qquad (7.61)$$

7.9 THE DOPPLER EFFECT

We have so far discussed wave motion in a homogenous medium travelling in different directions. You must have noted that in all such cases, wave speed and frequency were taken to be constant. Do you know of any situation where frequency of a wave either changes or appears to change?

You all must have heard the whistle of a train steaming in and steaming out from a railway platform. The pitch (frequency) of the whistle seems to rise when the train comes closer and fall when it goes away. While standing near a highway, you may have listened to the sound of a loaded truck. While approaching you, it makes a relatively high-pitched sound but as it recedes, the pitch drops abruptly and stays low. *This apparent change of frequency due to relative motion between the source and the listener is known as Doppler effect.*

In general, when the source approaches the listener or the listener approaches the source or both approach each other, the apparent frequency is higher than the actual frequency of sound produced by the source. But when the source moves away from the listener or when the listener moves away from the source, or when both move away from each other, the apparent frequency is lower than the actual frequency of the sound produced by the source.

Do you know that Doppler shift in ultra-sound waves reflected from moving body tissues is commonly used by obstetricians to detect foetal heart-beat. Do you know how it arises? As the heart muscle pulsates, the frequency of reflected ultra-sound waves is different from the frequency of incident waves. Similarly, sonar makes use of the Doppler effect in determining the velocity of a submarine relative to a ship.

The electromagnetic waves also exhibit Doppler effect. In aircraft navigation, radar works by measuring the Doppler shift of high frequency radio waves reflected from moving aeroplanes. The Doppler shift of star-light allows us to study stellar motion. When we examine light from stars in a spectrograph, we observe several spectral lines. These lines are slightly shifted as compared to the corresponding lines from the same elements on the earth. This shift is generally towards the *red-end*, which implies that frequency of light reflected by the star increases. This observation is interpreted as if the star is receding, and the universe is expanding.

To study Doppler effect, we generally consider the following situations:

- the source or the listener or both are in motion;
- the motion is along the line joining the source and the listener;
- the direction of motion of the medium is along or opposite to the direction of propagation of sound; and
- the speed of source is greater or smaller than the speed of sound produced by it.

We now consider some of these possibilities with particular reference to sound waves.

7.9.1 Source Stationary and Listener in Motion

Let us suppose that a stationary source S is producing sound of frequency f and wavelength $\lambda = (v/f)$. The waves emitted by the source spread out as spherical wavefronts of sound (Fig. 7.19a). It means that the length of the block of waves passing a stationary listener per second is v and contains f waves (Fig. 7.19b). But when a listener moves away from the source at a velocity v_0, he will be at O' after one second and find that the length of the block of waves passing him in one second is $v - v_0$. That is, the sound waves will appear to the moving listener to have a speed $v' = v - v_0$. However, the distance between two successive wave maxima in the listener's moving frame of reference remains the same as for the stationary frame of the source, equal to λ (Fig. 7.19b).

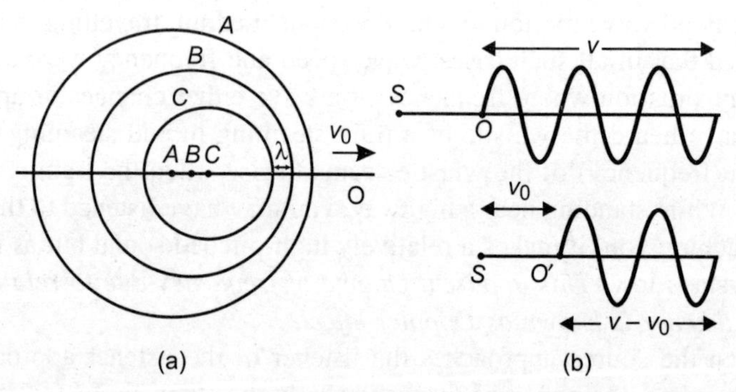

Fig. 7.19 (a) Waves leave the stationary source with speed v but appear to be moving with speed $v - v_0$ to the listener, and (b) representation of waves received by the listener in motion at intervals of one second.

The frequency heard by the listener is given by

$$f' = \frac{v'}{\lambda}.$$

We know that $v' = v - v_0$ and $\lambda = v/f$. Therefore, we find that for a listener moving away from a source, the apparent frequency is given by

$$f' = f\left(\frac{v - v_0}{v}\right). \tag{7.62a}$$

If the listener approaches the source, v_0 is to be regarded as negative. Then apparent frequency is given by

$$f' = f\left(\frac{v + v_0}{v}\right). \tag{7.62b}$$

From Eqs. (7.62a, b) we find that the frequency heard by the listener is lower than the source frequency when the listener is moving away from the stationary source. But the perceived frequency is greater than the emitted frequency when the listener approaches a stationary source. Note that the difference in perceived and emitted frequencies lasts only as long as there is relative motion between the source and the listener.

7.9.2 Source in Motion and Observer Stationary

Let us now consider the situation that a source S, producing sound of frequency f and wavelength λ, moves with a speed v_s ($< v$) towards a listener, who is at rest in a stationary medium. The wavefronts emitted by the moving source at several positions are shown in Fig. 7.20. As may be noted, these are not concentric; more closely spaced in the direction of motion of the source and widely separated on the opposite side. To a listener at rest on either side, this corresponds, respectively to a shorter and larger effective wavelength.

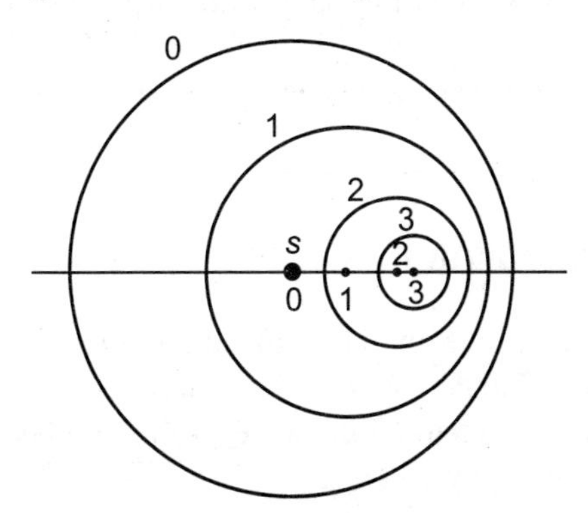

Fig. 7.20 Wavefronts emitted by a moving source.

To represent the same situation in terms of waves, we recall that for a stationary source, f waves occupy a length v in one second. When the source moves a distance v_s towards the stationary listener in one second, these f waves are bunched together in the direction of motion in a length $v - v_s$, as shown in Fig. 7.21. It means that the distance between successive crests decreases from $\lambda = v/f$ to $\lambda' = (v - v_s)/f$. That is, the wavelength of waves perceived by the listener decreases and manifests as apparent change in frequency of sound.

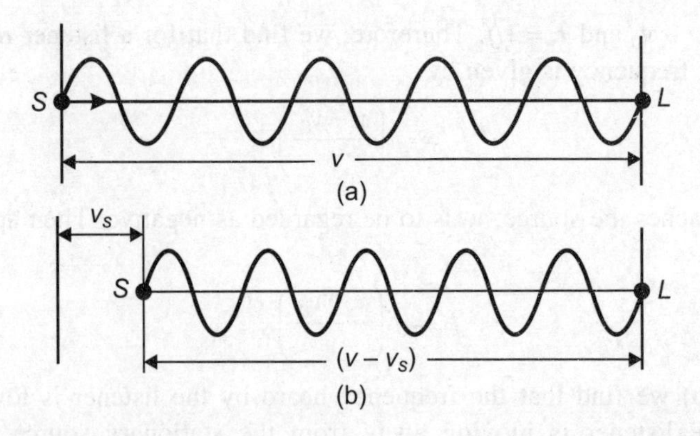

Fig. 7.21 Crowding of waves when a source moves towards a stationary listener.

The apparent frequency f' is given by

$$f' = \left(\frac{v}{\lambda'}\right) = f\left(\frac{v}{v - v_s}\right). \tag{7.63a}$$

If, however, the source moves away from the listener, v_s will be negative and Eq. (7.63a) takes the form

$$f' = f\frac{v}{v + v_s}. \tag{7.63b}$$

To fix up the ideas discussed in this section so far, you may like to work on Practice Exercise 7.6.

***Practice Exercise* 7.6** A person is standing near a railway track. A train approaches him with a speed of 54 kmh^{-1}. The apparent frequency of the whistle heard by the person is 660 Hz. Calculate the actual frequency of the whistle. Take the speed of sound in air as 330 ms^{-1}.

[***Ans.*** 630 Hz]

7.9.3 Source and Listener both in Motion

When both source and listener move in the same direction, we have to combine the results contained in Eqs. (7.62a) and (7.63a). When the source is in motion, it causes a change in wavelength. The listener in motion results in change in the number of waves received. In such a case, apparent frequency f' is given by

$$f' = \frac{\text{Length of block of waves received}}{\text{Reduced wavelength}}$$

$$= \left(\frac{v - v_0}{v - v_s}\right) f. \tag{7.64}$$

From this result we note that the magnitude of apparent frequency with respect to emitted frequency will be determined by the values of v_0 and v_s. If the listener recedes faster, f' will be less than f, but if the source approaches faster, f' will be greater than f.

You may now ask: Is there any difference in the apparent frequency when the source approaches stationary listener or the listener approaches stationary source with the same velocity? Equation (7.64) tells us that the apparent frequencies will be different in these two cases.

The Doppler effect in optics is very similar to that in sound. To change over from sound to optics, one has simply 'to replace the velocity of sound v by the velocity of light c in each expression. In optics, the 'listener' becomes the 'observer'.

REVIEW EXERCISES

7.1 Express the wave equation $y = a \sin \frac{2\pi}{\lambda}(x - vt)$ in its equivalent forms:

(i) $y = a \sin (kx - \omega t)$ (ii) $y = a \sin k(x - vt)$

(iii) $y = a \sin 2\pi\left(\frac{x}{\lambda} - \frac{t}{T}\right)$ (iv) $y = a \sin \omega\left(\frac{x}{v} - t\right)$

7.2 (a) Determine the frequency range corresponding to

(i) the visible range in the spectrum ranging from λ_V (= 400 nm) to λ_R (= 400 nm).

(ii) X-rays in the range 0.1 Å to 50 Å.

(b) Calculate the wavelength range for radio waves having frequencies in the range 1.5 MHz to 300 MHz.

[*Ans.* (a) (i) 4.26×10^{14} Hz to 7.5×10^{14} Hz, (ii) 6×10^{16} Hz to 3×10^{19} Hz, (b) 1 m to 200 m]

7.3 A plane progressive wave is represented by the equation

$$y = 5 \sin \pi (0.01x - 4.0t),$$

where y and x are measured in cm and time is measured in seconds. Express the amplitude, time period, wavelength and velocity of the wave in SI units.

[*Ans.* $a = 0.05$ m, $\lambda = 2$ m, $T = 0.5$ s, $v = 4$ ms^{-1}]

7.4 A plane progressive travelling along negative x-axis is characterised by frequency 500 Hz, amplitude 0.02 m and phase velocity 400 ms^{-1}. Write down its equation.

[*Ans.* $y = 0.02 \sin \pi (2.5 + 1000t)$ m]

7.5 The velocity and frequency of a wave are 200 ms^{-1} and 400 Hz, respectively. Determine

(a) how far are points which have phase difference of 45°?

(b) the phase difference between (i) two points separated by 7.5 m at a particular instant and (ii) at a point at an interval of 10^{-1} s.

[*Ans.* (a) $x = 0.025$ m, (b) (i) π (ii) zero]

7.6 Two sound waves of same frequency, one in air and the other in water, have same intensity. (i) Calculate the ratio of pressure amplitude of the wave in water to that in air. (ii) If their pressure amplitudes were equal, determine the ration of their intensities. Use the data $\rho_{air} = 43$ in cgs units and $\rho_{water} = 1.5 \times 10^5$ in cgs units.

[*Ans.* (i) p_{water}: $p_{air} = 60$, (ii) $I_{air} : I_{water} = 3.5 \times 10^3$]

7.7 A uniform circular loop of radius a is rotating clockwise with an angular speed ω. Calculate the velocity of the wave.

[*Ans.* $v = a\omega$]

7.8 Two waves having same frequency, 500 Hz, amplitudes 3 cm and 4 cm and differing in phase by 90° combine and move at a speed of 10 ms^{-1}. Write down the equation of motion of the resultant wave.

[*Ans.* $y = 5 \sin \pi (x - 1000t)$ cm]

7.9 A 2.0 W source emits spherical waves in an isotropic medium. By ignoring the absorption coefficient of the medium, calculate the intensity of the wave at a distance of 50 cm from the source.

[*Ans.* $I = (2/\pi)$ Wm^{-2}]

7.10 A train moving with speed 72 kmh^{-1} emits a whistle of frequency 600 Hz. Calculate the apparent frequency heard by a stationary observer, if the train is (i) approaching near, (ii) receding away.

[*Ans.* (i) 765 Hz, (ii) 680 Hz]

8

Reflection and Refraction of Waves

EXPECTED LEARNING OUTCOMES

In this Chapter, you will acquire capability to:

- correlate impedance to wave motion;
- obtain expressions for impedances offered by transverse and longitudinal waves;
- derive expressions for reflection and transmission amplitude coefficients;
- explain formation of stationary waves and beats;
- explain formation of wave groups and derive expressions for group velocity and wave velocity;
- solve numerical examples based on these concepts.

8.1 INTRODUCTION

In Chapter 7, we discussed propagation of waves on strings in solid rods, and in fluids (gases and liquids) with particular reference to sound waves. You may now ask: What happens to a wave when it encounters a rigid barrier, as in the case of a string whose one end is tied to a rigid wall or on a stringed musical instrument. In Physics laboratory, a sonometer or Melde's experiment are familiar examples of stringed devices. You may logically ask: What happens to the energy of a wave at an interface? If you think that the wave turns around and bounces back along the string, you are thinking logically since energy carried by the wave can not be destroyed at an interface. We then say that the wave has been *reflected*.

You must have heard echoes in large halls or in a valley. This is due to reflection of sound. You may have also observed reflection of water waves from the sea shore. In case of light waves, reflection from silvered surfaces, say a mirror, is the most common optical effect. The reflection of ultrasound waves forms the operating principle of *sonar* in depth-ranging, navigation, prospecting for oil and mineral deposits. The same principle is used in medical diagnostics to locate abnormalities in the abdomen. The reflection of electromagnetic waves in air navigation constitutes the working principle of radar. Reflection of radio waves by the ionosphere makes signal (information) transmission possible. In the knowledge era, it is vital for data transmission and retrieval, knowledge management and audio/video communications. However, in this Chapter, we shall confine ourselves only to sound waves.

When a wave is incident on a boundary between two different media, and if the boundary is not very rigid, as for example, when two strings of different mass per unit lengths are tied, the wave (energy) is partly transmitted into the second medium and is partly reflected back in the first. The phenomenon of partial reflection and transmission at the junction of strings has its analogue in the behaviour of all waves at an interface between two different media. Shallow water waves are partially reflected if water depth changes suddenly. Light incident on our atmosphere from the sun undergoes partial reflection because of changes in the density of atmosphere with height. Partial reflection of ultrasound waves at the interfaces of body tissues with different densities makes ultrasound a valuable diagnostic tool, particularly for detection of kidney stones and abdominal disorders. The phenomena of partial reflection and partial refraction are theoretically explained in terms of reflection and transmission amplitude and power coefficients. This discussion is facilitated by introducing the concept of *impedance*; the resistance offered to wave motion by a medium. These form the subject matter of discussion of Sections 8.2–8.4.

Stationary waves are formed due to superposition of two identical (same frequency and same amplitude) sound waves travelling in opposite directions. Unlike progressive waves, energy in stationary waves is confined to a limited region of space. This aspect finds useful manifestation in the form of music—one of the most charming aspects of life—using wind pipes and stringed instruments. As a child, you must have been charmed by ektara and flute. Similarly, sitar, violin, harp or clarinet must have attracted your attention at a musical concert or on a radio/TV. Did you ever observe a violinist in an orchestra tie up or loosen the pegs of his instrument while tuning with other musicians? This helps in changing the frequency and depending on the tension, thickness (area of cross section) and length of string, different notes are generated. Similar situation is obtained in a flute pipe and a reed pipe. You will learn the principle of superposition and its applications to formation of stationary waves in Sections 8.5 and 8.6.

When the frequencies of superposing (sound) waves are slightly different, we hear waxing and waning of sound. These are known as *beats*. You will learn about these in Section 8.7.

Wave groups arise when waves of slightly different frequencies are made to superpose. These are also referred to as *wave packets*. The concept of wave packet is of great importance in the study of quantum mechanics and communication engineering. These are discussed in Section 8.8.

8.2 WAVE MOTION AND IMPEDANCE

When a wave propagates in a medium, it experiences some opposition to its motion. This is analogous to the opposition offered to flow of current in a circuit or flow of a liquid in a capillary tube. *The resistance to wave motion offered by medium particles is called wave impedance.* The impedance offered to transverse waves travelling on stretched strings is termed as *characteristic impedance*, whereas the impedance offered to sound waves in air is called *acoustic impedance*.

You may now ask: Why impedance arises and what factors determine it? To know answer to this question, we recall that during wave propagation, each particle in the medium vibrates about its respective mean position and attempts to make the succeeding particle vibrate by transferring its own energy. On the other hand, each particle at rest tends to drag or slow down the particle in motion. That is, a vibrating particle experiences a dragging force which is similar to the viscous force experienced by a liquid while flowing in a capillary tube. According to Newton's third law of motion, this viscous drag is equal to the driving force F. From Chapter 4, you will recall that in small oscillation approximation, we can model the viscous force on Stokes' law:

$$\mathbf{F} = Z\mathbf{v}. \tag{8.1}$$

Here Z is constant of proportionality and is known as *wave impedance*. From this equation, it is clear that the impedance is numerically equal to the driving force which produces unit velocity.

You must have visited a building with big glass doors. Did you at any time fail to notice a door of glass and bang into it. This happens because glass transmits most of the light falling on it. Physically it means that glass offers very low impedance for transmission of light waves. This fact is used by geologists in determining the refractive index of smaller irregular specimen of transparent minerals. The specimen is immersed in a mixture of two miscible liquids whose refractive indices are different and at least one of them has refractive index greater than that of the specimen. The proportion of the liquids is adjusted so that the mineral specimen becomes almost invisible. The refractive index of this mixture corresponds to that of the mineral.

Let us consider another situation. Suppose that the walls of a room are covered with glass and a person claps/shouts/cries. Will you hear any sound? No, you will be surprised that no voice is heard because glass offers very high impedance to sound waves. This property is used in designing a sound recording studio. If you get an opportunity to visit any radio station or the A/V production centre, you will see that the chamber of the technician and his recording equipment are separated from the chamber of the artist/presenter/performer by a thick glass sheet. The glass does not allow any sound produced in the artist's room to reach technician's chamber and ensures disturbance free recording.

Let us now learn to calculate impedances produced by different types of waves.

8.2.1 Impedance Offered by Stretched Strings: Transverse Waves

Consider a stretched string and choose x-axis along its length (Fig. 8.1). We apply a harmonic force at the end $x = 0$ of the string to generate transverse waves. Let the instantaneous magnitude of force be given by $F = F_0 \cos \omega t$. The displacement of the particles of the string at position x and time t can be expressed as

$$y(x, t) = a \sin (\omega t - kx). \tag{8.2}$$

Let us obtain an expression for the impedance offered by the string to transverse waves. On physical considerations, we expect the amplitude of impressed force to gradually decrease. Therefore, we define characteristic impedance as

$$Z = \frac{\text{Amplitude of the applied force}}{\text{Transverse velocity amplitude of the wave}} \tag{8.3}$$

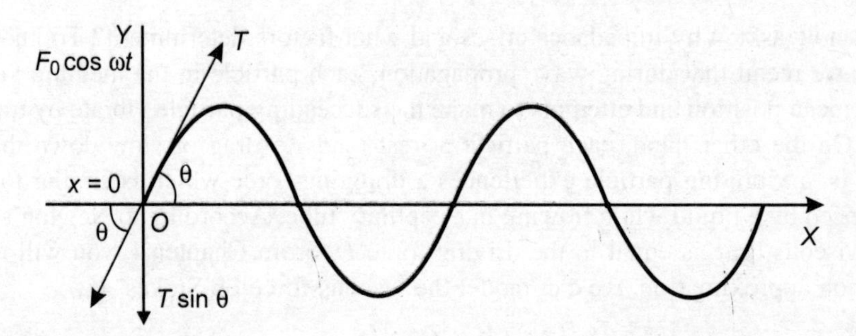

Fig. 8.1 A string vibrating under the harmonic force of magnitude $F = F_0 \cos \omega t$.

> The displacement of a particle in terms of x and t is given by
> $$y(x, t) = a \sin(\omega t - kx).$$
> Differentiating with respect to x and t, we get
> $$\frac{dy}{dx} = a\cos(\omega t - kx)(-k)$$
> and
> $$\frac{dy}{dt} = a\cos(\omega t - kx)(\omega).$$
> Hence
> $$\frac{dy}{dx} \cdot \frac{dt}{dy} = -\frac{k}{\omega}.$$
> $$\Rightarrow \quad \frac{dt}{dx} = -\frac{k}{\omega}$$

To proceed further, we denote the tension in the string as T. At $x = 0$, the vertical component of tension T along the negative y-direction will be equal to the applied transverse force (to give zero resultant displacement):

$$F_0 \cos \omega t = -T \sin \theta.$$

For small values of θ (< 4°), $\sin \theta = \tan \theta$ so that we can write

$$F_0 \cos \omega t = -T \tan \theta. \tag{8.4}$$

We know that the tangent (or slope) is defined at the $x = 0$ end of the string. Using Eq. (8.2), we can relate (dy/dx) to (dy/dt):

$$\frac{dy}{dx} = \frac{dy}{dt} \cdot \frac{dt}{dx} = -\frac{k}{\omega}\frac{dy}{dt}.$$

On substituting this result in Eq. (8.4), we get

$$F_0 \cos \omega t = \frac{kT}{\omega}\left(\frac{dy}{dt}\right)_{x=0}. \tag{8.5}$$

Since $(dy/dt)_{x=0} = a\omega \cos \omega t$, we can rewrite it as

$$F_0 \cos \omega t = \frac{Tk}{\omega} a\omega \cos \omega t = \frac{T}{v} a\omega \cos \omega t,$$

where $v = \omega/k$ is speed of the wave.
Writing $a\omega = v_0$ as *velocity amplitude* of the wave, the above equation takes the form

$$F_0 \cos \omega t = \frac{Tv_0}{v} \cos \omega t.$$

From this relation it readily follows that amplitude of applied harmonic force is given by

$$F_0 = \frac{Tv_0}{v}$$

so that

$$\frac{F_0}{v_0} = \frac{T}{v}. \tag{8.6}$$

This result shows that the ratio of the amplitude of the applied force to the amplitude of particle velocity for transverse waves is equal to the ratio of tension in the string and particle velocity. On combining Eqs. (8.3) and (8.6), we can write

$$Z = \frac{F_0}{v_0} = \frac{T}{v}. \tag{8.7}$$

From this result we may conclude that *characteristic impedance will be low for waves travelling with high speed.* The units of characteristic impedance are Nsm^{-1} and its dimension is MT^{-1}.

From Chapter 7, you may recall that velocity of a wave excited on a string can be expressed in terms of tension and mass per unit length through the relation $v = \sqrt{T/m}$, where m is mass per unit length of the string. On using this expression in Eq. (8.7), we get

$$Z = \frac{T}{v} = \sqrt{Tm}. \tag{8.8}$$

This result shows that *characteristic impedance is directly proportional to the square root of the product of mass per unit length of the string and tension in it.* It means that in a wired instrument, the impedance depends on the area of cross-section of the wire, its material and how tightly it is tied.

Sometimes it may be more convenient to express impedance in terms of wave velocity and thereby relate it to medium characteristics. So, we eliminate T from Eq. (8.7), by combining it with Eq. (8.8):

$$Z = \frac{v^2 m}{v} = mv. \tag{8.9}$$

This result tells us that *impedance is directly proportional to the velocity of the wave and hence depends on inertia as well as elasticity of the medium.*
You may now like to solve Practice Exercise 8.1.

***Practice Exercise* 8.1** Calculate the characteristic impedance offered by a sonometer wire stretched by a force of 20 N. It weighs 2 g per metre.

[***Ans.*** 0.2 Nsm^{-1}]

8.2.2 Impedance Offered by Gases: Sound Waves

You now know that when a sound wave travels in a gaseous medium, it generates some excess pressure. Its role is analogous to that of applied force in case of a transverse wave. So following Eq. (8.3), we define *acoustic impedance* as

$$Z = \frac{\text{Excess pressure due to a sound wave}}{\text{Particle velocity}} = \frac{\Delta p}{\Delta\psi/\Delta t},$$

where ψ signifies particle displacement for a longitudinal wave. Obviously, acoustic impedance Z has dimensions of the ratio of force per unit area to velocity, i.e., $ML^{-2}T^{-1}$.
The excess pressure generated by a longitudinal wave is given by

$$\Delta p = -E\frac{\partial\psi}{\partial x}, \tag{8.10}$$

where E is the bulk modulus of elasticity of the medium. This means that to know Z, we have to determine $(\partial\psi/\partial t)$ and $(\partial\psi/\partial x)$. To do so, we resort to the expression for particle displacement. For a longitudinal wave travelling along $+x$-direction, we can write

$$\psi(x, t) = a \sin\left[\frac{2\pi}{\lambda}(vt - x)\right]. \tag{8.11}$$

On differentiating $\psi(x, t)$ with respect to x, we get

$$\frac{\partial\psi}{\partial x} = -a\frac{2\pi}{\lambda}\cos\left[\frac{2\pi}{\lambda}(vt - x)\right]. \tag{8.12a}$$

Similarly, by differentiating $\psi(x, t)$ with respect to t, we obtain

$$\frac{\partial\psi}{\partial t} = a\left(\frac{2\pi v}{\lambda}\right)\cos\left[\frac{2\pi}{\lambda}(vt - x)\right]. \tag{8.12b}$$

On substituting the value of $(\partial\psi/\partial x)$ from Eq. (8.12a) in Eq. (8.10), we get the expression for excess pressure:

$$\Delta p = Ea\left(\frac{2\pi}{\lambda}\right)\cos\left[\frac{2\pi}{\lambda}(vt - x)\right]. \tag{8.13}$$

Now, substitute for Δp and $\partial\psi/\partial t$ from Eqs. (8.13) and (8.12b) in the expression defining acoustic impedance. On simplifying the resultant expression, you will find that

$$Z = \frac{\Delta p}{\partial\psi/\partial t} = \frac{E}{v}, \tag{8.14}$$

where v is wave velocity. This result shows that the units of acoustic impedance are $Nm^{-3}s$. In the preceding chapter you learnt that wave velocity in a fluid medium is given by

$$v = \sqrt{\frac{E}{\rho}},$$

where ρ is density of the medium. Using this expression for v in Eq. (8.14), we can express acoustic impedance in terms of medium density and wave velocity:

$$Z = \frac{E}{v} = \sqrt{E\rho} = \rho v. \tag{8.15}$$

This result shows that *acoustic impedance is product of wave velocity and the density of the medium in which wave propagates.* It means that a denser medium will offer greater impedance and we expect the speed of the wave to decrease. But we know that sound moves faster in liquids than gases. This apparent contradiction is resolved by noting that increase in elasticity of the medium overcomes the changes due to density. To convince yourself, you should solve the following Practice Exercise.

***Practice Exercise* 8.2** Calculate acoustic impedance of air and water at STP. Use $\rho_{air} = 1.29$ kgm^{-3}, $v_{air} = 332$ ms^{-1}, $\rho_{water} = 10^3$ kgm^{-3} and $v_{water} = 1500$ ms^{-1}.

[***Ans.*** 428.3 Nsm^{-3}, 1.5×10^6 Nsm^{-3}]

We now apply the results obtained in Section 8.2 to derive expressions for reflection and transmission amplitude and energy coefficients for a wave incident on the interface separating two media.

8.3 REFLECTION AND TRANSMISSION AMPLITUDE COEFFICIENTS

You now know that different media offer different impedances to waves travelling through them. These impedances depend on the characteristic properties of the medium. How do waves respond to an abrupt change in impedance as encountered at the interface separating two media? Let us now discover answer to this interesting question. For simplicity, we shall first consider transverse waves.

8.3.1 Transverse Waves

Refer to Fig. 8.2. Two strings *AO* and *OB* are joined together at *O* and are stretched by the same tension *T*. We choose *x*-axis along the length of the string *AB*. Suppose that the strings *AO* and *OB* offer characteristic impedances Z_1 and Z_2, respectively. From Eq. (8.8), you may recall that the impedance offered to a wave in this case is given by $\sqrt{Tm}$, where *m* is mass per unit length of the string. Suppose that end *B* is not clamped and the waves are generated by moving end *A* up and down with frequency *f*.

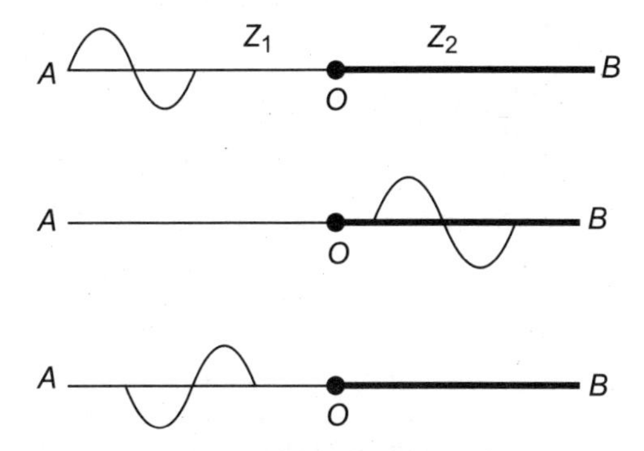

Fig. 8.2 Transverse waves in strings of different mass per unit lengths.

Consider a wave excited in string *AO* along the positive *x*-direction. We expect it to be partly reflected and partly transmitted at point *O*, which joins the strings. Since the frequency of the wave is *f*, its angular frequency $\omega = 2\pi f$. The particle displacements due to the incident, reflected and transmitted waves can be written as

$$y_i(x, t) = a_i \sin(\omega t - k_1 x), \tag{8.16}$$

$$y_r(x, t) = a_r \sin(\omega t + k_1 x), \tag{8.17}$$

and

$$y_t(x, t) = a_t \sin(\omega t - k_2 x), \tag{8.18}$$

where the subscripts *i, r* and *t* on the displacements and the amplitudes refer to the incident, reflected and the transmitted waves, respectively. Note that

- The angular frequency has remained unchanged.
- The propagation constant k_1 for the incident and the reflected waves is the same but differs from that for the transmitted wave. This is because the wave speed depends on the density of the medium.

To give physical meaning to the reflection and transmission coefficients, we have to consider *boundary conditions*. In this case, total displacement and restoring force or total transverse component of tension on one side of the boundary (section *AO* of the string in the instant case) arises due to combination of incident and reflected waves. However, continuity demands that total displacement and tension must be same immediately on both sides. So the boundary conditions can be written as:

- The particle displacements immediately to the left and immediately to the right of the boundary (at $x = 0$) must be the same, since the string is unbroken at O. This implies that the particle velocity $\partial y(x, t)/\partial t$ will also be the same at that point.
- The transverse components of tension $-T[\partial y(x, t)/\partial x]$ must also be equal at the boundary when approached from opposite directions. Otherwise, there will be a net force at the point $x = 0$. Physically it means that this point will have infinite acceleration since the mass of a single point is zero.

These conditions can be expressed as

$$y_i(x,t)\Big|_{x=0} + y_r(x,t)\Big|_{x=0} = y_t(x,t)\Big|_{x=0} \tag{8.19a}$$

and

$$-T\frac{\partial y_i}{\partial x}\bigg|_{x=0} - T\frac{\partial y_r}{\partial x}\bigg|_{x=0} = -T\frac{\partial y_t}{\partial x}\bigg|_{x=0}. \tag{8.19b}$$

Putting $x = 0$ in Eqs. (8.16) to (8.18) and combining these with Eq. (8.19a), the first boundary condition takes the mathematical form

$$a_i \sin \omega t + a_r \sin \omega t = a_t \sin \omega t.$$

On cancelling $\sin \omega t$ from both sides, we get

$$a_i + a_r = a_t. \tag{8.20}$$

The condition expressed by Eq. (8.19b) gives

$$a_i k_1 T \cos \omega t - a_r k_1 T \cos \omega t = a_t k_2 T \cos \omega t.$$

On simplification, it reduces to

$$k_1 T(a_i - a_r) = k_2 T a_t. \tag{8.21}$$

We know that

$$k_1 T = \frac{2\pi}{\lambda_1} T = \frac{2\pi f}{v_i} T = 2\pi f m_1 v_i = 2\pi f Z_1,$$

where m_1 is mass per unit length of string *AO*, which offers impedance Z_1 and v_i is speed of incident waves of frequency f. Note that in arriving at the final expression, we have used Eq. (8.9) which relates impedance with wave speed and mass per unit length of the wire.

Similarly, we can write

$$k_2 T = 2\pi f Z_2,$$

where Z_2 is the impedance offered by string *OB*.

Using these results in Eq. (8.21), we find that the impedances of two media are related as

$$Z_1(a_i - a_r) = Z_2 a_t. \tag{8.22}$$

By combining Eqs. (8.20) and (8.22), you can easily calculate the fraction of the incident amplitude that is reflected or transmitted at the boundary. Mathematically, these are given by the expressions for a_r/a_i and a_t/a_i. These ratios are usually called the *reflection* and *transmission amplitude coefficients*. Let us denote these by symbols R_{12} and T_{12}, respectively. On substituting the value of a_t from Eq. (8.20) into Eq. (8.22), you will get

$$Z_1(a_i - a_r) = Z_2(a_i + a_r)$$

or

$$(Z_1 - Z_2)a_i = (Z_1 + Z_2)a_r,$$

so that

$$R_{12} = \frac{a_r}{a_i} = \frac{Z_1 - Z_2}{Z_1 + Z_2}. \tag{8.23}$$

Similarly, by eliminating a_r from Eqs. (8.20) and (8.22), you will get

$$T_{12} = \frac{a_t}{a_i} = \frac{2Z_1}{Z_1 + Z_2}. \tag{8.24}$$

Let us now pause for a while and ponder as to what have we achieved so far. We note that

- The reflection and transmission amplitude coefficients depend only on the impedances of the two media. The frequency of the wave has no effect whatsoever.
- If string *OB* is rigidly fixed to a wall, i.e., the second medium is extremely dense, Z_2 will be almost infinite. In such a case, $R_{12} = -1$ and $T_{12} = 0$. This suggests that $a_r = -a_i$ and $a_t = 0$. We may, therefore, conclude that the amplitude of the reflected wave is equal to the amplitude of the incident wave with just a reversal of sign and there is no transmitted wave. This means that *when a wave undergoes reflection at an interface with denser medium, it suffers a phase change* of π.
- As long as $Z_2 > Z_1$, i.e., second medium is denser (but not infinitely dense) than the first, R_{12} will be negative implying a phase change of π on reflection. In this case, however, *the incident wave will be partly reflected and partly transmitted.*
- For $Z_2 < Z_1$, R_{12} will be positive. It means that no change of phase will take place when a wave is reflected at the interface with a rarer medium. Both transmitted and reflected waves will exist in this case also.
- When $Z_1 = Z_2$, $R_{12} = 0$. It means that there is no reflected wave. Since energy is to be conserved, we will have only transmitted wave, i.e., $T_{12} = 1$, which gives $a_t = a_i$ indicating that the wave proceeds undisturbed.

To summarise, when *a wave travelling in a medium of lower impedance is incident on the boundary of a medium of higher impedance (air to water), the reflected wave undergoes a phase change* of π. If, however, a wave travelling through a medium of higher impedance reaches the boundary of a medium of lower impedance (water to air), no change of phase takes place on reflection. Moreover, T_{12} is always positive indicating that there is no change of phase for the transmitted wave in any case. These results are depicted in Fig. 8.3.

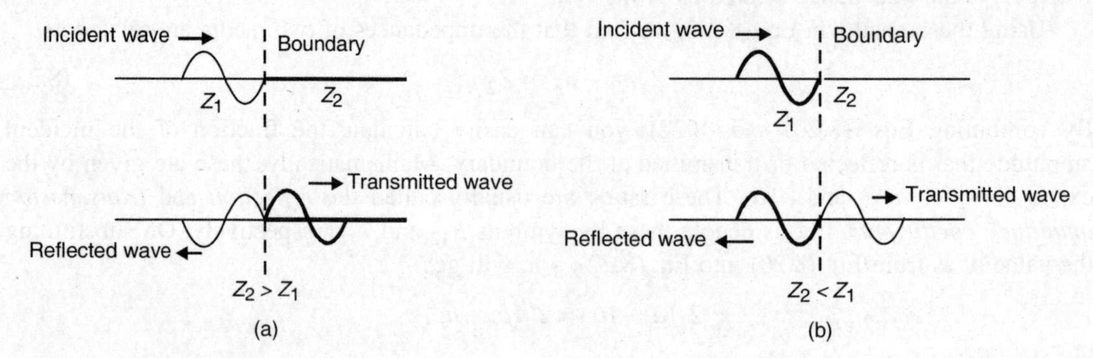

Fig. 8.3 Reflection and transmission of waves incident from (a) medium of lower impedance to medium of higher impedance; and (b) medium of higher impedance to medium of lower impedance.

To fix up these ideas, you should read Solved Examples 8.1 and 8.2. Then do Practice Exercise 8.3.

EXAMPLE 8.1

A sound wave travelling through air falls normally on the surface of water. The ratio of amplitude of sound wave that enters the second medium to the amplitude of incident wave is 6×10^{-4}. Calculate the speed of sound in water using the data $\rho_{air} = 1.29$ kgm^{-3}, $\rho_{water} = 10^3$ kgm^{-3}, and speed of sound in air = 350 ms^{-1}.

Solution: Here $\rho_{air} = 1.29$ kgm^{-3}, $\rho_{water} = 1000$ kgm^{-3}, and $v_{air} = 350$ ms^{-1}.
Since sound waves are longitudinal, from Eq. (8.24), we have

$$\frac{a_t}{a_i} = \frac{2Z_1}{Z_1 + Z_2} = \frac{2(Z_1/Z_2)}{1 + (Z_1/Z_2)} = r, \text{ say.} \qquad \text{(i)}$$

Since $Z = \rho v$, we can write

$$\frac{Z_1}{Z_2} = \frac{\rho_{air} v_{air}}{\rho_{water} v_2}$$

$$= \frac{(1.29 \text{ kg m}^{-3}) \times (350 \text{ ms}^{-1})}{(1000 \text{ kg m}^{-3}) \times v_2} = \frac{0.4515}{v_2} = x, \text{ say.}$$

Using this result in Eq. (i), we get

$$\frac{a_t}{a_i} = \frac{2x}{1 + x} = r.$$

$$\therefore \qquad (2 - r)\, x = r$$

or

$$x = \frac{r}{2 - r}.$$

Using the value of r in this expression, we get

$$x = 3 \times 10^{-4}.$$

Hence
$$v_2 = \frac{0.4515}{x} = \frac{0.4515}{3 \times 10^{-4}} = 1505 \text{ ms}^{-1}.$$

EXAMPLE 8.2

A rope is made up of four identical strands twisted together. At one point, the condition of the rope becomes such that only a single strand continues. The rope is held under tension and a wave of amplitude 1.0 cm is sent from the single strand. Calculate the amplitude of the wave reflected back along the single strand.

Solution: From Eq. (8.23), you can write

$$R_{12} = \frac{a_r}{a_i} = \frac{Z_1 - Z_2}{Z_1 + Z_2}.$$

By dividing the numerator and denominator by Z_2, we can rewrite the expression for amplitude reflection coefficient as

$$R_{12} = \frac{(Z_1/Z_2) - 1}{(Z_1/Z_2) + 1}. \tag{i}$$

For a string under tension we know that impedance is directly proportional to the square root of mass per unit length, i.e., $Z \propto \sqrt{m}$. So we can write

$$\frac{Z_1}{Z_2} = \sqrt{\frac{m_1}{m_2}}. \tag{ii}$$

On combining Eqs. (i) and (ii), we get

$$R_{12} = \frac{\sqrt{\dfrac{m_1}{m_2}} - 1}{\sqrt{\dfrac{m_1}{m_2}} + 1}.$$

In terms of the number of strands in the first portion (n, say), we can express R_{12} as

$$R_{12} = \frac{\sqrt{n} - 1}{\sqrt{n} + 1} = \frac{2 - 1}{2 + 1} = \frac{1}{3}.$$

$$\therefore \qquad a_r = \frac{a_i}{3} = 0.33 \text{ cm}.$$

***Practice Exercise* 8.3** Two strings are joined together and stretched under the same tension. For transverse waves, calculate the reflection and transmission amplitude coefficients when the ratio of their linear densities is 1 : 9.

$$\left[\textbf{\textit{Ans.}} \quad R_{12} = -\frac{1}{2}, T_{12} = \frac{3}{4}\right]$$

8.3.2 Longitudinal Waves

To analyse reflection and transmission of longitudinal waves, we follow the same procedure as outlined for transverse waves. Let us choose the boundary separating two media of acoustic impedances Z_1 and Z_2 as origin ($x = 0$) of the co-ordinate system. Now, consider a longitudinal wave incident on such a boundary. As in the case of transverse waves, we can represent the particle displacements for the incident, reflected and transmitted waves by Eqs. (8.16), (8.17) and (8.18):

$$\psi_i(x, t) = a_i \sin(\omega t - k_1 x),$$

$$\psi_r(x, t) = a_r \sin(\omega t + k_1 x)$$

and

$$\psi_t(x, t) = a_t \sin(\omega t - k_2 x).$$

The boundary conditions in this case are:

- The particle displacement $\psi(x, t)$ is continuous at the boundary. That is, it has the same value immediately to the left and right of the boundary at $x = 0$.
- The excess pressure in the immediate vicinity of the boundary on its two sides is the same.

As before, the first boundary condition leads us to the relation

$$a_i + a_r = a_t. \tag{8.25}$$

We know that for a longitudinal wave, excess pressure is given by the relation $\Delta p = -E(\partial\psi/\partial x)$, where E is bulk modulus of elasticity. You may recall that $E = \gamma p_0$, where $\gamma = C_p/C_V$ is ratio of heat capacities at constant pressure and constant volume and p_0 is equilibrium pressure. Therefore, the second boundary condition implies that p_0 cancels out from both sides and the resultant expression takes the form

$$\frac{\partial\psi_i}{\partial x} + \frac{\partial\psi_r}{\partial x} = \frac{\partial\psi_t}{\partial x}.$$

On inserting the values of ψ_i, ψ_r, and ψ_t in this expression, we get

$$-a_i k_1 \cos\omega t + a_r k_1 \cos\omega t = -a_t k_2 \cos\omega t.$$

This expression readily simplifies to

$$k_1(a_i - a_r) = k_2 a_t. \tag{8.26}$$

To express this result in terms of impedance, recall that wave vector $k_1 = (\omega/v_1)$, where v_1 is wave velocity in medium 1. By multiplying and dividing the RHS by $\rho_1 v_1$, we can rewrite the expression for wave vector as

$$k_1 = \frac{\omega}{\rho_1 v_1^2}\rho_1 v_1 = \frac{\omega Z_1}{\gamma p_0},$$

where we have used the relation $Z_1 = \rho_1 v_1$ and $v_1 = \sqrt{\gamma p_0/\rho}$.

Similarly, you can show that

$$k_2 = \frac{\omega Z_2}{\gamma p_0}.$$

Using these results in Eq. (8.26), we get

$$\frac{\omega Z_1}{\gamma p_0}(a_i - a_r) = \frac{\omega Z_2}{\gamma p_0} a_t$$

or

$$Z_1(a_i - a_r) = Z_2 a_t. \tag{8.27}$$

Since Eqs. (8.25) and (8.27) relating the incident, reflected and transmitted amplitudes are exactly the same as in the transverse case, we can safely say that the reflection and transmission amplitude coefficients will be given by Eqs. (8.23) and (8.24), respectively.

You now know that progressive waves provide us a very useful mechanism for transferring energy from one region to another region of space. You may, therefore, like to know as to what happens to the energy of a wave when it encounters a boundary between two media of different impedances. This information is usually contained in reflection and refraction power coefficients. You will now learn to calculate these for transverse as well as longitudinal waves.

8.4 REFLECTION AND TRANSMISSION POWER COEFFICIENTS

From Section 7.7, you may recall that when a string of mass per unit length m vibrates with amplitude a and angular frequency ω, the total energy per unit length is given by

$$E = \left(\frac{1}{2}\right) m a^2 \omega^2. \tag{8.28}$$

Suppose that the wave is travelling with speed v. Then the rate at which energy is carried per unit length along the string, i.e., power is obtained by multiplying the expression for energy with the speed v of the wave. That is, it is given $(1/2)ma^2\omega^2 v$. For transverse waves discussed in the preceding section, the rate at which energy reaches the boundary along with the incident wave is given by

$$P_i = \frac{1}{2} m_1 a_i^2 \omega^2 v = \frac{1}{2} Z_1 \omega^2 a_i^2, \tag{8.29}$$

since $Z_1 = m_1 v$.

Similarly, the rates at which the energy leaves the boundary along with the reflected and the transmitted waves, respectively, are

$$P_r = \frac{1}{2} Z_1 \omega^2 a_r^2 \tag{8.30}$$

and

$$P_t = \frac{1}{2} Z_2 \omega^2 a_t^2. \tag{8.31}$$

Recall that Eqs. (8.23) and (8.24) express a_r and a_t in terms of a_i. Substituting these expressions in Eqs. (8.30) and (8.31), we get

$$P_r = \frac{1}{2} Z_1 \omega^2 \left(\frac{Z_1 - Z_2}{Z_1 + Z_2}\right)^2 a_i^2 = \left(\frac{Z_1 - Z_2}{Z_1 + Z_2}\right)^2 P_i \tag{8.32}$$

and

$$P_t = \frac{1}{2} Z_2 \omega^2 \left(\frac{2Z_1}{Z_1 + Z_2} \right)^2 a_i^2 = \frac{4Z_1 Z_2}{(Z_1 + Z_2)^2} P_i. \tag{8.33}$$

These results can be used to obtain the reflection and transmission power coefficients R_p and T_p. The reflection coefficient is defined as the ratio of the rate at which energy is reflected at the interface to the rate at which energy is incident on it. Hence, from Eq. (8.32), we can write

$$R_p = \frac{P_r}{P_i} = \left(\frac{Z_1 - Z_2}{Z_1 + Z_2} \right)^2 \tag{8.34}$$

From this we note that the rate at which energy is reflected at an interface depends on the nature of media and the impedance it offers. This explains why almost half of solar energy is reflected by the earth. However, if $Z_1 = Z_2$ (which is also possible if we have $T_1 m_1 = T_2 m_2$), $R_p = 0$. That is, no energy is reflected back when the impedances match. Such an impedance matching plays a very important role in electricity transmission. Long distance cables carrying power are matched so accurately at the joints that no energy is wasted due to reflection. We also need impedance matching when we wish to transfer sound energy using a loudspeaker. Similarly, when light waves travel from air into a glass lens or a slab, we wish to minimise reflection (as it reduces intensity).

The transmission power coefficient is defined as the ratio of the rate at which energy is transmitted at the interface into the second medium to the rate at which energy is incident on it. Hence, from Eq. (8.33), we can write

$$T_p = \frac{P_t}{P_i} = \frac{4Z_1 Z_2}{(Z_1 + Z_2)^2}. \tag{8.35}$$

To appreciate the importance of transmission power coefficient in practical applications, study Example 8.3 carefully.

EXAMPLE 8.3

Two strings having mass per unit lengths 0.1 kgm^{-1} and 0.4 kgm^{-1} are joined together. If tension in the strings is 250 N, calculate the fraction of power transmitted to the second string.

Solution: To calculate the power transmission coefficient in the second string, we recall from Eq. (8.8) that the impedance offered to a stretched string is given by $Z = \sqrt{Tm}$. On substituting the given values, we get

$$Z_1 = \sqrt{(250 \text{ N}) \times (0.1 \text{ kg m}^{-1})}$$

$$= 5 \text{ kgs}^{-1}$$

and

$$Z_2 = \sqrt{(250 \text{ N}) \times (0.4 \text{ kg m}^{-1})}$$

$$= 10 \text{ kgs}^{-1}.$$

Hence

$$T_p = \frac{4Z_1 Z_2}{(Z_1 + Z_2)^2}$$

$$= \frac{4 \times (5 \text{ kg s}^{-1}) \times (10 \text{ kg s}^{-1})}{15 \text{ kg s}^{-1} \times 15 \text{ kg s}^{-1}}$$

$$= \frac{8}{9} \approx 0.89 \text{ kgs}^{-1}.$$

It means that nearly 89 percent power is transmitted.
You may now like to do a Practice Exercise 8.4.

***Practice Exercise* 8.4** Show that energy is conserved when a transverse wave meets the boundary between two media of characteristic impedances Z_1 and Z_2.

For longitudinal waves, it is customary to calculate energy transfer in terms of their intensity. In Chapter 7, you learnt to derive an expression for the intensity of sound waves in a gas. It is given by

$$I = \frac{1}{2}\rho a^2 \omega^2 v$$

$$= 2\pi^2 f^2 a^2 Z, \tag{8.36}$$

where $Z = \rho v$ is impedance offered by the medium to a longitudinal wave. Hence the incident, reflected and transmitted wave intensities can be written as

$$I_i = 2\pi^2 f^2 a_i^2 Z_1, \tag{8.37a}$$

$$I_r = 2\pi^2 f^2 a_r^2 Z_1 \tag{8.37b}$$

and

$$I_t = 2\pi^2 f^2 a_t^2 Z_2. \tag{8.37c}$$

If we define reflection power coefficients as the ratio of reflected to incident intensities $[R_p = (I_r/I_i)]$ and transmission power coefficient as the ratio of transmitted to incident intensities at the interface $[T_p = (I_t/I_i)]$, these equations lead us to exactly the same expressions for R_p and T_p as the ones for transverse waves. This suggests that same conclusions hold for transverse and longitudinal waves.

8.5 PRINCIPLE OF SUPERPOSITION OF WAVES

In Chapter 3, you learnt that when a particle is acted upon by two or more simple harmonic oscillations simultaneously, its resultant displacement at any time is given by the algebraic sum of individual displacements. You may now ask: Can this result be extended to waves? If your answer is affirmative, you are thinking logically. When two or more waves travel the same path, independent of one another, the resultant displacement of a particle at a given time is found to be equal to the algebraic sum of its displacements due to individual waves. In other words, we can say that *the resultant displacement of a particle can be determined by algebraically adding the displacements of individual waves*. This is known as *principle of superposition of waves*. The genesis of the superposition principle for waves lies in individuality of waves. A vivid demonstration of the individualistic behaviour of waves is seen in radio broadcast and TV transmission, wherein we tune to a particular frequency/radio station/channel out of the many waves propagating in the atmosphere.

You can demonstrate the principle of superposition by considering two identical pulses travelling on a rope in opposite directions. Refer to Fig. 8.4a, which depicts two pulses having positive amplitude. When these pulses cross each other, the resultant amplitude increases. However, they continue to move in their respective original directions. It means that each pulse behaves as if it has not been influenced at all by the other, i.e., they act completely independently. Similarly, Fig. 8.4b shows two pulses which are approaching each other but their amplitudes are oppositely directed. When these pass each other, the entire string is straightened momentarily, and they appear to cancel out the effect of each other. Thereafter, the pulses reappear and keep moving individually. Each pulse carries energy and at the time of crossing, though their resultant displacement is zero, their energies add up. The principle of conservation of energy suggests that energy cannot be destroyed; it reappears in the form of pulses. These observations can be explained on the basis of the principle of superposition.

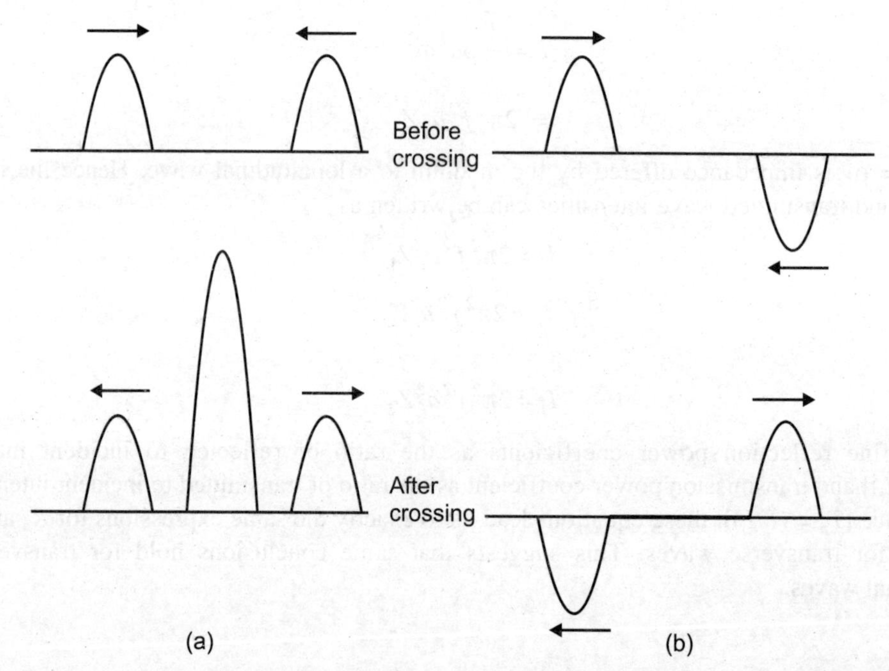

Fig. 8.4 Superposition of two pulses travelling in opposite directions having (a) positive amplitudes, and (b) oppositely directed amplitudes.

In Chapter 3, you learnt that mathematical basis of superposition of oscillations lies in the linearity of the equation of motion. Suppose that $y_1(x, t)$ and $y_2(x, t)$ denote the displacements of a particle at space print x and time t due to propagation of two waves in a medium. Its resultant displacement can be mathematically written as

$$y(x, t) = y_1(x, t) + y_2(x, t). \qquad (8.38)$$

In Chapter 7, you have learnt that a wave is characterised by its amplitude, angular frequency, wave vector and phase. Therefore, depending on which of these components is same or different, superposition of waves gives rise to significantly different but very interesting phenomena. These find applications in diverse fields. We now discuss some of these.

8.5.1 Superposition of In-phase Waves of Different Amplitudes

Consider two in-phase waves having same angular frequency, wave vector and phase but different amplitudes propagating along positive x-axis. We can mathematically describe these as

$$y_1(x, t) = a_1 \sin(\omega t - kx)$$

and

$$y_2(x, t) = a_2 \sin(\omega t - kx)\,. \tag{8.39}$$

To determine the amplitude of resultant wave, we add the individual displacements algebraically and obtain

$$\begin{aligned} y(x, t) &= a_1 \sin(\omega t - kx) + a_2 \sin(\omega t - kx) \\ &= (a_1 + a_2) \sin(\omega t - kx). \end{aligned} \tag{8.40}$$

This equation implies that amplitude of the resultant wave is equal to the sum of the amplitudes of individual waves. However, the frequency and phase remain unchanged. It is graphically shown in Fig. 8.5.

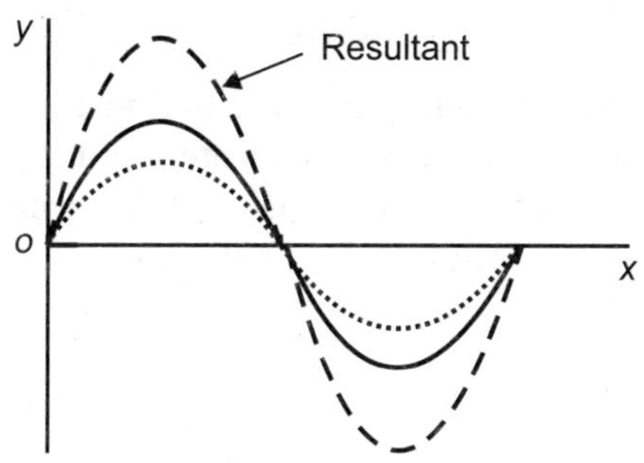

Fig. 8.5 Superposition of two in-phase waves of unequal amplitudes.

8.5.2 Superposition of Identical Out-of-phase Waves

As the next possibility, let us consider that two identical but out-of-phase waves moving in the same direction along positive x-axis are made to superpose. Mathematically, we write

$$y_1(x, t) = a \sin(\omega t - kx)$$

and

$$y_2(x, t) = a \sin(\omega t - kx + \phi). \tag{8.41}$$

When such waves superpose, depending on their phase difference, the amplitude of resultant wave can vary from zero to $2a$. This leads to redistribution of energy in space giving rise to the phenomenon of *interference*. You may recall from your school Physics that interference is more vividly seen with light waves in the form of dark and bright fringes. You will learn about it in detail in Wave Optics.

8.5.3 Superposition of identical Waves Moving in Opposite Directions

Let us now consider superposition of two identical waves propagating in opposite directions in a medium having rigid boundaries. For the wave propagating along the positive direction of x-axis, we write

$$y_1(x, t) = a \sin(\omega t - kx), \quad (8.42a)$$

whereas for the wave propagating along the negative direction of x-axis, we write

$$y_2(x, t) = a \sin(\omega t + kx). \quad (8.42b)$$

Since these waves are identical and superpose in a bound region of space, they give rise to *stationary* or *standing waves*. These are responsible for production of music and help us understand the working of musical instruments. You will learn about formation of stationary waves in detail in Section 8.6.

8.5.4 Superposition of Waves of Slightly Different Frequencies

When two waves having slightly different frequencies but equal amplitudes move in the same direction, we can write

$$y_1(x, t) = a \sin(\omega_1 t - kx)$$

and

$$y_2(x, t) = a \sin(\omega_2 t - kx). \quad (8.43)$$

Suppose that these waves are made to superpose. We find that irrespective of phase difference, their superposition leads to production of *beats*, which are heard alternately in the form of waxing and waning of sound. If many waves of slightly different frequencies are allowed to superpose, they give rise to *wave groups* or *wave packets*. You will learn about these in Sections 8.7 and 8.8, respectively.

8.6 STATIONARY WAVES

You now know that stationary waves are generated when two identical waves travelling in opposite directions are made/allowed to superpose. You may now ask: How to generate such waves? In practice, it is inconvenient, if not difficult, to excite identical waves of exactly the same amplitude and wavelength. But a convenient and easy way is to allow a wave to reflect from a *boundary*. The reflection can take place at a fixed boundary such as a wall, a peg in a stringed instrument, and closed end of an organ pipe or a free boundary like the open end of an organ pipe. In Section 8.3, you have learnt that when a wave travelling in a rarer medium is reflected at a rigid boundary, the displacement $y(x, t)$ is zero at the interface and a phase change of π takes place. However, no such change of phase takes place at a free boundary.

Let us first consider reflection at a rigid boundary. In this case, the resultant displacement is given by

$$y(x, t) = a \sin(\omega t - kx) + a \sin(\omega t + kx).$$

Using the trigonometric formula $\sin A + \sin B = 2 \sin\left(\frac{A+B}{2}\right)\cos\left(\frac{A-B}{2}\right)$, we can write

$$
\begin{aligned}
y(x, t) &= 2a \sin \omega t \cos kx \\
&= (2a \cos kx) \sin \omega t. \qquad (8.44)
\end{aligned}
$$

This result shows that the resultant wave is sinusoidal and for a given t, its amplitude varies in space as $2a \cos kx$; changing from $+ 2a$ to $-2a$. However, the resultant motion has the same frequency and wave vector as the individual waves.

Let us pause for a moment and ask: What does Eq. (8.44) represent? It does not represent a travelling wave since the dependence of resultant motion on x and t is not through the combination of spatial and temporal components $(x \pm vt)$. Note that we started with two progressive waves moving in opposite directions, but we have ended up with a wave confined between two points (the source and the reflector). That is why such a wave is called a *stationary wave*. Since stationary waves remain confined in space, they *transport no energy* from one point to another.
Equation (8.44) implies that displacement $y(x, t)$ will take maximum value for

$$\cos kx = \cos\frac{2\pi}{\lambda}x = 1,$$

or

$$\frac{2\pi}{\lambda}x = m\pi; \quad m = 0, 1, 2, \ldots \qquad (8.45)$$

That is, points of maximum displacement occur at $x = 0, \lambda/2, \lambda, \ldots, m\lambda$. These are called *antinodes*. Similarly, displacement $y(x, t)$ will take minimum value when

$$\cos kx = \cos\frac{2\pi}{\lambda}x = 0,$$

or

$$\frac{2\pi}{\lambda}x = \left(\frac{2m+1}{2}\right)\pi; \quad m = 0, 1, 2, \ldots \qquad (8.46)$$

This equation shows that points of no disturbance occur at $\lambda/4, 3\lambda/4, \ldots, (2m+1)\lambda/4$ and are called *nodes*. Figure 8.6 shows the envelop of stationary waves. The nodes and antinodes are also depicted here.

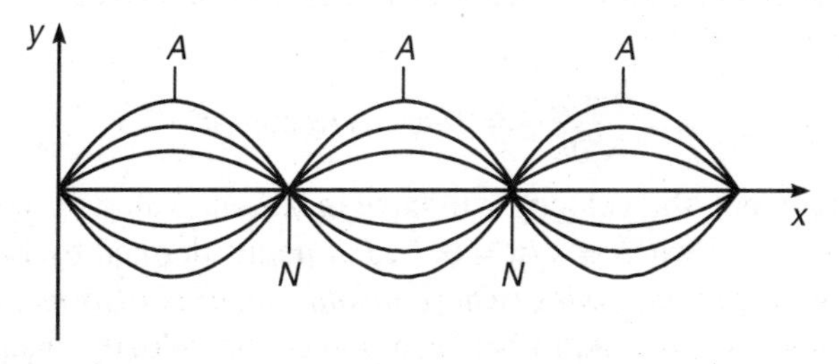

Fig. 8.6 Envelop of a stationary wave. Nodes (*N*) and antinodes (*A*) respectively correspond to zero and maximum displacements.

Note that distance between any two successive nodes or antinodes is $\lambda/2$. But the distance between a node and an immediately next antinode is $\lambda/4$.
Now, study Example 8.4 carefully.

EXAMPLE 8.4

Show that in a stationary wave, all particles between any two consecutive nodes are in phase but they are in opposite phase with the particles between the next pair of nodes.

Solution: From Eq. (8.44), we recall that instantaneous displacement of a standing wave is given by

$$y(x, t) = 2a \cos kx \sin \omega t.$$

For any two points $x = x_1$ and $x = x_2$ between consecutive nodes, the phase is same, equal to ωt. But if $x_2 = x_1 + (\lambda/2)$, i.e., the points belong to adjacent loops (which means that they lie between consecutive pairs of nodes). The corresponding displacements are

$$y_1 = 2a \cos kx_1 \sin \omega t$$

and $$y_2 = 2a \cos kx_2 \sin \omega t = 2a \cos k\left(x_1 + \frac{\lambda}{2}\right) \sin \omega t = 2a \cos\left(kx_1 + \frac{k\lambda}{2}\right) \sin \omega t$$

$$= 2a \cos (kx_1 + \pi) \sin \omega t = -2a \cos kx_1 \sin \omega t = 2a \cos kx_1 \sin (\omega t + \pi).$$

This result shows that the phases of y_1 and y_2 differ by π.
You may now like to solve Practice Exercise 8.5.

***Practice Exercise* 8.5** Depict the amplitudes of various particles in the region of the standing wave on a string fixed at both ends.

Let us now calculate the velocity of a particle at any point in a stationary wave.

8.6.1 Velocity of a Particle in a Stationary Wave

We know that velocity of a particle is defined as the rate of change of displacement with time. So we can obtain the expression for velocity of a particle in a stationary wave by differentiating the resultant displacement $y(x, t)$ of a particle with respect to time at a fixed point. From Eq. (8.44) for fixed x, we get

$$v = \frac{\partial y(x, t)}{\partial t} = 2a\omega \cos kx \cos \omega t. \tag{8.47}$$

As in the case of displacement, the velocity will have maximum value for $\cos kx = 1$, i.e., at points defined by Eq. (8.45) and zero when $\cos kx = 0$, i.e., at points defined by Eq. (8.46). It means that the *velocity is maximum at the antinodes (where displacement is also maximum)* and *zero at the nodes (where displacement is zero)*. At in-between points, the velocity changes from maximum to zero.

To determine spatial variation of displacement in a stationary wave, we differentiate the expression for resultant displacement with respect to x at a given time t. Then Eq. (8.44) leads us to

$$\left.\frac{\partial y(x, t)}{\partial x}\right|_{t = \text{constant}} = -2ak \sin kx \sin \omega t. \tag{8.48}$$

It is important to realise here that *particles at the antinodes experience minimum strain,* whereas those *at nodes experience maximum strain.* Physically, we can say that particles immediately to the left and to the right of the nodes experience pull by particles moving in the opposite direction. Again by referring to Fig. 8.6, you can see that particles at the antinodes always move along with the particles at their sides, not causing much strain on particles at the antinodes.

Now, refer to Fig. 8.7 which depicts the shapes of the string at different times. Note that all particles in a particular segment reach their respective extreme positions simultaneously. Also, they pass through the mean positions at the same time. This is because all particles have the same time period but different velocities. The particles which have to cover greater distances have greater velocities and vice versa.

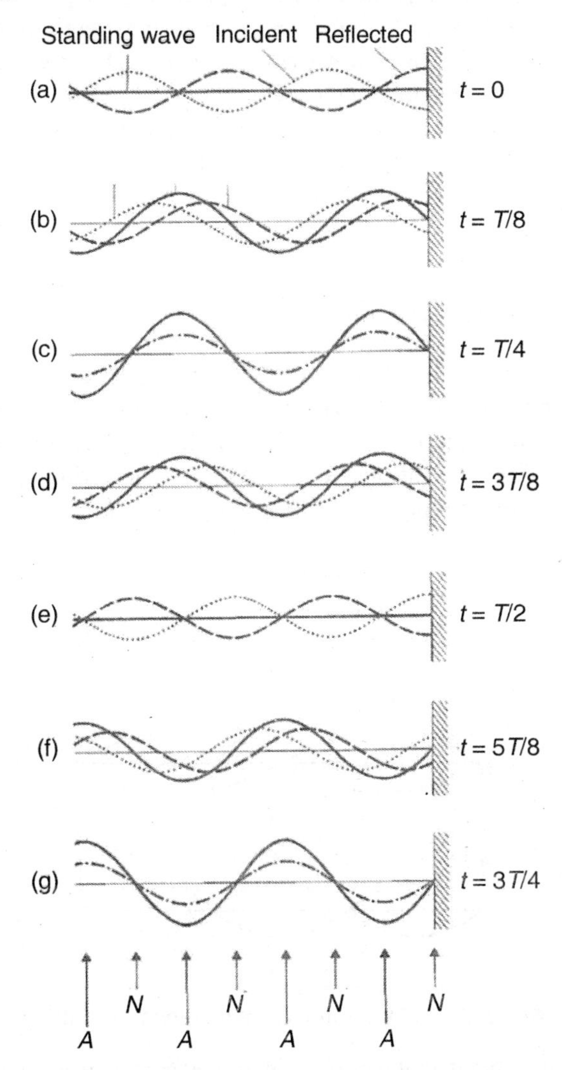

Fig. 8.7 Shapes of a string, fixed at both ends, at different times for a stationary wave.

Let us now consider individual particles. To know the time at which the velocity of a particle will be maximum, we rewrite Eq. (8.47) as

$$v = 4\pi af \cos\left(\frac{2\pi}{\lambda}x\right)\cos(2\pi ft)$$

$$= \frac{4\pi a}{T}\cos\left(\frac{2\pi}{\lambda}x\right)\cos\left(\frac{2\pi}{T}t\right). \tag{8.48}$$

You can easily convince yourself that particle velocity will be zero for $t = T/4$, $3T/4$, ..., $(2n + 1)T/4$; $n = 0, 1, 2, 3,...$, and maximum for $t = 0$, $T/2\ T$, ..., $nT/2$; $n = 0, 1, 2, 3, ...$. Thus, in one cycle, the *velocity of particles of the medium becomes maximum when they pass through the mean position, and zero when they are at the extreme positions*. It means that every particle attains zero and maximum velocities twice in one cycle.

In the introduction we pointed out that musicians handling string instruments in an orchestra are seen adjusting the length of string or tension in the string to generate frequencies which enable them to be in unison with each other. This is facilitated by the formation of stationary waves in the string. Let us learn about these now.

8.6.2 Harmonics in Stationary Waves

You now know that all stringed musical instruments produce sound due to generation of stationary waves. Since ends of a stretched string are fixed, these necessarily act as nodes. The three modes of vibration which satisfy this condition are shown in Fig. 8.8. The simplest way in which a string fixed at both ends can vibrate is as just one segment. The string is then said to vibrate in *fundamental mode*. The corresponding lowest frequency is given by $f = v/\lambda$.

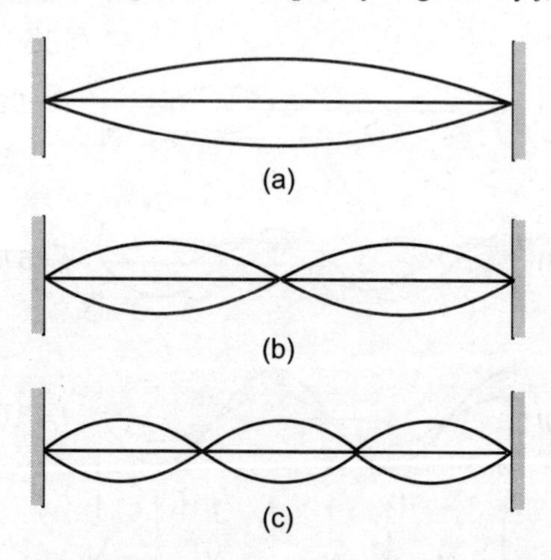

Fig. 8.8 Different modes of vibration of a string.

From Chapter 7, you may recall that velocity of transverse waves on a string is given by the relation $v = \sqrt{T/\mu}$, where T is tension in the string and μ is its mass per unit length. Hence the fundamental frequency of vibration of a string fixed at both ends is given by

$$f = \frac{v}{\lambda} = \frac{1}{\lambda}\sqrt{\frac{T}{\mu}}. \tag{8.49}$$

When the string vibrates in the fundamental mode, the length l of the string is equal to one-half of the wavelength, i.e., $l = \lambda/2$. Hence, the fundamental mode frequency in terms of the physically measurably quantities is given by

$$f_1 = \frac{1}{2l}\sqrt{\frac{T}{\mu}}. \qquad (8.50)$$

Note that the fundamental frequency is the minimum frequency with which a string can vibrate. For this reason, the fundamental frequency is also called the *first harmonic*.

If the string vibrates in n number of segments, the corresponding frequency of vibration f_n is given by

$$f_n = \frac{n}{2l}\sqrt{\frac{T}{\mu}}, \quad n = 1, 2, 3, \ldots \qquad (8.51)$$

The frequencies f_n $(n \geq 2)$ are called *higher harmonics* or *overtones*. These are actually excited with the fundamental mode. Note that f_n is an integral multiple of f_1. Does this result hold for pipes as well? You will discover answer to this question in the section 8.6.3.

If you place a bucket/bottle under a running tap, you will realise that the sound produced by the water continuously changes. Do you know why? It is essentially because of change in the length of air column which produces different frequencies. Similar situation is obtained when you play a flute. These essentially arise due to formation of stationary waves. Let us learn more about these now.

8.6.3 Waves in Pipes

Organ pipes are generally classified into two categories:

- *Open-end* organ pipes, where both ends are open; and
- *Closed-end* organ pipes, where one end is closed.

When air is blown in a pipe, stationary waves are produced due to superposition of longitudinal waves.

A closed-end organ pipe has an *antinode at the open end and a node at the closed end*, whereas *an open end organ pipe has antinodes at both ends*. The formation of waves in such pipes is shown in Fig. 8.9. Note that in a closed-end organ pipe, even number harmonics are absent. To understand this, we note that in the fundamental mode of vibration, the length of the column in such a pipe will be one-fourth of the wavelength, i.e., $l = \lambda_1/4$ and frequency of vibration will be $f_1 = v/4l$. In the second harmonic (first over tone), there will be two nodes and two antinodes so that $l = 3\lambda/4$ and frequency of vibration $f_2 = 3v/4l$. Similarly, in the third harmonic $l = 5\lambda/4$ and $f_3 = 5v/4l$. Hence $f_1 : f_2 : f_3 :: 1: 3: 5$. That is, only odd harmonics are excited in a closed-end organ pipe and even number harmonics are absent.

You may now like to do Practice Exercise 8.6.

***Practice Exercise* 8.6** A piano string has a length of one metre. It is fixed at both ends. Its mass per unit length is 0.015 kgm^{-1}. It is used to excite fundamental note of frequency 220 Hz. Calculate the tension in the string. Compare its frequency with that of the fundamental mode in a closed-end organ pipe of length 1.0 m. Take velocity of sound $v = 350$ ms^{-1}.

[***Ans.*** $T = 2904$ N; 40% (apx.)]

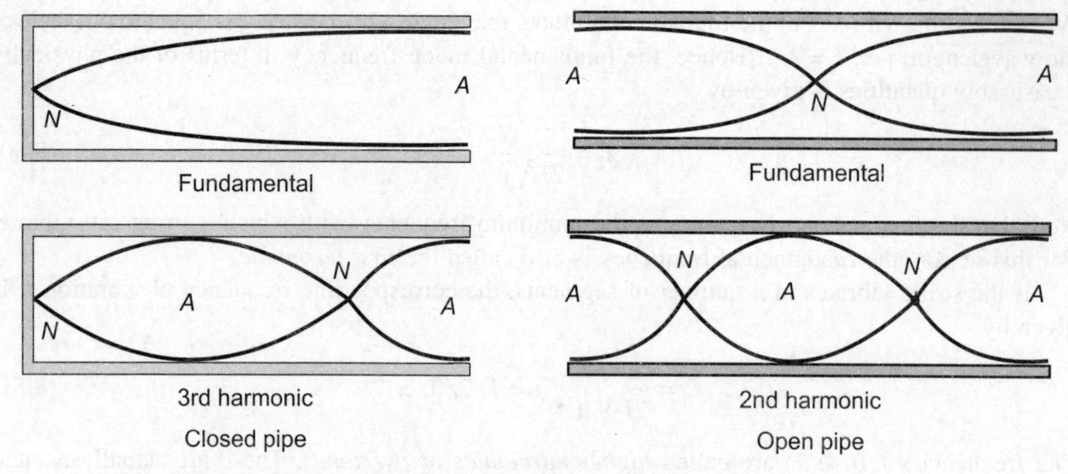

Fig. 8.9 Modes of vibration in (a) closed end; and (b) open end organ pipes.

From experience we know that quality of sound produced by an instrument depends on the source of vibration, the shape and size of cavity, size and position of finger holes, presence and richness of overtones and musical expertise of the player. A normal organ pipe always has an antinode at the mouth and a node or anti-node depending on whether it is closed or open at the top. A flute is essentially an organ pipe, but in reed instruments, the end with the reed acts as a closed end. The air in these instruments is stored under pressure in the wind chest and the various combinations of notes are played with fingers.

Having understood the genesis of generation of sound of different frequencies by string based and organ pipe based instruments, it is instructive to learn the characteristics of sound.

8.6.4 Musical Sound and Noise

Sound is classified as *musical* and *noise*. A musical sound leaves a pleasing effect on our ears, brain and mind and is generated by periodic vibrations with a sinusoidal waveform. Different musical instruments produce sounds corresponding to different frequencies and waveforms. In Table 8.1, we have listed the frequencies of a musical scale used in western music (C, D, ...) and Indian music (Sa, Re, ...).

Table 8.1 Musical notes and corresponding frequencies

Note	*Frequency (in Hz)*
C : Sa	256
D : Re	288
E : Ga	320
F : Ma	340
G : Pa	384
A : Dha	427
B : Ni	480
C′ : Sa′	512

A noise produces an unpleasant effect in the ears. It results from non-periodic irregular waveforms (Fig. 8.10a). Honking by the drivers, bursting of crackers or machines in a factory generate a lot of noise. In recent years, noise pollution in metropolitan cities is fast emerging as the source of hypertension and blood pressure due to unsound sleep. You should, therefore, avoid unnecessary honking, bursting crackers, and playing loud music to minimise noise population.

Musical sound is produced by superposition of vibrations of the fundamental mode with those of higher harmonics. The amplitude of particles corresponding to the fundament mode is large, whereas the amplitude of vibrations representing higher harmonics is small. The waveform of resultant oscillation thus depends on the number and relative amplitudes of higher harmonic vibrations (Fig. 8.10b).

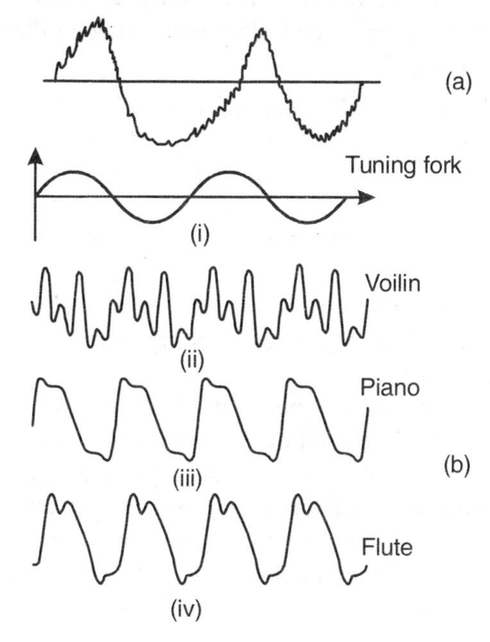

Fig. 8.10 (a) An irregular waveform leads to noise; and (b) a regular waveform produces music.

Sound is generally characterised by (i) loudness, (ii) pitch and (iii) timbre. From Section 7.6, you may recall that intensity (or loudness) of sound depends on the amplitude of wave and the distance between the listener and the source. Loudness increases with the square of the amplitude and decreases as the inverse of the square of the distance between the source and the observer.

The pitch of sound is determined by the frequency of the vibrating body. Sounds of same loudness can be distinguished by their pitch.

The timbre of sound depends on the shape of the waveform, which, in turn, depends on the richness of harmonics.

Having considered superposition of two identical waves travelling in opposite directions, let us now learn about production of beats and concepts of wave groups and group velocity due to superposition of waves of slightly different angular frequencies travelling in the same direction. (Section 8.5.4).

8.7 BEATS

Consider two sound waves of slightly different angular frequencies ω_1 and ω_2, but equal amplitude a, propagating in the same direction. Mathematically, we can write

$$y_1(x, t) = a \sin(\omega_1 t - k_1 x)$$

and

$$y_2(x, t) = a \sin(\omega_2 t - k_2 x). \tag{8.52}$$

For simplicity, we fix the spatial co-ordinate x in these equations, say, at $x = 0$. It represents the situation where an observer is standing at one place and watching the resultant wave passing by. The resultant waveform is given by

$$y(x, t) = y(0, t) = a \sin \omega_1 t + a \sin \omega_2 t$$

$$= 2a \cos \frac{\Delta\omega}{2} t \sin \omega_{av} t, \tag{8.53}$$

where $\omega_{av} = (\omega_1 + \omega_2)/2$ is average angular frequency and $\Delta\omega = \omega_1 - \omega_2$ signifies the frequency difference.

This result suggests that amplitude of resultant wave at a given point is not constant; it varies with angular frequency $(\Delta\omega/2)$ and its magnitude varies between $2a$ and zero. The sin $(\omega_{av} t)$ term acts as envelope of the resultant waveform.

If ω_1 and ω_2 are nearly equal, $\Delta\omega$ will be small. Then, amplitude of the resultant wave will vary rather slowly. The periodic rise and fall of this wave leads to generation of beats in the form of waxing (loudness) and waning (shrillness) of sound at regular intervals of time.

Beats are heard at the maxima of amplitude (Fig. 8.11). Mathematically, these correspond to the condition cos $(\Delta\omega/2)t = \pm 1$. This is because the intensity of sound is directly proportional to the square of the amplitude. Since the maximum amplitude occurs twice in one time period associated with the angular frequency $(\Delta\omega/2)$, the frequency of beats equals the difference of the two component frequencies.

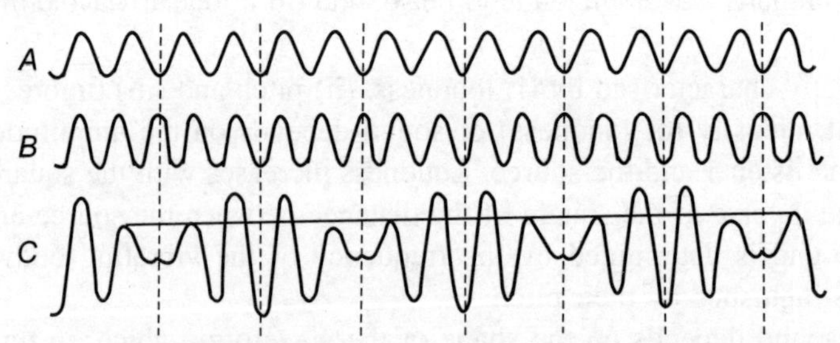

Fig. 8.11 Formation of beats due to superposition of two waves of nearly the same frequency.

In terms of frequencies f_1 and f_2, *beat frequency* $\Delta f = f_1 - f_2 = (\Delta\omega/2\pi)$. The time elapsed between any two consecutive beats, called the beat period $= 1/\Delta f$.

You may now like to answer Practice Exercise 8.7.

***Practice Exercise* 8.7** A note of a piano is sounded with a tuning fork of frequency 512 Hz. If six beats are heard every second, calculate the frequency of the note.

[***Ans.*** 506 Hz, 518 Hz]

8.8 WAVE GROUPS AND GROUP VELOCITY

When two waves of same amplitude but slightly different frequencies ω_1 and ω_2 travelling in the same direction superpose, the displacement of the resultant wave is given by

$$y(x, t) = a \sin(\omega_1 t - k_1 x) + a \sin(\omega_2 t - k_2 x)$$

$$= 2a \sin\left[\frac{(\omega_1 + \omega_2)t - (k_1 + k_2)x}{2}\right] \cos\left[\frac{(\omega_1 - \omega_2)t - (k_1 - k_2)x}{2}\right]. \tag{8.54}$$

Note that we have taken the amplitudes of the two waves to be equal for mathematical convenience. If the angular frequencies and wave vectors characterising original waves are only slightly different, we can write

$$\omega_1 - \omega_2 = \Delta\omega,$$

and

$$k_1 - k_2 = \Delta k.$$

Further, if we define average angular frequency and average wave vector as

$$\omega_{av} = \frac{\omega_1 + \omega_2}{2},$$

and

$$k_{av} = \frac{k_1 + k_2}{2},$$

Eq. (8.54) takes the form

$$y(x, t) = 2a \cos\left(\frac{\Delta\omega}{2}t - \frac{\Delta k}{2}x\right) \sin(\omega_{av}t - k_{av}x)$$

$$= a_m \sin(\omega_{av}t - k_{av}x), \tag{8.55}$$

where

$$a_m = 2a \cos\left(\frac{\Delta\omega}{2}t - \frac{\Delta k}{2}x\right). \tag{8.56}$$

Let us pause for a while and first see how the new waveform represented by Eq. (8.55) looks like. Note that it represents a non-harmonic wave as its amplitude a_m varies both in space (x) and time (t). The modulated amplitude of the resultant wave gives rise to a wave envelope, as shown in Fig. 8.12. Furthermore, the resultant wave is made up of two parts: (i) the sinusoidal part, which varies rapidly with a frequency that is mean of the frequencies of the component waves; and (ii) the cosine part which varies (very slowly) with a frequency that is half of the difference of the two frequencies. The propagation vector of the slowly varying part of the resultant wave is $\Delta k/2$.

As you can see from Fig. 8.12, the superposition of waves of slightly different frequencies results in the formation of groups, called the *wave groups* or the *wave packets*. A wave group can travel with a velocity which may be different from that of the individual waves or of the resultant wave. The velocity of the wave group is termed as *group velocity*. We denote it by the symbol v_g. From Eq. (8.55), it readily follows that

$$v_g = \frac{\Delta\omega/2}{\Delta k/2} = \frac{\Delta\omega}{\Delta k}. \tag{8.57}$$

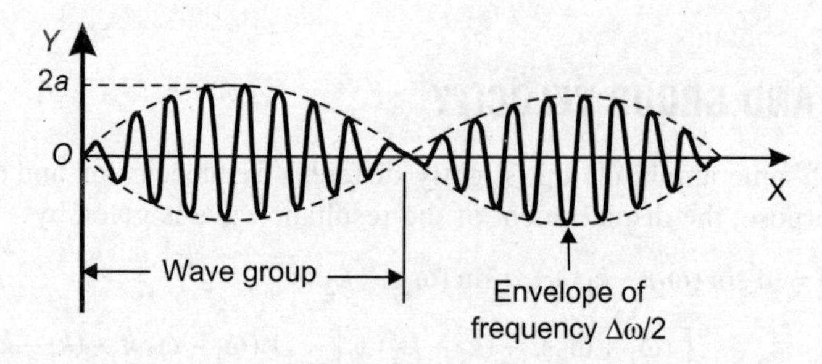

Fig. 8.12 Wave envelope of the resultant arising due to superposition of two waves of slightly different frequencies.

If a group consists of a number of component waves with angular frequencies lying between ω_1 and ω_2 (with $\omega_1 \approx \omega_2$), and wave vectors between $\mathbf{k_1}$ and $\mathbf{k_2}$ (with $|\mathbf{k_1}| \approx |\mathbf{k_2}|$), the group velocity v_g can be written as

$$v_g = \frac{\Delta\omega}{\Delta k} = \frac{d\omega}{dk}. \tag{8.58}$$

Here $d\omega$ and dk represent the spread of or difference between the maximum and the minimum angular frequencies and wave vectors of the component waves that constitute the wave group.

The velocity of the resultant wave is called the *phase velocity* v_p. From Eq. (8.55), it follows that

$$v_p = \frac{\omega_{\text{av}}}{k_{\text{av}}}. \tag{8.59}$$

If wave velocities of individual waves are equal, i.e.

$$\frac{\omega_1}{k_1} = \frac{\omega_2}{k_2} = v \text{ (say)},$$

then

$$v_p = \frac{\omega_1 + \omega_2}{k_1 + k_2} = \frac{k_1 v + k_2 v}{k_1 + k_2} = v.$$

And

$$\begin{aligned} v_g = \frac{\Delta\omega}{\Delta k} &= \frac{\omega_1 - \omega_2}{k_1 - k_2} \\ &= \frac{k_1 v - k_2 v}{k_1 - k_2} \\ &= v = v_p. \end{aligned} \tag{8.60}$$

That is, if the individual wave velocities are equal, the group velocity is equal to the phase velocity.

We have said earlier that the resultant wave travels with phase velocity. But energy transfer is governed by group velocity. So we can say that group velocity is a more fundamental quantity in physics.

The *phase velocity* and the *group velocity* are connected through the relation

$$v_g = \frac{d\omega}{dk} = \frac{d}{dk}(kv).$$

On simplification, we obtain

$$v_g = v + k\frac{dv}{dk}. \tag{8.61}$$

Using the relation $k = 2\pi/\lambda$, we can express v_g in terms of λ:

$$\frac{dv}{dk} = \frac{dv}{d\lambda}\cdot\frac{d\lambda}{dk} \tag{8.62}$$

Since $k = 2\pi/\lambda$, it readily follows that

$$\frac{dk}{d\lambda} = -\frac{2\pi}{\lambda^2}$$

Inserting this result in Eq. (8.62), you will obtain

$$\frac{dv}{dk} = -\frac{\lambda^2}{2\pi}\frac{dv}{d\lambda}$$

Hence,

$$v_g = v + \frac{2\pi}{\lambda}\left(-\frac{\lambda^2}{2\pi}\right)\frac{dv}{d\lambda}.$$

$$= v - \lambda\frac{dv}{d\lambda}. \tag{8.63}$$

The wavelength of the enveloping wave is given by

$$\lambda_c = \frac{2\pi}{\Delta k/2} = \frac{4\pi}{\Delta k}.$$

Since Δk is very small compared to k, $\lambda_c >> \lambda$.

If λ_1 and λ_2 denote the wavelengths of the component waves, you can easily verify that

$$\frac{\lambda_c}{2} = \frac{\lambda_1\lambda_2}{\lambda_2 - \lambda_1}. \tag{8.64}$$

This gives the length (or the extent) of the wave group. We can see from Fig. 8.13 that the length of the wave group is half of the wavelength of the enveloping wave, i.e., it is equal to $\lambda_c/2$.

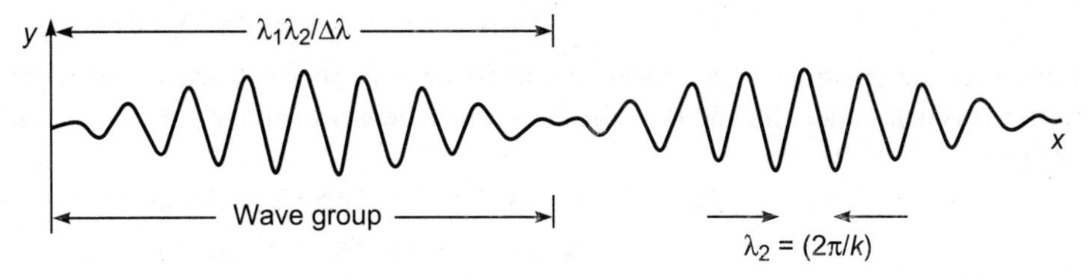

Fig. 8.13 Wave group and its extent.

To further illustrate the difference between phase and group velocities, we consider *gravity waves*. These waves are strongly dispersive and their phase velocity is known to be proportional to the square root of wavelength:

$$v_p = C\lambda^{1/2}. \tag{8.65}$$

In terms of wave vector

$$v_p = \frac{A}{\sqrt{k}},$$

where $A = C\sqrt{2\pi}$. Since

$$v_p = \frac{\omega}{k},$$

we note that angular frequency of gravity waves can be expressed as

$$\omega = kv_p = Ak^{1/2}.$$

On differentiating ω with respect to k, we get

$$v_g = \frac{d\omega}{dk} = \frac{1}{2}\frac{A}{\sqrt{k}} = \frac{1}{2}v_p. \tag{8.66}$$

That is, *group velocity for gravity waves is just half of their phase velocity.* In other words, for these waves, the component wave crests move faster through the group as a whole.
You may now like to answer Practice Exercise 8.8.

***Practice Exercise* 8.8** Establish the relation between group velocity and phase velocity for a dispersive medium.

REVIEW EXERCISES

8.1 The equation of a stationary wave is given by $y = 2 \sin \pi x \cos 100\pi t$, where x and y are measured in metres and t is in seconds. Calculate the amplitude, wavelength, frequency of component waves whose superposition generated stationary waves. Also determine the phase difference between two points 1 m apart.

[***Ans.*** $a = 2.5$ m; $\lambda = 2$ m; $f = 50$ Hz; phase difference = π]

8.2 Two identical plane progressive waves characterised by $a = 2.5$ cm, $f = 20$ Hz and $\lambda = 6$ cm are moving in opposite directions. Write down the equation of the stationary waves generated by their superposition. What is the velocity of a particle of the medium at a distance 1.5 cm from the source when $t = 1.125$ s.

[***Ans.*** $y(x, t) = 5 \sin (\pi x/3) \cos 40\pi t$; v = zero]

8.3 Determine the position of the nodes and antinodes of stationary waves for (i) reflection from a less dense medium and (ii) reflection from a denser medium. The wavelength of the original waves is 10 cm.

[***Ans.*** (i) Nodes: 2.5, 7.5, 12.5,... Antinodes: 0, 5, 10, 15,...

(ii) Nodes: 0, 5, 10, 15,... Antinodes: 2.5, 7.5, 12.5,...]

8.4 Two identical transverse harmonic waves travel in opposite directions along a string. These are characterised by $a = 2$ cm, $\lambda = 4$ cm, $v = 0.33$ cms^{-1}. Starting from $t = 0$, depict the shape of the string at $t = 1.5$ s, 3.0 s, 4.5 s, and 6.0 s.

8.5 A plane progressive wave is represented by the equation $y(x, t) = 10 \cos (\pi/2)(0.004x - 6.0t)$. x and y are measured in centimetres and t is measured in seconds. Write down the equation of the wave which on interfering with this wave produces a stationary wave. Also determine the positions of nodes and antinodes of the stationary wave.

[**Ans.** $y_1(x, t) = 10 \cos (\pi/2)(0.004x + 6.0t)$; Nodes: 2.5 cm, 7.5 cm, 12.5 cm; Antinodes: 5 cm, 10 cm, 15 cm]

8.6 (a) In a wave guide, the wavelength is related to frequency through the relation

$$\lambda = \frac{c}{\sqrt{f^2 - f_0^2}}.$$

Calculate the group velocity of the waves.

(b) For surface waves in shallow water, the frequency and wavelength are connected through the relation

$$f = \sqrt{\frac{2\pi S}{\rho \lambda^3}},$$

where S and ρ respectively denote surface tension and density of water. Calculate the group velocity of waves. How is it related to phase velocity?

$$\left[\textbf{Ans.} \quad c\sqrt{1 - \left(\frac{f_0}{f}\right)^2}; \; v_g = \frac{3}{2} v_p\right]$$

8.7 For a stationary wave, show that whole medium attains maximum displacement and zero displacement simultaneously twice in each cycle. Describe energy distribution at these times.

8.8 The instantaneous displacement for a plane progressive wave is given by

$$y_1 = a \sin (bx + ct),$$

where a, b, c are positive constants. The wave is reflected at $x = 0$ and the intensity of the reflected wave is 0.36 times that of the incident wave.

(a) Calculate the wavelength and period of the incident wave.
(b) Obtain the expression for instantaneous displacement of the reflected wave.
(c) For the resultant wave obtain the values of minimum and maximum particle speeds.
(d) Express the resultant wave as a superposition of stationary wave and a progressive wave.
(e) What are the positions of nodes and antinodes?
(f) What is the direction of propagation of the travelling wave?

[**Ans.** (a) $\lambda = (2\pi/b), f = (c/2\pi)$; (b) $y_2 = -0.6a \sin (bx - ct)$; (c) $v_{max} = 1.6ac$, $v_{min} = 0$; (d) $y_s = 1.2a \sin bx \cos ct$, $y_t = 0.4a \sin (bx + ct)$; (e) $x = (n\pi/b)$, $x = (2n + 1)\pi/2b$; (f) Along negative x-axis]

9

Vibrations of Strings

Expected Learning Outcomes

In this Chapter, you will acquire capability to:

- write wave equation for transverse vibrations of a stretched string and solve it using the method of separation of variables;
- calculate instantaneous displacement of a stretched string for given initial and boundary conditions;
- obtain eigenvalues and corresponding eigenfunctions for a stretched string executing transverse oscillations;
- derive an expression for the energy of a vibrating string executing transverse oscillations;
- express displacement of a plucked string in terms of different harmonics;
- derive expressions for displacement of struck and a bowed strings; and
- highlight salient features of plucked, struck and bowed stringed instruments.

9.1 INTRODUCTION

In Chapter 6, you learnt that longitudinal as well as transverse oscillations of coupled masses lead to energy exchange. Moreover, when the number of oscillators is infinitely large, the energy exchange results in wave motion. You may now ask: Where and when can the number of oscillators be so large? The answer to this question is: We can regard a string, and for that matter any medium, as made up of a large number of identical masses. The study of vibrations of strings is important to understand how music is produced by instruments such as mandoline, harp, piano, sitar, guitar, violin, ektara, esraj, tanpura, sarangi, sarod, veena, flute and banjo.

Suppose that a guitar or a sitar string is displaced from its equilibrium position by plucking, piano by striking and sarangi or violin by bowing at a particular point. You will observe that waves travel along the string back and forth due to reflection at fixed points. In the preceding chapter, you have learnt that the superposition of incident and reflected waves travelling along a string in opposite directions gives rise to stationary waves. And the string may vibrate in one or more segments giving out a musical note of definite pitch and quality.

You will appreciate that a string constitutes a remarkable class by itself in that it is both a vibrator and medium for wave motion. According to Lord Rayleigh, "a string is a perfectly uniform and perfectly flexible filament of solid matter stretched between two fixed ends ...". In actual practice, however, a string is neither perfectly uniform nor perfectly flexible. (The non-uniform density along the length of the string implies non-uniform tension in the string.) Moreover, a string always possesses some rigidity whose effect diminishes with increase in length-to-diameter ratio. The transverse vibrations of an ideal string depend on tension rather than rigidity.

In Section 9.2, you will learn about transverse waves on stretched strings and obtain expression for displacement of medium particles by solving the wave equation, subject to boundary and initial conditions. You will be introduced to the concept of eigenvalues and eigenfunctions and learn to derive expressions for these and relate them to the corresponding characteristic frequencies. This knowledge will be used to discuss how wave in a string can be modelled as a stationary wave. We also discuss the experiment which can be used to visualise such an arrangement. In Section 9.3, we shall consider a plucked string and obtain expression for its displacement. This is followed by discussion of struck and bowed strings in Sections 9.4 and 9.5, respectively. The basic differences in the waves excited by different methods are discussed in detail. In subsequent chapters, we discuss stationary waves in solids and gases.

9.2 TRANSVERSE WAVES ON A STRETCHED STRING

In Chapter 7, you learnt that 1-D transverse motion of a stretched string fixed at both ends is described by the wave equation:

$$\frac{\partial^2 y}{\partial t^2} = v^2 \frac{\partial^2 y}{\partial x^2}, \tag{9.1}$$

where $v = \sqrt{T/m}$ is velocity of waves.

Note that Eq. (9.1) has no term containing mixed partial derivatives such as $\partial^2 y/\partial x \partial t$ or $\partial^2 y/\partial t \partial x$. This is because Eq. (9.1) has been derived under the assumption that the transverse displacement of the string is very small, i.e., the string is displaced only slightly from its equilibrium position. From Chapter 7, you may recall that we assumed a sinusoidal solution of the wave equation. We now discuss mathematical basis of this assumption.

In Chapter 1, you have seen that a function of the form $vt - x$ or $vt + x$ is a solution of Eq. (9.1). Therefore, its general solution can be written as superposition of individual solutions:

$$y(x, t) = f_1(vt - x) + f_2(vt + x), \tag{9.2}$$

where f_1 and f_2 are arbitrary functions. The function f_1 represents a plane transverse wave progressing along the positive x-direction with wave velocity v, and the function f_2 represents a plane transverse wave progressing along the negative x-direction with the same velocity. Therefore, the general solution represents two plane transverse waves moving in the opposite directions.
In general, waves may not have the same waveform since f_1 and f_2 can be different functions. In the simplest form, f_1 and f_2 are taken as sinusoidal and represent plane progressive harmonic waves. Mathematically, we can write

> There is no loss of generality in expressing the argeements of the cosine and sine functions in terms of $(x \pm vt)$ instead of $(vt \pm x)$.

$$y(x, t) = y_m \cos k(x \pm vt).$$

or

$$y(x, t) = y_m \sin k(x \pm vt).$$

Recall that cosine and sine functions are respectively the real and imaginary parts of an exponential function. Therefore, for mathematical convenience, you can work with complex functions and express $y(x, t)$ as

$$\begin{aligned} y(x, t) &= y_m \exp jk(x \pm vt) = y_m \exp j(kx \pm \omega t) \\ &= y_m \exp\{\pm j(\omega t)\} \exp(jkx). \end{aligned} \tag{9.3}$$

Note that $y(x, t)$ is a function of two variables, x and t. Therefore, we can solve Eq. (9.1) using the method of separation of variables. This is the simplest and most widely used method for solving partial differential equations. You would recall that $(kx \pm \omega t)$ represents the phase of the wave, which is a quantity varying with space and time. On separation the phase becomes only time-variant and the amplitude becomes a function of x, which is a feature of a stationary wave. This happens under certain specific condition which you will discover through the next section.

9.2.1 Method of Separation of Variables

In Chapter 1, you have seen that in the method of separation of variables, we express the solution of an equation as a product of as many independent functions as is the number of independent variables. This method helps us to reduce a linear partial differential equation in two or more ordinary differential equations, which are cast in easily solvable form. In the instant case, there are only two independent variables (x and t). So we express $y(x, t)$ in the form of a product as

$$y(x, t) = X(x)\ T(t), \tag{9.4}$$

where X depends on position co-ordinate x only and T is a function of time t only. Physically it means that dependence of the unknown function on one variable is in no way affected by its dependence on the other variable. Does this mean that there is no connection at all between X and T? No, it only means that the function X does not depend on t in any way whatsoever and there is no dependence of the function T on x.

Now, we calculate second order partial derivatives of $y(x, t)$ by differentiating Eq. (9.4) twice with respect to space co-ordinate x. This gives

$$\frac{\partial y}{\partial x} = \left(\frac{dX}{dx}\right) T$$

and

$$\frac{\partial^2 y}{\partial x^2} = \left(\frac{d^2 X}{dx^2}\right) T. \tag{9.5a}$$

Note that the derivative on the right hand side is total derivative of X. This is because X is a function of only one variable.

Similarly, on differentiating Eq. (9.4) twice with respect to time, you will get

$$\frac{\partial^2 y}{\partial t^2} = \left(\frac{d^2 T}{dt^2}\right) X. \tag{9.5b}$$

On substituting Eqs. (9.5a), and (9.5b) in Eq. (9.1), you will get

$$v^2 \frac{d^2X}{dx^2} T = \frac{d^2T}{dt^2} X$$

Dividing both sides by XT, we get

$$v^2 \frac{1}{X} \frac{d^2X}{dx^2} = \frac{1}{T} \frac{d^2T}{dt^2}. \tag{9.6}$$

Note that in this equation, the left-hand side involves functions which depend only on x, whereas the right-hand side depends only on t. Thus, if we vary t and keep x fixed, the LHS will not change. This means that LHS and RHS are independent. Therefore, two sides must be equal to a constant. This constant is called the *separation constant*. Note that the value of separation constant must be negative as otherwise y will become non-periodic and change (increase or decrease) indefinitely with x or t. Let us take this constant as $-\omega^2$. Then we can rewrite Eq. (9.6) as a set of two linear ordinary differential equations.

> Writing the constant equal to $-\omega^2$ is not a matter of choice. It will become further evident through Eq. (9.8) that the constant would indeed be equal to the –ve of the square of the angular frequency. A careful reader of the text should not have any problem about this judgement.

$$v^2 \frac{1}{X} \frac{d^2X}{dx^2} = -\omega^2$$

or

$$\frac{d^2X}{dx^2} + \frac{\omega^2}{v^2} X = 0 \tag{9.7}$$

and

$$\frac{1}{T} \cdot \frac{d^2T}{dt^2} = -\omega^2$$

or

$$\frac{d^2T}{dt^2} + \omega^2 T = 0. \tag{9.8}$$

Recall that Eq. (9.7) is a standard differential equation for simple harmonic motion. So we can say that X varies simple harmonically with x. And the general solution of this equation can be written as

$$X(x) = A_1 \cos\left(\frac{\omega}{v}\right)x + B_1 \sin\left(\frac{\omega}{v}\right)x$$

or

$$= A_1 \cos kx + B_1 \sin kx, \tag{9.9}$$

where we have put $k = \omega/v$. A_1 and B_1 are arbitrary constants, which can be determined by specifying boundary conditions.

Similarly, the general solution of Eq. (9.8) can be written as

$$T(t) = C_1 \cos \omega t + D_1 \sin \omega t. \tag{9.10}$$

The values of the constants C_1 and D_1 are determined using the initial conditions. Thus, the general solution of the wave equation for the stretched string may be written as

$$y(x, t) = X(x)\,T(t) = (A_1 \cos kx + B_1 \sin kx)(C_1 \cos \omega t + D_1 \sin \omega t). \tag{9.11}$$

Let us pause for a while and ask: What does Eq. (9.11) imply? Note that in Eq. (9.11), there is no restriction on the value of ω. It implies that an infinite number of solutions are possible and progressive wave is one of these. But its form is not immediately apparent from this solution. To determine this, we specify initial and boundary conditions.

9.2.2 Displacement of a Stretched String

Suppose that a stretched string, fixed rigidly at both ends, is of length l. The boundary conditions to be satisfied in this case are:

(i) $y(0, t) = 0$ for all values of t, and

(ii) $y(l, t) = 0$ for all values of t.

Now, using the boundary condition (i) in Eq. (9.11), we get

$$0 = A_1(C_1 \cos \omega t + D_1 \sin \omega t). \tag{9.12}$$

Since the term within the parentheses cannot be zero for all values of t, this relation shall be satisfied if and only if $A_1 = 0$.

Similarly, using the boundary condition (ii), we get

$$0 = B_1 \sin kl\,(C_1 \cos \omega t + D_1 \sin \omega t), \tag{9.13}$$

where $k = \omega/v$. On substituting this value of k in Eq. (9.13), we get

$$0 = B_1 \sin \frac{\omega l}{v}(C_1 \cos \omega t + D_1 \sin \omega t). \tag{9.14}$$

Note that $B_1 = 0$ will lead to a trivial solution. Therefore, for this relation to hold for all values of t, $\sin \omega l/v$ must be equal to zero. This is possible when

$$\frac{\omega l}{v} = n\pi, \quad n = 1, 2, 3, \ldots \tag{9.15}$$

From this relation we note that an infinte set of angular frequencies is permissible: $\omega_1 = \pi v/l$, $\omega_2 = 2\pi v/l$, $\omega_3 = 3\pi v/l$, ... Note that we have not considered the term corresponding to $n = 0$ as it leads to a trivial solution. Moreover, the solutions corresponding to any negative integer ($n = -m$) value will be linearly related to the solution corresponding to the positive integer ($n = m$). That is why negative integers have also not been included.

By using these results in Eq. (9.11), we find that solutions of 1-D wave equation for a string fixed at both ends at all times can be written as

$$y_n = B_{1n} \sin \frac{\omega_n x}{v}(C_{1n} \cos \omega_n t + D_{1n} \sin \omega_n t). \tag{9.16}$$

The subscript n denotes the nth mode of vibration having a characteristic angular frequency

$$\omega_n = \frac{n\pi v}{l}. \tag{9.17}$$

In general, the values of coefficients and displacements for different modes are different. For brevity, we write $a_n = B_{1n}C_{1n}$ and $b_n = B_{1n}D_{1n}$. Then Eq. (9.16) takes the form

$$\begin{aligned} y_n(x, t) &= \sin \frac{\omega_n x}{v}(a_n \cos \omega_n t + b_n \sin \omega_n t) \\ &= \sin \frac{n\pi x}{l}\left(a_n \cos \frac{n\pi v t}{l} + b_n \sin \frac{n\pi v t}{l}\right). \end{aligned} \tag{9.18}$$

Since there is a solution for each integral value of n, and the wave equation is linear and homogeneous, the most general solution of Eq. (9.1) is obtained by superposition of all the solutions:

$$y(x,t) = \sum_{n=1}^{\infty} \sin\frac{n\pi x}{l}\left(a_n \cos\frac{n\pi vt}{l} + b_n \sin\frac{n\pi vt}{l}\right). \tag{9.19}$$

We now introduce a change of variable by defining $a_n = c_n \cos\phi_n$ and $b_n = c_n \sin\phi_n$. Then the term within the brackets in Eq. (9.19) can be expressed as $c_n \cos[(n\pi v/l)t - \phi_n]$, where $c_n^2 = a_n^2 + b_n^2$ and $\tan\phi_n = b_n/a_n$. Hence Eq. (9.19) may be rewritten as

$$y(x,t) = \sum_{n=1}^{\infty} c_n \sin\frac{n\pi x}{l} \cos\left(\frac{n\pi v}{l}t - \phi_n\right). \tag{9.20}$$

From Eq. (9.20), you will observe that in the nth mode, the string vibrates simple harmonically with the angular frequency ω_n equal to $n\pi v/l$ and amplitude $c_n \sin n\pi x/l$.

The frequencies with which a string stretched between two rigid supports vibrates are given by Eq. (9.17). Hence, we can express the frequency of the nth mode of vibration of the string in terms of tension and mass per unit length as

$$f_n = \frac{\omega_n}{2\pi} = \frac{nv}{2l} = \frac{n}{2l}\sqrt{\frac{T}{m}}. \tag{9.21}$$

The fundamental frequency, which corresponds to $n = 1$, is the lowest frequency. It is given by

$$f_1 = \frac{1}{2l}\sqrt{\frac{T}{m}}. \tag{9.22}$$

Since $\lambda_1 = v/f_1$, we find that for the fundamental mode, $\lambda_1 = 2l$. By comparing Eqs. (9.21) and (9.22) you can easily conclude that frequencies of higher harmonics are integral multiples of the fundamental frequency f_1.

The corresponding wavelengths are n-times shorter. This is illustrated in Fig. 9.1. Note that in the fundamental mode of vibration, the string between the fixed ends vibrates in a single loop. From Chapter 8 you may recall that the points on the string where the displacement is always zero are called nodes, whereas the points of maximum displacement are known as antinodes. These are denoted by N and A, respectively in Fig. 9.1.

Let us pause for a while and ask: What have we achieved so far? You now know that when a string of length l and linear density m is made to vibrate under tension T, in the fundamental mode ($n = 1$), it forms one segment and the fundamental frequency is given by Eq. (9.22). In your school, you have learnt the laws of vibration of stretched strings. These readily follow from Eq. (9.22):

(i) *Law of length:* The frequency of vibration is inversely proportional to the vibrating length, provided tension and mass per unit length remain constant:

$$f_1 \propto \frac{1}{l}.$$

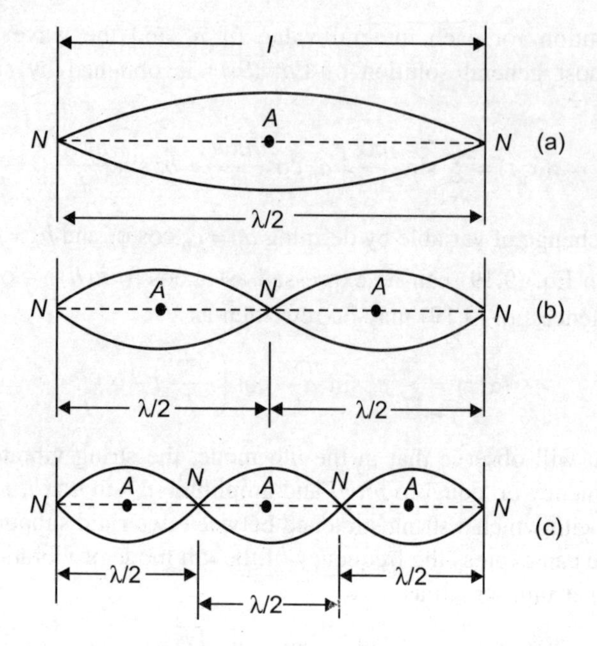

Fig. 9.1 Modes of vibration of a string fixed at both ends.

(ii) *Law of tension:* The frequency of vibration is directly proportional to the square root of tension when the length and mass per unit length of string remain constant:

$$f_1 \propto \sqrt{T}.$$

(iii) *Law of mass:* The frequency of vibration is inversely proportional to the square root of mass per unit length, provided the length of string and tension in the string do not change:

$$f_1 \propto \sqrt{\frac{1}{m}}.$$

You should now study Example 9.1.

EXAMPLE 9.1

A veena string is 2 m long and has mass per unit length 0.012 kgm^{-1}. If tension in it is 500 N, calculate the frequency of fundamental mode.

Solution: From Eq. (9.22), we recall that frequency of the fundamental mode is given by

$$f = \frac{1}{2l}\sqrt{\frac{T}{m}}.$$

On substituting the given data, we get

$$f = \frac{1}{2 \times 2}\sqrt{\frac{500}{0.012}} = \frac{1}{4}\sqrt{\frac{50}{12} \times 10^4} = 51 \text{ Hz}.$$

You may now like to answer Practice Exercise 9.1.

***Practice Exercise* 9.1** For a string of circular cross-section, linear density is defined as the product of area of cross-section and density. Discuss the dependence of fundamental frequency on diameter and density.

$$\left[\textbf{Ans.} \quad f \propto \frac{1}{d}; f \propto \frac{1}{\sqrt{\rho}}\right]$$

9.2.3 Eigenvalues and Eigenfunctions

When an operator $\hat{O}$ acts on a function $f(x)$ and gives $O_s f(x)$, where O_s is a constant, the function $f(x)$ is said to be the eigenfunction of $\hat{O}$ with eigenvalue O_s. For example,

$$\frac{d}{dx}(e^{mx}) = me^{mx}.$$

So e^{mx} is an eigenfunction of d/dx with eigenvalue m.

Recall that for the space part of the wave equation for a stretched string [Eq. (9.7)], $A_1 \cos kx + B_1 \sin kx$ is an acceptable solution for any value of k. But on applying boundary conditions, we find that the permissible values of k are restricted. These values of k are known as *eigenvalues*.
Since $k = \omega/v$ using Eq. (9.15), we can write

$$k_n = \frac{n\pi}{l}, \tag{9.23}$$

where $n = 1, 2, 3, \ldots$
From Eq. (9.18), you will note that space part of displacement of a stretched string fixed at both ends is given by the function

$$P_n(x) = \sin\frac{n\pi x}{l}.$$

For integral values of n, it is called *eigenfunction*.

By closely examining Eq. (9.7), we can say that when operator d^2/dx^2 acts on the eigenfunction $\sin n\pi x/l$, it gives

$$\frac{d^2}{dx^2}\left(\sin\frac{n\pi x}{l}\right) = -\left(\frac{n\pi}{l}\right)^2 \sin\frac{n\pi x}{l}.$$

That is, the eigenvalue of the operator d^2/dx^2 is $-n^2\pi^2/l^2$. The corresponding frequencies given by $\omega_n = n\pi v/l$, are known as *eigenfrequencies*. Note that eigenfunctions are undetermined with respect to multiplier B_n. The values of B_n will be different for different n.

Let us now check whether or not the eigenfunctions of a stretched string are orthogonal. You may recall that two vectors **A** and **B** are said to be orthogonal if their scalar product vanishes. Does a similar condition hold for the above eigenfunctions as well? If you think so, your thinking is logical. The eigenfunctions are said to be orthogonal when the definite integral of the product of any two eigenfunctions vanishes over the admissible range of the independent variable. We can write from the knowledge of Eq. (1.60i)

$$\int_0^l \sin\frac{n\pi x}{l} \sin\frac{m\pi x}{l}\, dx = \frac{l}{2}\delta_{mn}, \tag{9.24}$$

where δ represents the Kronecker delta function:

$$\delta_{mn} = 1 \quad \text{for } m = n,$$
$$= 0 \quad \text{for } m \neq n.$$

Equation (9.24) gives the orthogonality condition of the eigenfunctions.

The other important property which eigenfunctions of a stretched string must satisfy is the property of *completeness*. A set of eigenfunctions is said to be complete if an arbitrary function $f(x)$, which satisfies the same boundary conditions as the given set, can be expressed in terms of the given eigenfunctions as

$$f(x) = \sum_{n=1}^{\infty} a_n P_n(x). \tag{9.25}$$

Here $P_n(x)$ is the nth eigenfunction and a_n is a constant coefficient.

9.2.4 Mathematical Modelling of a String as a Stationary Wave

From Chapter 8, you will recall that genesis of stationary waves is in superposition of two identical waves moving in opposite directions. In the preceding section, you have learnt that when a stretched string fixed at both ends is disturbed at an in-between point, the harmonics excited in it resemble stationary waves. These solutions (of the wave equation) were obtained using the method of separation of variables. In this section, we will write the general solution of wave equation and establish that both approaches lead to the identical result.

As before, consider a stretched string of length l fixed between two ends. When any portion of the string is displaced from its equilibrium state in a direction perpendicular to its length, a disturbance travels along the string, which can be described by Eq. (9.1). You have learnt through Eq. (9.2) that the general solution of a linear second order partial differential equation is of the form

$$y(x, t) = y_1 \sin (kx + \omega t) + y_2 \sin (kx - \omega t), \tag{9.26}$$

where y_1 and y_2 denote the amplitudes of the waves moving in opposite directions. As before, by applying the boundary conditions at $x = 0$ and $x = l$, you will get

$$y_1 = y_2 = \frac{1}{2} C \text{ (say)} \tag{9.27}$$

and

$$y = C \sin kl \cos \omega t = 0.$$

For this to hold for all values of t, we must have

$$\sin kl = 0$$

or

$$k = \frac{n\pi}{l}, \; n = 1, 2, 3, \ldots$$

and

$$\omega = \frac{n\pi v}{l}. \tag{9.28}$$

Note that the value of k depends on n. But boundary conditions do not allow waves of all frequencies. In order to study that we write C as C_n and y as y_n following Eq. (9.28). Hence, for a string fixed at both ends, the displacement of the nth mode of transverse oscillations at point x and time t is given by

$$y_n = C_n \sin \frac{n\pi x}{l} \cos \omega t. \tag{9.29}$$

We, therefore, find that amplitude varies with x and becomes maximum when $n\pi x/l = \pi/2$ or $x = l/2n$, and a minimum when $n\pi x/l = \pi$ or $x = l/n$. These points correspond to antinodes and nodes, respectively. In the nth mode, the string vibrates in n segments and neighbouring segments are in opposite phases, i.e., when one segment moves up, the neighbouring segments move down (see Example 8.4).

We now discuss the experimental arrangement used to demonstrate formation of stationary waves.

9.2.5 Resonant Vibrations of Stretched Strings: Melde's Experiment

Melde's experiment helps to demonstrate resonant vibrations of stretched strings in a simple way. Refer to Fig. 9.2, which shows a very thin wire/string whose one end is attached to one prong of an electrically maintained tuning fork. The other end of the string passes over a pulley. The string is stretched by adding known weights in a pan. When the fork is excited, the string is set in motion giving rise to stationary waves. By adjusting the distance between the fixed ends, we can make the string to resonate with the tuning fork for a given load. This manifests in the form of points of maximum and minimum (zero) displacements. And the number of loops in the string depends on the load and the manner of vibration of the fork.

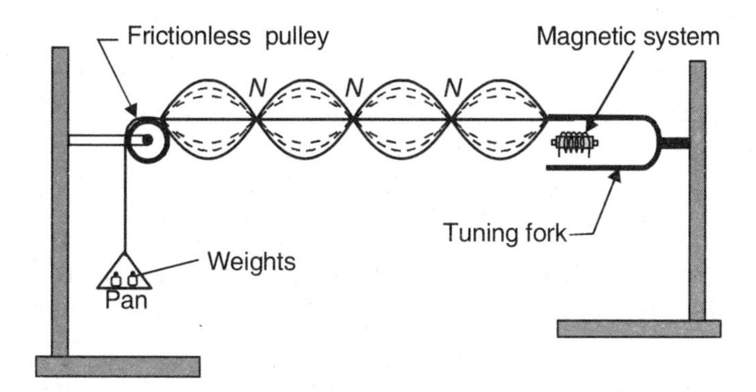

Fig. 9.2 Melde's apparatus: Transverse arrangement.

The experiment can be performed by arranging the fork with respect to the string in two ways: (i) perpendicular to the string and (ii) along the string. Accordingly, we can excite two types of waves:

- Transverse waves and
- Longitudinal waves

In transverse arrangement, the tip of the prong vibrates at right angles to the length of the string. In Section 8.6, you have learnt that if a string of length l resonates with a tuning fork when n harmonics are excited in it under tension T, the frequency of the string is given by

$$f = \frac{n}{2l}\sqrt{\frac{T}{m}}, \tag{9.30}$$

where m is its linear density.

Since f, l, and m are constants, we find that the load required to make a string vibrate in a number of loops is inversely proportional to the square of the corresponding number:

$$Tn^2 = \text{Constant}. \tag{9.31}$$

This result shows that the same length of the wire will vibrate in unison with a given fork in two segments, if tension is reduced by one-fourth of the value required for one segment. To give you an idea of numerical value, if a load (tension) of 100 g makes the given length of a string resonate in one segment, a load of 25 g will make it resonate in two segments with the same tuning fork.

The arrangement for producing longitudinal waves is shown in Fig. 9.3. In this arrangement, to-and-fro motion of the prong of the tuning fork is along the length of the string.

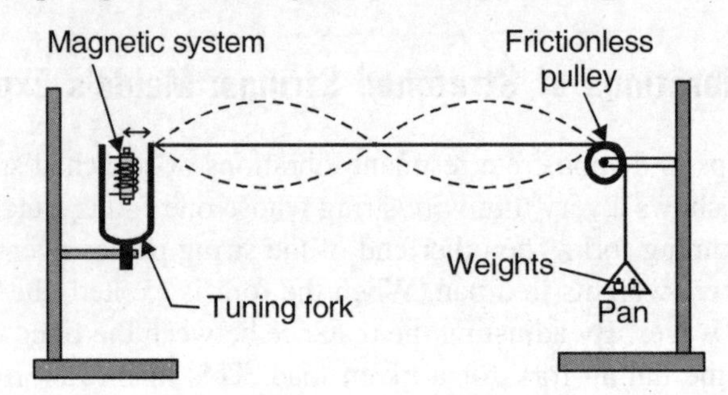

Fig. 9.3 Melde's apparatus for producing longitudinal waves.

You will note that during one complete vibration of the fork, the string executes half a vibration. So we can write

$$\text{Frequency of the fork, } f = 2 \times \text{Frequency of the string}$$

or

$$f = 2 \times \frac{n}{2l}\sqrt{\frac{T}{m}} = \frac{n}{l}\sqrt{\frac{T}{m}}, \tag{9.32}$$

where n is the number of loops. In other words, the frequency of the string is half that of the fork.

If you compare Eqs. (9.30) and (9.32), you will note that tension required to make a string execute longitudinal vibrations in a particular number of loops is one-fourth of that required for transverse motion. Furthermore, if the same length of a string is under the same tension, the number of loops generated in longitudinal and transverse arrangements will be different. In fact, if the number of loops in the transverse motion is 2, 4, 6, ..., the corresponding number in the longitudinal motion will be 1, 2, 3, ...

EXAMPLE 9.2

A stretched string is vibrating in fundamental mode. A weight of 3.5 kg is added on it. Its frequency increases in the ratio 4 : 3. Calculate the initial tension in the string.

Solution: We know that fundamental frequency is given by the relation

$$f_1 = \frac{1}{2l}\sqrt{\frac{T}{m}},$$

where $T = Mg$. Then we can write

$$f_1 = \frac{1}{2l}\sqrt{\frac{Mg}{m}}.$$

When tension is increased by 3.5 kg, let the new frequency be

$$f_2 = \frac{1}{2l}\sqrt{\frac{(M+3.5)g}{m}}.$$

On combining the relations for the frequencies in the two cases, we can write

$$\frac{f_2}{f_1} = \sqrt{\frac{(M+3.5)}{M}} = \frac{4}{3}.$$

By squaring both sides and simplifying the resultant expression, you will obtain

$$M = 4.5 \text{ kg}.$$

EXAMPLE 9.3

A load of 9.0 kg is put on a sonometer wire of length 1.0 m and mass per unit length 1.0×10^{-3} kgm^{-1}. If it is plucked at its mid-point, calculate the wavelength and frequency of the emitted note. Take g = 10 ms^{-2}.

Solution: If the wire vibrates in fundamental mode and emits frequency f, which corresponds to wavelength λ, we have

$$l = \frac{\lambda}{2},$$

where l is the length of the wire. On inserting the given data, we get

$$\lambda = 2l = 2 \times 1.0 = 2 \text{ m}.$$

When a wire of length l and mass per unit length m is under tension T, the fundamental frequency f emitted by it is given by

$$f = \frac{1}{2l}\sqrt{\frac{T}{m}}$$

$$= \frac{1}{2\times 1}\sqrt{\frac{90}{1.0\times 10^{-3}}} = 3.0 \times 10^2 \text{ Hz}.$$

Note that we have been able to express displacement of a wave excited on a stretched string in terms of eigenfunctions. Let us now discuss energy conservation.

9.2.6 Energy of a Vibrating String

The transverse displacement $y(x, t)$ at any point x and time t for a stretched string fixed at both ends is given by Eq. (9.20):

$$y(x,t) = \sum_{n=1}^{\infty} c_n \sin\frac{n\pi x}{l}\cdot\cos\left(\frac{n\pi v}{l}t - \phi_n\right).$$

For mathematical ease and convenience, we put

$$\xi_n(t) = c_n \cos(\omega_n t - \phi_n),$$

so that the expression for displacement $y(x, t)$ takes a very compact form:

$$y(x,t) = \sum_{n=1}^{\infty} \xi_n(t) \sin\frac{n\pi x}{l}. \tag{9.33}$$

On differentiating this expression with respect to t, keeping x fixed, we get

$$\dot{y} = \left(\frac{\partial y}{\partial t}\right) = \sum_{n=1}^{\infty} \dot{\xi}_n(t) \sin\frac{n\pi x}{l},$$

where dot over $\xi_n(t)$ as well as $y(x, t)$ signifies the time derivative. Therefore, the total kinetic energy of the string at any time t is given by

$$\text{KE} = \frac{1}{2}\int_0^l m\dot{y}^2\, dx.$$

On substituting the value of $(\partial y/\partial t)$ in this expression, we get

$$\text{KE} = \frac{m}{2}\int_0^l \left(\sum_{n=1}^{\infty} \dot{\xi}_n(t) \sin\frac{n\pi x}{l}\right)^2 dx$$

$$= \frac{m}{2}\int_0^l \left(\sum_{n=1}^{\infty} \dot{\xi}_n^2 \sin^2\frac{n\pi x}{l} + 2\sum_{n\neq m} \dot{\xi}_n \dot{\xi}_m \sin\frac{n\pi x}{l} \sin\frac{m\pi x}{l}\right) dx. \tag{9.34}$$

Now, from the orthogonality property of sine functions, we can write

$$\int_0^l \sin\frac{n\pi x}{l} \sin\frac{m\pi x}{l}\, dx = 0, \qquad \text{for } n \neq m$$

$$= l/2, \qquad \text{for } n = m.$$

Using this result in Eq. (9.34), we can write the expression for kinetic energy as

$$\text{KE} = \frac{ml}{4}\sum_{n=1}^{\infty} \dot{\xi}_n^2$$

$$= \frac{ml}{4}\frac{\pi^2 v^2}{l^2}\sum_{n=1}^{\infty} n^2 c_n^2 \sin^2\left(\frac{n\pi v}{l}t - \phi_n\right).$$

We know that $M = ml$ denotes the mass of the string and $\omega_n^2 = \pi^2 v^2 n^2/l^2$ signifies the square of the angular frequency of the string. Then expression for kinetic energy can be written as

$$\text{KE} = \frac{M}{4}\sum_{n=1}^{\infty} \omega_n^2 c_n^2 \sin^2\left(\frac{n\pi v}{l}t - \phi_n\right). \tag{9.35}$$

We know that the total energy of a mechanical wave is made up of kinetic and potential energies. To calculate potential energy of the string, we consider an element dx of the string and note that it is equal to the work done on this element in stretching it to length ds:

$$ds = \sqrt{dy^2 + dx^2}$$

$$= \sqrt{1 + \left(\frac{\partial y}{\partial x}\right)^2}\, dx.$$

For constant tension (force), the work done is equal to the product of force and distance. In the instant case, it is given $T(ds - dx)$.

On substituting for ds, we can write the expression for potential energy of the infinitesimal element of the string as:

$$dU = T\left[\sqrt{1+\left(\frac{\partial y}{\partial x}\right)^2} - 1\right]dx. \tag{9.36}$$

In practice, the slope of the string at any position is small. So $\partial y/\partial x << 1$, and we can use binomial expansion and retain terms only upto second order in $\partial y/\partial x$:

$$\sqrt{1+\left(\frac{\partial y}{\partial x}\right)^2} \approx 1+\frac{1}{2}\left(\frac{\partial y}{\partial x}\right)^2.$$

Using this result in Eq. (9.36), we get

$$dU = T\left[1+\frac{1}{2}\left(\frac{\partial y}{\partial x}\right)^2 - 1\right]dx = \frac{T}{2}\left(\frac{\partial y}{\partial x}\right)^2 dx. \tag{9.37}$$

Now, you can easily calculate the value of $\partial y/\partial x$ from Eq. (9.33):

$$\frac{\partial y}{\partial x} = \frac{\pi}{l}\sum_{n=1}^{\infty} n\xi_n(t)\cos\frac{n\pi x}{l}.$$

On substituting this result in Eq. (9.37) and integrating over x from 0 to l, we obtain the expression for potential energy of the string at time t:

$$U = \frac{\pi^2}{l^2}\frac{T}{2}\int_0^l \left(\sum n\xi_n(t)\cos\frac{n\pi x}{l}\right)^2 dx$$

$$= \frac{\pi^2 T}{2l^2}\int_0^l \left(\sum n^2\xi_n^2(t)\cos^2\frac{n\pi x}{l} + \sum_{n\neq m} nm\xi_n\xi_m \cos\frac{n\pi x}{l}\cos\frac{m\pi x}{l}\right). \tag{9.38}$$

You can easily convince yourself that

$$\int_0^l \cos^2\frac{n\pi x}{l}dx = \frac{l}{2}$$

and

$$\int_0^l \cos\frac{n\pi x}{l}\cos\frac{m\pi x}{l}dx = 0, \quad \text{for } n \neq m.$$

Using these results in Eq. (9.38), we get

$$U = \frac{\pi^2 T}{2l^2}\cdot\frac{l}{2}\sum_{n=1}^{\infty}\xi_n^2(t)n^2$$

$$= \frac{\pi^2 T}{2l^2}\cdot\frac{l}{2}\sum_{n=1}^{\infty} c_n^2\cdot n^2\cos^2\left(\frac{n\pi v}{l}t - \phi_n\right)$$

$$= \frac{\pi^2 v^2 M}{4l^2}\sum_{n=1}^{\infty} n^2 c_n^2\cos^2\left(\frac{n\pi v}{l}t - \phi_n\right),$$

where we have used the relation

$$v = \sqrt{\frac{T}{m}}.$$

This can be rewritten as

$$U = \frac{M}{4} \sum_{n=1}^{\infty} \omega_n^2 c_n^2 \cos^2(\omega_n t - \phi_n). \tag{9.39}$$

The total energy of the vibrating string is the sum of the kinetic and potential energies and is obtained by adding Eqs. (9.35) and (9.39):

$$E = \text{KE} + U = \frac{M}{4} \sum_{n=1}^{\infty} \omega_n^2 c_n^2. \tag{9.40}$$

Note that the total energy of the string is independent of spatial co-ordinate x and temporal co-ordinate t. This expression comprises an infinite sum of terms, each of which corresponds to one mode only. Moreover, for a particular mode, the energy of vibration is proportional to the square of the amplitude of vibration as well as the square of the eigenfrequency corresponding to that mode.

We can also rewrite Eq. (9.35) as

$$\text{KE} = \frac{M}{4} \sum \dot{\xi}_n^2$$

and Eq. (9.39) as

$$U = \frac{M}{4} \sum_{n=1}^{\infty} \omega_n^2 \xi_n^2$$

so that

$$E = \text{KE} + U$$

$$= \frac{M}{4} \sum \dot{\xi}_n^2 + \frac{M}{4} \sum_{n=1}^{\infty} \omega_n^2 \xi_n^2. \tag{9.41}$$

Let us now calculate the time averages of kinetic energy and potential energy of the string. Since average value of $\sin^2(\omega_n t - \phi_n)$ and $\cos^2(\omega_n t - \phi_n)$ over one period of oscillation is one-half, the time averages of the kinetic and potential energies are equal:

$$\langle \text{KE} \rangle = \langle U \rangle = \frac{M}{8} \sum_{n=1}^{\infty} \omega_n^2 c_n^2. \tag{9.42}$$

In section 9.2.2, you learnt to calculate displacement of a string rigidly fixed at both ends and discovered that the coefficients A_1 and B_1 appearing in Eq. (9.11) can be determined by specifying the initial displacements. The coefficients a_n and b_n can be similarly determined by specifying initial velocities at fixed points of the string. Since no new physics is involved, we leave it as an exercise for you. This will help you to get more familiarised with mathematical steps.

***Practice Exercise* 9.2** Consider a string fixed at both ends. The most general solution of the wave equation for transverse motion is given by Eq. (9.19). Calculate the coefficients a_n if velocity is zero at $t = 0$ and $t = T$.

$$\left[\textbf{Ans.} \quad a_n = \frac{2}{l} \int_0^l y(x, 0) \sin \frac{n\pi x}{l} dx \right]$$

You now know that stringed musical instruments produce music when the string(s) is (are) plucked, struck, or bowed at a single point by a sharp object. You may have noticed that the wires used in such instruments have different lengths, tension, and mass per unit length. The strings are fixed to screws mounted on a board and vibrations of the strings generate forced vibrations in the air inside and around the board. These vibrations are responsible for propagation of sound energy. In Section 9.3, we will discuss how plucked strings produce music.

9.3 VIBRATIONS OF A PLUCKED STRING

You now know that mandoline, harp, guitar, banjo, tanpura, veena, sarode and sitar are plucked stringed instruments. You must have enjoyed pleasing sitar vādan performances of Pt. Ravi Shankar. To understand how music is produced in sitar and such other stringed instruments, let us consider a string of length l stretched between two fixed supports P and Q. We choose x-axis along the length of the wire. Suppose that displacement of the string, when it is plucked at point A on the string at a distance a from point P, is h ($<< l$), as shown in Fig. 9.4. On being released, the string is set into motion. The initial configuration of the string can be described as

$$y_0 = \frac{h}{a}x \quad (0 \le x \le a)$$

$$= \frac{h(l-x)}{(l-a)}, (a \le x \le l), \tag{9.43}$$

where we have put the subscript 0 with y to signify $t = 0$.

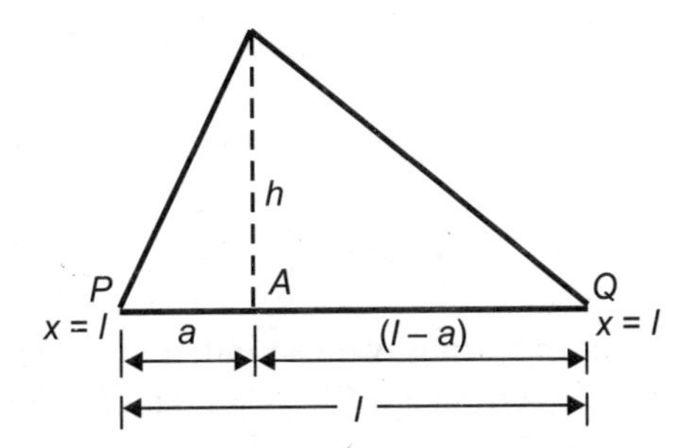

Fig. 9.4 A transversely plucked string.

Since the string is plucked from its equilibrium position of rest, its initial velocity is zero everywhere, i.e., $\dot{y}_0 = 0$. Note that the dot over y_0 denotes its first time-derivative. Recall that Eq. (9.19) describes the most general solution of wave equation for transverse motion. Therefore, the expression for initial displacement is obtained by taking $t = 0$ in this equation. This gives

$$y_0(x, 0) = \sum_{n=1}^{\infty} a_n \sin \frac{n\pi x}{l}. \tag{9.44}$$

Next we differentiate Eq. (9.19) with respect to time. This gives

$$\dot{y}(x,t) = \sum_{n=1}^{\infty} \sin\frac{n\pi x}{l}\left(-a_n \sin\frac{n\pi vt}{l} + b_n \cos\frac{n\pi vt}{l}\right)\frac{n\pi v}{l}.$$

Now, substitute $t = 0$ in this expression. This leads us to a compact expression for initial velocity of the string:

$$\dot{y}_0(x,0) = \sum_{n=1}^{\infty} \omega_n b_n \sin\frac{n\pi x}{l}, \tag{9.45}$$

where $\omega_n = n\pi v/l$.

As the initial velocity of the string is zero, for Eq. (9.45) to hold, the coefficients b_n must be zero. Using this result in Eq. (9.19), we get the desired expression for displacement of a plucked string:

$$y(x,t) = \sum_{n=1}^{\infty} a_n \sin\frac{n\pi x}{l} \cos\omega_n t. \tag{9.46}$$

Note that the coefficients a_n in Eq. (9.46) are still unknown. To determine these, we multiply Eq. (9.44) by $\sin(m\pi x/l)$, where m takes integral values and integrate the resultant expression over x from $x = 0$ to $x = l$. This gives

$$\int_0^l y_0 \sin\frac{n\pi x}{l}\,dx = \int_0^l \sum_{n=1}^{\infty} a_n \sin\frac{n\pi x}{l} \sin\frac{m\pi x}{l}\,dx.$$

Using orthogonality property of sine functions, we note that the integral on the right hand side is non-zero only for $n = m$. As before, we can write

$$\int_0^l y_0 \sin\frac{n\pi x}{l}\,dx = a_n \int_0^l \sum \sin^2\frac{n\pi x}{l}\,dx.$$

You can easily evaluate the integral on the right hand side to obtain

$$a_n = \frac{2}{l}\int_0^l y_0 \sin\frac{n\pi x}{l}\,dx. \tag{9.47}$$

To obtain the value of a_n for a plucked string, we use initial conditions [Eq. (9.43)] and split the integral in Eq. (9.47) into two parts: (i) from $x = 0$ to $x = a$ and (ii) from $x = a$ to $x = l$. This leads to

$$a_n = \frac{2}{l}\left\{\int_0^a \frac{h}{a} x \sin\frac{n\pi x}{l}\,dx + \int_a^l \frac{h(l-x)}{(l-a)} \sin\frac{n\pi x}{l}\,dx\right\}$$

$$= \frac{2}{l}\left\{\int_0^a \frac{h}{a} I + \int_a^l \frac{h}{(l-a)} J\right\},$$

where

$$I = \left\{\int_0^a x \sin\frac{n\pi x}{l}\,dx\right\}$$

and

$$J = \left\{ \int_a^l (l - x) \sin \frac{n\pi x}{l} dx \right\}.$$

You can evaluate these integrals using the method of integration by parts. Let us first consider the first integral, I, and for ease in writing, put $k = n\pi/l$, so that we can write

$$I = \left\{ \int_0^a x \sin k\, dx \right\} = -a \frac{\cos ka}{k} + \int_0^a \frac{\cos kx}{k} dx$$

$$= -\frac{a}{k} \cos ka + \frac{\sin ka}{k^2}.$$

Similarly, the integral J can be written as

$$J = \left\{ \int_a^l (l - x) \sin \frac{n\pi x}{l} dx \right\} = \left(\frac{l - a}{k} \right) \cos ka - \frac{1}{k^2} (\sin kl - \sin ka).$$

Now, inserting these values of I and J in the expression of a_n, we get

$$a_n = \frac{2}{l} \left[\frac{h}{a} \left\{ -\frac{a}{k} \cos ka + \frac{\sin ka}{k^2} \right\} + \frac{h}{l - a} \left(\left\{ \frac{l - a}{k} \right\} \cos ka - \frac{1}{k^2} \{\sin kl - \sin ka\} \right) \right]$$

$$= \frac{2}{l} \left[-\frac{h}{k} \cos ka + \frac{h}{ak^2} \sin ka + \frac{h}{l - a} \frac{l - a}{k} \cos ka - \frac{h}{k^2 (l - a)} (\sin kl - \sin ka) \right].$$

The first and third terms cancel out and if we note that $\sin kl = 0$ for all values of n, this expression simplifies to

$$a_n = \frac{2h \sin ka}{lk^2} \left(\frac{1}{a} + \frac{1}{l - a} \right)$$

$$= \frac{2hl^2}{an^2 \pi^2 (l - a)} \sin \frac{n\pi a}{l}.$$

Note that the coefficient a_n is inversely proportional to n^2. It means that the amplitudes of higher harmonics fall off very rapidly.

Proceeding further, the expression for $y(x, t)$ is obtained by substituting for a_n in Eq. (9.46):

$$y(x, t) = \frac{2hl^2}{a\pi^2 (l - a)} \left[\sum_{n=1}^{\infty} \frac{1}{n^2} \sin \frac{n\pi a}{l} \sin \frac{n\pi x}{l} \cos \frac{n\pi v}{l} t \right] \tag{9.48a}$$

In the expanded form, we rewrite this expression as

$$y(x, t) = \frac{2hl^2}{a\pi^2 (l - a)} \left[\sin \frac{\pi a}{l} \sin \frac{\pi x}{l} \cos \frac{\pi vt}{l} + \frac{1}{4} \sin \frac{2\pi a}{l} \sin \frac{2\pi x}{l} \cos \frac{2\pi vt}{l} \right.$$

$$\left. + \frac{1}{9} \sin \frac{3\pi a}{l} \sin \frac{3\pi x}{l} \cos \frac{3\pi ct}{l} + \cdots + \frac{1}{n^2} \sin \frac{n\pi a}{l} \sin \frac{n\pi x}{l} \cos \frac{n\pi ct}{l} \cdots \text{upto infinite number of terms} \right]. \tag{9.48b}$$

Note that the terms on the right hand side respectively give the displacement for the fundamental, second, third and the nth mode.

Let us now take $a = l/q$, where q is an integer. Then $\sin(n\pi a/l) = \sin(n\pi/q)$. So, if n is an integral multiple of q, i.e., $n = q, 2q, 3q, \ldots$, $\sin(n\pi/q)$ will become zero. For $q = 2$, which means that the string is plucked at the mid point, it readily follows that all even harmonics will be absent since n is an even number and all these harmonics have a node at $a = l/2$. We can generalise this result as: *For a plucked stretched string, harmonics which have a node at the point of plucking will be absent. This is known as Young's law.*

Note that if I_1 and I_3 are the intensities of the fundamental and the third harmonic, respectively, then Eq. (9.48b) implies that the ratio

$$\frac{I_1}{I_3} = 81\frac{\sin^2(\pi a/l)}{\sin^2 3\pi a/l}.$$

In writing this expression, we have used the result that intensities are proportional to the square of the amplitudes. That is, if the string is plucked at its mid point ($a = l/2$), the intensity of fundamental mode will become 81 times that of the third harmonic.

Let us pause for a while and see as to what is implied by Eq. (9.48b). We note that when a string in the mandoline is plucked at a single point by a plectrum—a sharp thin celluloid sheet, a high quality note is excited due to the presence of a large number of modes. On the other hand, if the string is plucked over a finite length by a round object like finger, fewer modes are excited and the sound is softer.

9.4 VIBRATIONS OF A STRUCK STRING

In the preceding section, you have learnt that when a string stretched between two fixed supports is plucked at a point, a high quality note is excited which produces music. Another way by which such a string can be made to vibrate is to strike it by a large force. A familiar example of a struck stringed musical instrument is a piano. This imparts a sudden impulse at the point where wire is struck. However, the initial velocity of all points, except the point of contact, on the string remains zero. Thus, the initial conditions in a struck string differ from those of a plucked string: A struck string has no initial displacement but an initial velocity at the point of strike. Therefore, if we say that the initial conditions of a plucked string are static while those of the struck string are dynamic, it will not be wrong.

In the case of a struck string, we have to take into consideration factors such as the striking velocity of hammer, the duration of contact between the hammer and string, relative masses of hammer and string, etc. For mathematical ease and convenience, we make a simplifying assumption in that the duration of contact between the striking hammer and the string is very small so that the impact ceases before the disturbance spreads out. The waves excited from the point of contact move in opposite directions, get reflected at the fixed ends and their superposition leads to generation of stationary waves.

Consider a string of length l supported between two fixed ends. We choose x-axis along the length of the wire. Let the positions of fixed ends be at $x = 0$ and $x = l$. Suppose that the string is struck in an infinitesimal region extending from $x = a$ to $x = a + dx$ and this portion acquires an initial velocity u. Hence, in this case, the initial conditions are:

- $y(x, 0) = 0$ for all x, and
- velocity is zero at all points, except over the region $x = a$ to $x = a + dx$. In this region, the velocity is u.

The general expression for displacement of a stretched string at any point x and time t is given by Eq. (9.19):

$$y(x,t) = \sum_{n=1}^{\infty} \sin\frac{n\pi x}{l}\left(a_n \cos\frac{n\pi vt}{l} + b_n \sin\frac{n\pi vt}{l}\right)$$

We have to solve it subject to given initial conditions. Taking $y = 0$ at $t = 0$ in above equation, we obtain

$$y(x,0) = \sum_{n=1}^{\infty} a_n \sin\frac{n\pi x}{l} = 0.$$

But in this expression sin $(n\pi x/l)$ cannot be zero for all values of x. Hence, a_n must be zero. Using this result in Eq. (9.19), we get

$$y(x,t) = \sum_{n=1}^{\infty} b_n \sin\frac{n\pi x}{l} \sin\frac{n\pi vt}{l}. \tag{9.49}$$

Proceeding further, we differentiate this expression with respect to time and put $t = 0$ in the resultant expression, as before. This gives

$$\dot{y}_0(x,0) = \left(\frac{\pi v}{l}\right)\sum_{n=1}^{\infty} nb_n \sin\frac{n\pi x}{l}. \tag{9.50}$$

To obtain the value of the coefficient b_n, we use the orthogonality property of sine functions. So we multiply Eq. (9.50) by sin $(m\pi x/l)$, where m is an integer, and integrate the resultant expression from $x = 0$ to $x = l$. Hence

$$\int_0^l \dot{y}_0 \sin\frac{m\pi x}{l}\, dx = \sum \int_0^l \omega_n b_n \sin\frac{n\pi x}{l} \sin\frac{m\pi x}{l}\, dx,$$

where we have put $\omega_n = (\pi n v/l)$.
Using orthogonality property, we can simplify the right hand side as

$$\int_0^l \dot{y}_0 \sin\frac{m\pi x}{l}\, dx = \omega_n b_n \int_0^l \sin^2\frac{n\pi x}{l}\, dx = \omega_n\frac{b_n l}{2} = \frac{\pi v n b_n}{2}. \tag{9.51}$$

Since $\dot{y}_0 = u$ for $x = a$ to $a + dx$ and zero elsewhere, the integral in Eq. (9.51) contributes only over a finite region. Therefore, we can write

$$\frac{\pi v n b_n}{2} = \omega_n \frac{b_n l}{2} = \int_a^{a+x} u \sin\frac{n\pi a}{l} \cdot dx$$

or

$$\frac{\pi v n b_n}{2} = \sin\frac{n\pi a}{l}\int_a^{a+x} u\, dx = I \sin\frac{n\pi a}{l},$$

where $I = \int_a^{a+x} u\,dx$. Hence,

$$b_n = \frac{2I}{\pi v n} \sin \frac{n\pi a}{l}. \tag{9.52}$$

Note that the coefficients b_n are inversely proportional to n. It implies that the amplitudes of higher harmonics will decrease rapidly as n increases.

On substituting this value of b_n in the expression for resultant displacement of a struck string [Eq. (9.49)], we get

$$y(x, t) = \frac{2I}{\pi v} \sum_{n=1}^{\infty} \frac{1}{n} \sin \frac{n\pi a}{l} \sin \frac{n\pi x}{l} \sin \frac{n\pi vt}{l}. \tag{9.53}$$

In the expanded form, this equation may be rewritten as

$$y(x, t) = \frac{2I}{\pi v} \left[\sin\frac{\pi a}{l} \sin\frac{\pi x}{l} \sin\frac{\pi vt}{l} + \frac{1}{2} \sin\frac{2\pi a}{l} \sin\frac{2\pi x}{l} \sin\frac{\pi vt}{l} + \frac{1}{3} \sin\frac{3\pi a}{l} \sin\frac{3\pi x}{l} \sin\frac{3\pi vt}{l} + \cdots + \frac{1}{n} \sin\frac{n\pi a}{l} \sin\frac{n\pi x}{l} \sin\frac{n\pi vt}{l} \right]. \tag{9.54}$$

upto infinite number of terms

In the preceding section, you have learnt that for a plucked string, the amplitude of higher harmonics falls off as $1/n^2$. It means that the amplitude of higher harmonics drops faster in a plucked stringed instrument than in a struck stringed instrment.

You may now like to answer Practice Exercise 9.3.

***Practice Exercise* 9.3** Show that Young's law holds for a struck string.

Another way in which a stretched string can be excited is by bowing it with a horse-hair resined bow, say. Let us now discover how harmonics are excited in a bowed string and music is produced in stringed instruments such as ektara, sarengi, violin and jesraj. You will also learn how these differ from those generated by plucked and struck strings.

9.5 VIBRATIONS OF A BOWED STRING

Consider a string of length l stretched between two supports. It is set into vibrations by drawing a horse-hair resined bow across it near one of the supports. Such a string is referred to as *bowed string*. The quality of the tone in a bowed musical instrument is found to depend, to some extent, on the position of the point of bowing, the speed of the bow and the pressure it exerts on the string. As the bow moves across the string, it drags the string along with it due to static friction between them. The amplitude of vibrations of the string and hence the intensity of sound produced depends on the amount of bow hairs in contact with the string as well as the velocity of the bow; increase in bow speed increases intensity. However, bow pressure plays little role in controlling intensity, though it modifies tone structure better than bow speed. Moreover, bow pressure tends to increase the intensity of higher partials. When the point of bowing is nearer to the bridge, the higher harmonics are relatively more prominent.

Let us now understand the physics of a bowed string. As the string is dragged by the bow from its equilibrium position of rest, a restoring force develops in the string. Once restoring force exceeds the frictionial force, the string tends to slip towards its initial position and in this process, due to inertia, it overshoots the equilibrium position. As soon as the string comes to rest, the moving bow seizes the string once again and drags it along. This two-stage process is repeated rapidly as long as the bow is in contact with the string. Note that during slipping, dynamic friction is operative. As static friction is greater than the dynamic friction, more work is done on the string than the work done by it. This difference maintains vibrations of the string.

Refer to Fig. 9.5a, which depicts the snapshot of the vibrations of a bowed string. The portion *CB* of the string represents that part of motion where the bow forces the string to move, whereas the portion *AC* represents the 'slipping part'. Let us denote the period of vibration by T. We can say that for time t_1, the point of observation on the string moves forward with a uniform velocity v_1 and for time $(T - t_1)$, the point moves with uniform velocity v_2 (Fig. 9.5b). It means that we have to evaluate coefficients a_n and b_n occurring in Eq. (9.19) subject to the following boundary conditions:

$$\begin{aligned} \dot{y} &= v_1 \quad \text{for } 0 \le t \le t_1 \\ &= -v_2 \quad \text{for } t_1 \le t \le T. \end{aligned} \tag{9.55}$$

For mathematical convenience, we rewrite Eq. (9.19) as

$$y(x, t) = \sum_{n=1}^{\infty} \sin\left(\frac{n\pi x}{l}\right)[a_n \cos \omega_n t + b_n \sin \omega_n t], \tag{9.56}$$

where $\omega_n = \dfrac{2\pi}{T} = \dfrac{n\pi v}{l}$.

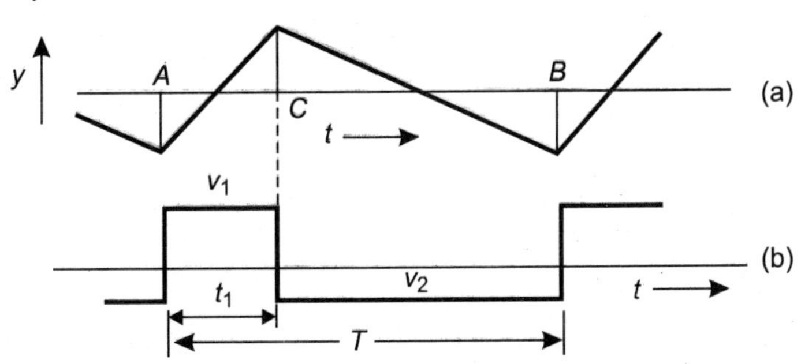

Fig. 9.5 (a) Snapshot of a bowed string; and (b).

The expression for velocity is obtained by differentiating the expression for displacement as

$$\dot{y}(x, t) = \sum_{n=1}^{\infty} \omega_n \sin\frac{n\pi x}{l}(-a_n \sin \omega_n t + b_n \cos \omega_n t).$$

Now, we multiply both sides of the resultant expression by $\sin \omega_m t$. Then, on integrating it from $t = 0$ to $t = T$ and using the orthogonality condition of sine functions, we get

$$\begin{aligned} \int_0^T \dot{y} \sin\frac{n\pi vt}{l}\, dt &= -a_n \omega_n \sin\frac{n\pi x}{l}\left(\frac{T}{2}\right) \\ &= -a_n n\pi \sin\frac{n\pi x}{l}. \end{aligned} \tag{9.57}$$

To determine the coefficients a_n, we use the initial conditions specified by Eq. (9.55) and substitute the values in the given ranges. This leads us to the equation

$$-a_n \sin \frac{n\pi x}{l} = \frac{1}{n\pi}\left[\int_0^{t_1} v_1 \sin \omega_n t dt + \int_{t_1}^{T} (-v_2) \sin \omega_n t \, dt\right].$$

The integrals can be readily evaluated, as before, to obtain

$$\left(-a_n \sin \frac{n\pi x}{l}\right) = \frac{1}{n\pi}\left[\frac{-v_1}{\omega_n}(\cos \omega_n t - 1) + \frac{v_2}{\omega_n}(1 - \cos \omega_n t_1)\right]$$

$$= \frac{T(v_1 + v_2)}{n^2 \pi^2} \sin^2 \frac{\omega_n t_1}{2} \tag{9.58a}$$

Following the same steps, we can evaluate coefficients b_n:

$$b_n \sin \frac{n\pi x}{l} = \frac{T(v_1 + v_2)}{n^2 \pi^2} \sin\left(\frac{\omega_n t_1}{2}\right) \cos\left(\frac{\omega_n t_1}{2}\right). \tag{9.58b}$$

On inserting Eqs. (9.58a) and (9.58b) in Eq. (9.56) and rearranging terms, we get

$$y(x, t) = \frac{T(v_1 + v_2)}{\pi^2} \sum_{n=1}^{\infty} \frac{1}{n^2} \sin\left(\frac{\omega_n t_1}{2}\right) \sin \omega_n \left(t - \frac{t_1}{2}\right). \tag{9.59}$$

From Eq. (9.19), we recall that displacement is zero at all times, if the argument of the sine function is an integral multiple of π. Obviously, this requirement must be satisfied by the solution given in Eq. (9.59) as well. Therefore, we must have

$$\frac{1}{2} \omega_n t_1 = \frac{n\pi x}{l}.$$

On rearranging the terms, we get

$$\frac{\omega_n t_1}{2n\pi} = \frac{x}{l}$$

or

$$\frac{t_1}{T} = \frac{x}{l}, \tag{9.60}$$

where we used the relation $T = (2n\pi/\omega_n)$. From this, you will note that at the midpoint ($x = l/2$), $t_1 = T/2$. This shows that the forward and backward motions take equal time to reach the midpoint. That is, for one-half period, particles acquire forward velocity v_1 and for the other half, the paricles possess backward velocity v_2.

Now, let us express displacement of the bowed string in terms of the amplitude of vibration, A. For this, we note that at the midpoint, velocities v_1 and v_2 are equal and may be taken as v. Then we can write

$$v_1 + v_2 = 2v = 2 \times \frac{2A}{T/2} = \frac{8A}{T}$$

On substituting this result in Eq. (9.59), we get

$$y(x, t) = \frac{8A}{\pi^2} \sum_{n=1}^{\infty} \frac{1}{n^2} \sin \frac{1}{2} \omega_n t_1 \sin \omega_n \left(t - \frac{t_1}{2} \right)$$

$$= \frac{8A}{\pi^2} \sum_{n=1}^{\infty} \frac{1}{n^2} \sin \frac{n\pi x}{l} \cdot \sin \omega_n \left(t - \frac{t_1}{2} \right). \tag{9.61}$$

From Eq. (9.61), we note that for $t = t_1/2$, $y(x, t)$ will be equal to zero for all values of x. It means that all points on the string pass through their equilibrium positions simultaneously.

If you now compare Eqs. (9.47) and Eq. (9.61) for plucked and bowed strings, respectively, you will note that at $t = 0$

$$\frac{hl^2}{a(l-a)} = \pm 4A. \tag{9.62}$$

So we can say that Eq. (9.61) represents a pair of two plucked strings whose co-ordinates (a, h) are given by Eq. (9.62). Obviously, a and h vary with time; a is given by the relation

$$\sin \frac{n\pi a}{l} = \pm \sin \omega_n t. \tag{9.63}$$

Equations (9.62) and (9.63) indicate that a projection of the point of intersection on the x-axis moves uniformly backward and forward between the supports at $x = 0$ and $x = l$. This is illustrated in Fig. 9.6. Note that the point of intersection K itself is situated on one or the other of two parabolic arcs defined by Eq. (9.63). The equilibrium position of the string is the common chord for these arcs.

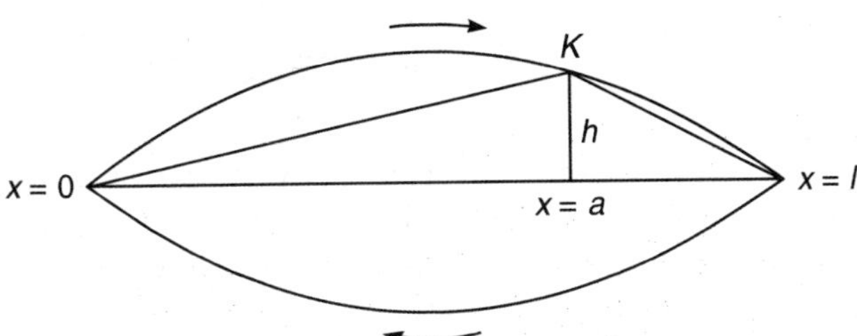

Fig. 9.6 The point of intersection K lies on one of the two parabolic arcs and the equilibrium position is the common chord.

We may now summarise by noting that the quality of tone excited by a bowed string is influenced by the location of the point of bowing, speed of the bow and pressure exerted by the bow on the string. The higher harmonics become relatively prominent if the point of bowing is closer to the bridge. The intensity of the emitted sound (or the amplitude of vibration of the string) depends on the speed of the bow; the intensity increases with the increase in the speed of the bow.

REVIEW EXERCISES

9.1 A string is fixed at two ends defined by $x = 0$ and $x = l$. The velocity of the wave in the string is c. Calculate the displacement $y(x, t)$ in each of the following cases. Also find the maximum speed attained by any point of the string far (a) and (b).

(a) $y(l/2, 0) = h$ and $\dot{y}(x, 0) = 0$

(b) $y(l, 0) = a \sin(3\pi x/l)$, $\dot{y} = 0$

(c) $y(x, 0) = a \sin(2\pi x/l) + b \sin(5\pi x/l)$; $\dot{y}(x, 0) = 0$

(d) $y(x, 0) = a \sin(\pi x/l)$; $\dot{y}(x, 0) = b \sin(\pi x/l)$

[**Ans.** (a) $y = \sum_{n=1}^{\infty} \frac{8h}{n^2\pi^2} \sin\frac{n\pi}{2} \sin\frac{n\pi x}{l} \cos\frac{n\pi ct}{l}$

(b) $y = a \cos\frac{3\pi ct}{l} \sin\frac{3\pi x}{l}$

(c) $y = a \cos\frac{2\pi ct}{l} \sin\frac{2\pi x}{l} + b \cos\frac{5\pi ct}{l} \sin\frac{5\pi x}{l}$

(d) $y = \sin\frac{\pi x}{l}\left(\frac{bl}{\pi c} \sin\frac{\pi ct}{l} + a \cos\frac{\pi ct}{l}\right)$

Maximum speeds:

(a) $\frac{8hc}{n\pi l} \sin\frac{n\pi}{2} \sin\frac{n\pi x}{l}$

(b) $\frac{3\pi ca}{l} \sin\frac{3\pi x}{l}$]

9.2 A string with ends at $x = 0$ and $x = l$ is initially at rest. It is set into vibrations by a transverse impulse at a distance b from the end $x = 0$. Obtain an expression for displacement $y(x, t)$ at any later time. The velocity of the wave is c.

[**Ans.** $\frac{2P}{\pi cm} \sum_{n=1}^{\infty} \frac{1}{n} \sin\frac{n\pi b}{l} \sin\frac{n\pi ct}{l} \sin\frac{n\pi x}{l}$]

9.3 A uniform string of length πl is fastened at its ends defined by $x = 0$ and $x = \pi l$. The point $x = \pi l/3$ is drawn upward through a distance b and then released. Show that

$$u(x, t) = \frac{9b}{\pi^2} \sum_{n=1}^{\infty} \frac{1}{n^2} \sin\frac{n\pi}{3} \sin\frac{n\pi}{l} \cos\left(\frac{n\pi}{l}\sqrt{\frac{T}{m}}\right),$$

where T denotes the tension in the string and m is mass per unit length.

9.4 In a Melde's experiment (transverse mode), the string vibrates in three loops, when the tension in the string is 1.92 N. What is the tension when the string vibrates in two loops in the longitudinal mode?

[**Ans.** 1.08 N]

9.5 Obtain a set of initial conditions for a string to vibrate with four nodes, excluding the end points. Are these initial conditions unique?

9.6 In Melde's experiment, the number of loops on the string decreases from 5 to 4 when the tension is increased by 0.4 N. Calculate the initial tension in the string.

[*Ans.* 0.7 N]

9.7 In Melde's experiment, 4.2 m long wire vibrates in 8 loops when the load is 0.25 kg. The frequencyof the tuning fork is 72 Hz. Calculate the mass of the wire.

[*Ans.* 16.5 g]

9.8 A string of length 1 m produces the same note as a tuning fork and gives 4 beats per second when the length is decreased by Δl. If the frequency of the tuning fork is 196 Hz, calculate Δl.

[*Ans.* 2 cm]

9.9 A tuning fork of frequency 224 Hz gives n beats in 12 s when sounded with a stretched string vibrating tranversely under a tension of either 102 N or 99 N. Determine the value of n.

[*Ans.* $n = 20$]

9.10 A steel string of length 0.63 m produces a tone with a frequency 320 Hz when it is stretched by a force of 100 N. Calculate the radius of the string. The density of steel $= 7.8 \times 10^3$ kgm^{-3}.

[*Ans.* 6.3×10^{-4} m]

10

Vibrations of Bars

EXPECTED LEARNING OUTCOMES

In this Chapter, you will acquire capability to:

- solve wave equation for longitudinal vibrations of a bar to determine instantaneous displacement for given initial conditions;
- derive an expression for the velocity of longitudinal waves in a bar;
- deduce expressions for frequency of fundamental and higher harmonics for longitudinal vibrations of a bar (i) free at both ends, (ii) clamped at one-end, and (iii) clamped at both ends.
- solve wave equation for transverse vibrations of a thin (i) fixed-fixed bar, (ii) free-free bar, and (iii) fixed-free bar and obtain expressions for frequencies of fundamental as well as higher harmonics; and
- explain vibrations of a tuning fork in terms of vibrations of bars.

10.1 INTRODUCTION

In the previous chapter, you have learnt that music in stringed instruments is generated when strings are plucked, struck or bowed under tension. You may recall that the frequency of fundamental mode of stationary waves set up due to superposition of longitudinal oscillations of a string is harmonically related to the frequencies of higher modes. In your school, you must have worked with a tuning fork. In this chapter, you will learn that it belongs to another category of primary vibrators-rectangular bars. When made to vibrate by striking one of its prongs, a tuning fork supports propagation of stationary waves. But natural frequencies of different modes are not harmonically related. Do you know that bars can support propagation of longitudinal as well as transverse waves?

The study of propagation of compressional waves in bars finds important scientific, engineering and industrial applications. For example, circular rods of fixed lengths serve as frequency standards. When longitudinal waves are excited in these, the frequency of vibrations is found to be inversely proportional to the length of the bar. Longitudinal vibrations in nickel tubes are used to make the diaphragm of a sonar transducer vibrate.

In Section 10.2, you will learn about longitudinal waves in different configurations of a bar. Section 10.3 is devoted to the study of transverse vibrations of a bar in different configurations.

10.2 LONGITUDINAL VIBRATIONS OF A BAR

We discussed propagation of longitudinal waves in a solid bar in Chapter 7. You may recall that for a solid bar of density ρ, cross-sectional area A and Young's modulus of elasticity Y, the wave equation is given by

$$\frac{\partial^2 y}{\partial t^2} = v^2 \frac{\partial^2 y}{\partial x^2}, \tag{10.1}$$

where

$$v = \sqrt{\frac{Y}{\rho}} \tag{10.2}$$

is wave velocity. In arriving at this result, we assumed that

- length of the bar is comparable to the wavelength of the longitudinal wave propagating in it; and
- the bar can be treated as thin compared to its length.

It means that a longitudinal stress applied across it produces the same displacement over a given cross-section.

Note that in Eq. (10.2), the Young's modulus of elasticity for metals is of the order of 10^{11} Pa and density ρ is of the order of 10^4 kgm^{-3}. Therefore, the velocity of sound in thin metallic bars is of the order of 10^3 ms^{-1}. In fact, in most metals, the velocity of sound is in the range $(3 - 5) \times 10^3$ ms^{-1}. Recall that this value is greater than that in liquids or gases.
You should now study Example 10.1.

EXAMPLE 10.1

Calculate the velocity and wavelength of longitudinal waves of frequency 500 Hz in a bar of material of density 10 gcm^{-3} and Young's modulus 16.0×10^{11} dynecm^{-2}.

Solution: (a) From Eq. (10.2), we know that the velocity of longitudinal wave is given by

$$v = \sqrt{\frac{Y}{\rho}},$$

where Y is the Young's modulus and ρ is the density of the material. We can write the given values as

$$Y = 16 \times 10^{11} \text{ dynecm}^{-2} = 16 \times 10^{11} \times \frac{1}{10^5} \times 10^4 \text{ Nm}^{-2}$$

and

$$\rho = 10 \text{ g cm}^{-3} = 10000 \text{ kgm}^{-3}.$$

$$\therefore \qquad v = \sqrt{\frac{16 \times 10^{10}}{10000}} = 4000 \text{ ms}^{-1}.$$

(b) Frequency $f = 500$ Hz

Therefore, the wavelength is determined using the relation

$$\lambda = \frac{v}{f} = \frac{4000\,\text{ms}^{-1}}{500\,\text{s}^{-1}}$$

$$= 8 \text{ m.}$$

To proceed further, we differentiate both sides of Eq. (10.1) partially with respect to time. This gives

$$\frac{\partial}{\partial t}\left(\frac{\partial^2 y}{\partial t^2}\right) = v^2 \frac{\partial}{\partial t}\left(\frac{\partial^2 y}{\partial x^2}\right). \tag{10.3}$$

Note that variables x and t are independent. So the order of differentiation with respect to x and t can be interchanged. Therefore, we can rewrite Eq. (10.3) as

$$\frac{\partial^2}{\partial t^2}\left(\frac{\partial y}{\partial t}\right) = v^2 \frac{\partial^2}{\partial x^2}\left(\frac{\partial y}{\partial t}\right). \tag{10.4}$$

If we define particle velocity $u = \partial y/\partial t$, then we can rewrite Eq. (10.4) as

$$\frac{\partial^2 u}{\partial t^2} = v^2 \frac{\partial^2 u}{\partial x^2}. \tag{10.5}$$

Note that Eqs. (10.1) and (10.5) are analogous. So we can say that particle velocity u obeys 1-D wave equation.

We can solve Eq. (10.1) by the method of separation of variables. So we assume that $y(x, t)$ can be expressed as a product of two functions $X(x)$ and $T(t)$:

$$y = X(x)\, T(t) \tag{10.6}$$

On substituting Eq. (10.6) in Eq. (10.1) and equating the spatial and temporal parts to a constant, as was done to solve Eq. (9.1), we can write the solutions of resultant equations as

$$X(x) = A_1 \cos \frac{\omega x}{v} + B_1 \sin \frac{\omega}{v} x \tag{10.7}$$

and

$$T(t) = C_1 \cos \omega t + D_1 \sin \omega t, \tag{10.8}$$

where A_1, B_1, C_1 and D_1 are arbitrary constants. These constants are determined using the initial and boundary conditions. Note that ω is related to the separation constant, as in case of vibrations of strings. You will discover that it is related to the properties of a wave.

On combining Eqs. (10.7) and (10.8) with Eq. (10.6), we can write

$$y(x, t) = \left(A_1 \cos \frac{\omega x}{v} + B_1 \sin \frac{\omega x}{v}\right)(C_1 \cos \omega t + D_1 \sin \omega t) \tag{10.9}$$

Let us now determine the form solution given by Eq. (10.9) takes for different bar configurations.

***Case* (i)** A bar resting lightly on two knife edges (Fig. 10.1) is a simple example of a *free-free bar*. If the length of the bar is l and we choose x-axis along its length, then we can say that ends of a free-free bar at $x = 0$ and $x = l$ can move freely. Suppose that tension in the bar is F. Since the area of cross section is α, the stress on the bar will be F/α. Then by definition of Young's modulus, we can write

$$\frac{F}{\alpha} = -Y\frac{\partial y}{\partial x},$$

where $\partial y/\partial x$ signifies longitudinal strain.

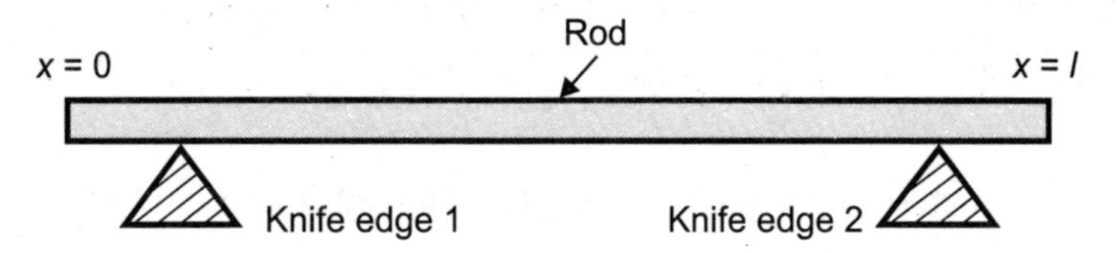

Fig. 10.1 A free-free bar.

We assume that strain at each end is zero.Then boundary conditions to be satisfied in this case are:

(i) At $x = 0$, $\partial y/\partial x = 0$ for all values of t; and

(ii) At $x = l$, $\partial y/\partial x = 0$ for all values of t.

To proceed further, we differentiate Eq. (10.9) with respect to x to obtain

$$\frac{\partial y}{\partial x} = \left(-A_1\frac{\omega}{v}\sin\frac{\omega x}{v} + B_1\frac{\omega}{v}\cos\frac{\omega x}{v}\right)(C_1\cos\omega t + D_1\sin\omega t). \tag{10.10}$$

On using the first boundary condition in this equation, we get

$$\left(B_1\frac{\omega}{v}\right)(C_1\cos\omega t + D_1\sin\omega t) = 0.$$

For this equation to hold for all values of t, B_1 must be equal to zero. Then Eq. (10.9) reduces to

$$y(x,t) = \left(A_1\cos\frac{\omega x}{v}\right)(C_1\cos\omega t + D_1\sin\omega t). \tag{10.11}$$

Next, we differentiate Eq. (10.11) with respect to x and use second boundary condition in the resultant expression. This gives

$$\left(-A_1\frac{\omega}{v}\right)\sin\left(\frac{\omega l}{v}\right)(C_1\cos\omega t + D_1\sin\omega t) = 0.$$

For it to hold for all values of t, we must have

$$\sin\frac{\omega l}{v} = 0.$$

This will hold for

$$\frac{\omega l}{v} = n\pi, \tag{10.12}$$

where n is an integer. It means that values of the constants A_1, C_1 and D_1 and hence displacement $y(x, t)$ depends on the value of n. For the nth mode, we can write

$$y_n(x, t) = A_{1n} \cos\frac{n\pi x}{l}(C_{1n} \cos \omega_n t + D_{1n} \sin \omega_n t)$$

$$= \cos\frac{n\pi x}{l}(a_n \cos \omega_n t + b_n \sin \omega_n t), \tag{10.13}$$

where a_n and b_n are two new constants.

From Chapter 9, you may recall that this equation is analogous to Eq. (9.18) for a stretched string. Since $y(x, t)$ satisfies a linear wave equation, the most general solution is obtained by linear superposition of all solutions of the form given by Eq. (10.13):

$$y(x, t) = \sum_{n=1}^{\infty} \cos\frac{n\pi x}{l}(a_n \cos \omega_n t + b_n \sin \omega_n t). \tag{10.14}$$

The expression for characteristic frequencies may be obtained from Eq. (10.12):

$$f_n = \frac{\omega_n}{2\pi} = \frac{nv}{2l} = \frac{n}{2l}\sqrt{\frac{Y}{\rho}} \tag{10.15}$$

and wavelength for the nth mode, λ_n may be written as follows:

$$\lambda_n = \frac{v}{f_n} = \frac{2l}{n}. \tag{10.16}$$

The frequency and wavelength of fundamental mode is given by

$$\text{Fundamental frequency } f_1 = \frac{1}{2l}\sqrt{\frac{Y}{\rho}} \tag{10.17}$$

and

$$\text{Fundamental wavelength } \lambda_1 = 2l.$$

You may now like to do Practice Exercise 10.1.

***Practice Exercise* 10.1** Consider a bar fixed at the centre and free at both ends. Show that in such a bar only odd harmonics are excited.

***Case* (ii)** A bar rigidly fixed at one end and free to move at the other end is said to be a *fixed-free* bar (Fig. 10.2).

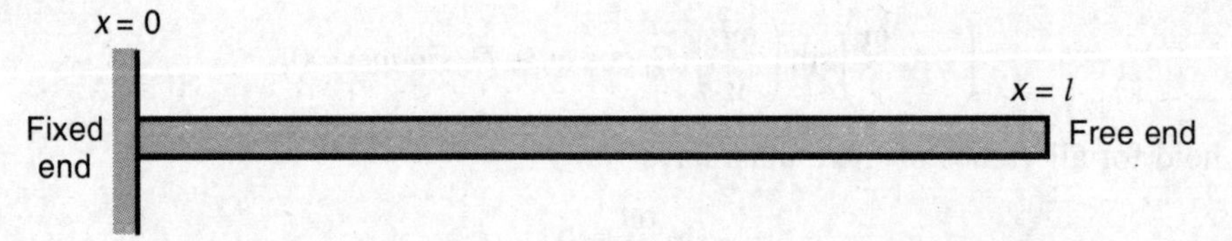

Fig. 10.2 A bar fixed at one end and free at the other end.

The boundary conditions in this case are

(i) At $x = 0$, $y = 0$ for all values of t and

(ii) At $x = l$, $\partial y/\partial x = 0$ for all values of t. (10.18)

As before, use of boundary condition (i) in Eq. (10.9), leads to the result $A_1 = 0$. Then Eq. (10.9) reduces to

$$y(x,t) = \left(B_1 \sin\frac{\omega x}{v}\right)(C_1 \cos \omega t + D_1 \sin \omega t).$$

Next, we use boundary condition (ii) in Eq. (10.10) with $A_1 = 0$. This gives

$$\left(B_1 \frac{\omega}{v}\right)\cos\left(\frac{\omega l}{v}\right)(C_1 \cos \omega t + D_1 \sin \omega t) = 0.$$

For it to hold for all values of t, we must have

$$\cos\frac{\omega l}{v} = 0$$

or

$$\frac{\omega l}{v} = (2n-1)\frac{\pi}{2}, \quad n = 1, 2, 3, \ldots \tag{10.19}$$

Hence, the characteristic frequencies of a free-fixed bar are given by

$$\omega_n = (2n-1)\frac{\pi v}{2l}$$

or

$$f_n = \frac{\omega_n}{2\pi} = (2n-1)\frac{v}{4l} = \frac{(2n-1)}{4l}\sqrt{\frac{Y}{\rho}}. \tag{10.20}$$

Note that the fundamental frequency and the corresponding fundamental wavelength are $f_1 = v/4l$ and $\lambda_1 = 4l$, respectively. The first higher harmonic corresponds to $n = 2$ and has frequency $f_2 = 3v/4l = 3f_1$. This indicates that only odd harmonics will be excited in a fixed-free bar.

The general expression for displacement $y(x, t)$ of a fixed-free bar executing transverse vibrations is given by

$$y(x,t) = \sum_{n=1}^{\infty} \sin\frac{(2n-1)\pi x}{2l}(a_n \cos \omega_n t + b_n \sin \omega_n t). \tag{10.21}$$

***Case* (iii)** A bar fixed at both ends is said to be *fixed-fixed bar* (Fig. 10.3). This case is similar to that of a stretched string fixed at both ends discussed in Chapter 9. Therefore, if you think that proceeding along the steps outlined for a stretched string, you can obtain an expression for displacement, you are thinking logically.

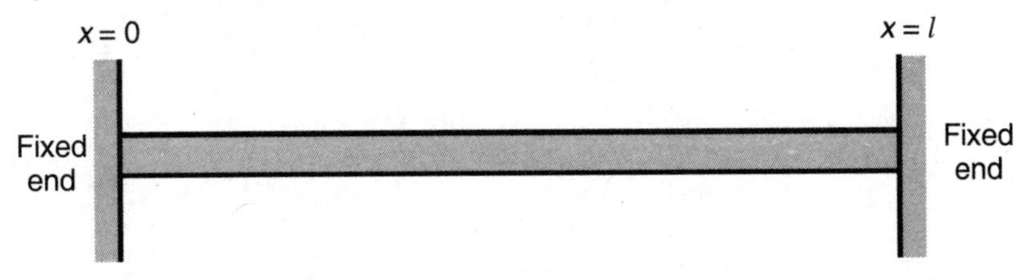

Fig.10.3 A bar fixed at both ends.

The boundary conditions for a fixed-fixed bar are:

(i) At $x = 0$, $y = 0$ for all values of t.

(ii) At $x = l$, $y = 0$ for all values of t. (10.22)

Using boundary condition (i) in Eq. (10.9), we get

$$A_1 (C_1 \cos \omega t + D_1 \sin \omega t) = 0. \tag{10.23}$$

This equation will be satisfied for all values of t only if $A_1 = 0$. Hence, Eq. (10.9) simplifies to

$$y(x, t) = \left(B_1 \sin \frac{\omega x}{v} \right) (C_1 \cos \omega t + D_1 \sin \omega t).$$

Now, using boundary condition (ii) in this equation, we can write

$$\left(B_1 \sin \frac{\omega l}{v} \right) (C_1 \cos \omega t + D_1 \sin \omega t) = 0. \tag{10.24}$$

For Eq. (10.24) to hold for all values of t, $\sin \omega l/v$ must be equal to zero. Hence, we must have

$$\frac{\omega l}{v} = n\pi, \quad n = 1, 2, 3, \ldots$$

or

$$\omega = \frac{n\pi v}{l}. \tag{10.25}$$

Note that we have not taken the value $n = 0$ as it will lead to a trivial solution. Physically, it will correspond to a non-vibrating rod.

The characteristic frequencies of vibrations of a fixed-fixed bar are given by

$$\omega_1 = \frac{\pi v}{l}, \omega_2 = \frac{2\pi v}{l}, \omega_3 = \frac{3\pi v}{l} \cdots$$

That is, a fixed-fixed bar has an infinite set of frequencies and we can write

$$\omega_n = n\omega_1 = \frac{n\pi v}{l}.$$

The solution corresponding to the nth mode can be written as

$$y_n(x, t) = \sin \frac{n\pi v}{l} (a_n \cos \omega_n t + b_n \sin \omega_n t), \tag{10.26}$$

where we have put $a_n = A_n C_n$ and $b_n = A_n D_n$.

Since solution exists for each value of n and wave equation is linear and homogeneous, as before, its general solution will be obtained by superposition of all these solutions. Hence, we can write

$$y(x, t) = \sum_{n=1}^{\infty} \sin \frac{n\pi x}{l} (a_n \cos \omega_n t + b_n \sin \omega_n t)$$

$$= \sum_{n=1}^{\infty} c_n \sin \frac{n\pi x}{l} \cos (\omega_n t - \phi_n). \tag{10.27}$$

From Eq. (10.27), we may conclude that an element dx of the bar vibrates simple harmonically with amplitude $c_n \sin (n\pi x/l)$ and angular frequency ω_n. The frequency of the nth mode of vibration and the corresponding wavelength are given by

$$f_n = \frac{\omega_n}{2\pi} = \frac{n\pi v}{l \cdot 2\pi} = \frac{nv}{2l} = \frac{n}{2l}\sqrt{\frac{Y}{\rho}}. \tag{10.28}$$

and

$$\lambda_n = \frac{v}{f_n} = \frac{2l}{nv}v = \frac{2l}{n}. \tag{10.29}$$

Note that all harmonics are present in a fixed-fixed bar, as in case of the free-free bar. The fundamental frequency in this case is given by

$$f_1 = \frac{1}{2l}\sqrt{\frac{Y}{\rho}}. \tag{10.30}$$

The frequencies of higher modes are simple multiples of the fundamental frequency.

Another situation that you may encounter is that of a *free-fixed-free bar*. In this case, the bar is free at both ends, but fixed (clamped) in the middle. Therefore, we can say that in this case stationary waves will be formed with a node at the middle point and antinodes at the ends.

Like case (i), the solution is given by Eq. (10.14) with characteristic frequencies given by Eq. (10.12).

Since the bar is clamped at the middle in this case, i.e., $y = 0$ at $x = l/2$, we can say that $\cos n\pi x/l$ should be equal to zero at $x = l/2$. That is

$$\cos \frac{n\pi}{l}\frac{l}{2} = 0$$

or

$$\frac{n\pi}{2} = (2p+1)\cdot\frac{\pi}{2}, \quad p = 0, 1, 2, 3, \ldots$$

so that

$$n = (2p + 1) = 1, 3, 5, \ldots \tag{10.31}$$

This result shows that frequencies of various harmonics excited in a bar fixed at the middle are proportional to the odd number and it can be said that all even harmonics will be absent. Therefore, we can write

$$f_n = \frac{n}{2l}\sqrt{\frac{Y}{\rho}}, \tag{10.32}$$

where $n = 1, 3, 5, \ldots$

It may be pointed out here that as in vibrating strings, a fixed-fixed or free-free vibrating bar can also be modelled as a stationary wave. We will not repeat the mathematical analysis as it is exactly similar to that given for vibrating strings. However, we may argue physically as follows:

The wave equation [Eq. (10.1)] admits plane wave solutions. These waves are reflected at the fixed ends or free ends, depending on the configuration of the bar (fixed-fixed or free-free). When reflection is perfect and there is no loss of energy, superposition of incident and reflected waves gives rise to displacement nodes and antinodes.

Now, study Example 10.2.

EXAMPLE 10.2

A brass bar of length 2.0 m is clamped at the centre. It emits a note of frequency 600 Hz, when it vibrates longitudinally. If the density of brass is 8.3 gcm^{-3}, calculate its Young's modulus.

Solution: The frequency for fundamental mode of vibration is given by

$$f = \frac{1}{2l}\sqrt{\frac{Y}{\rho}}$$

We are given that l = 200 cm, f = 600 Hz, and ρ = 8.3 gcm^{-3}. On substituting these values in the expression for frequency, we get

$$600 = \frac{1}{400}\sqrt{\frac{Y}{(8.3)}}$$

$$(240000)^2 = \frac{Y}{(8.3)}$$

$$\therefore \qquad Y = 8.3 \times (24)^2 \times 10^8 \text{ dynes cm}^{-2}$$

$$= 4.8 \times 10^{11} \text{ dynes cm}^{-2}$$

10.3 TRANSVERSE VIBRATIONS OF A BAR

In Chapter 9, you learnt that when a stretched string fixed at both ends is displaced transversely, stationary waves are generated along the length of the string. And in case of longitudinal oscillations of bars, we discovered that depending on the configuration, fundamental and other odd and/or even harmonics may be excited. You may now like to know: What happens in case of transverse vibrations of bars? You will learn that the resultant motion is a sum of harmonic components, each of which has a different velocity.

Recall that in a string, tension acts as the restoring force. But in a thin bar or a rod, this role is played by stiffness. When a bar is made to execute transverse vibrations, the forces that come into play are primarily those that arise in the bending of a beam; the strain is in the form of an elongation and a contraction on the two sides of the neutral axis. Refer to Fig. 10.4, which shows a straight thin bar subject to a force F acting normal to the neutral axis. We choose x-axis along the length of the bar such that it coincides with the neutral axis.

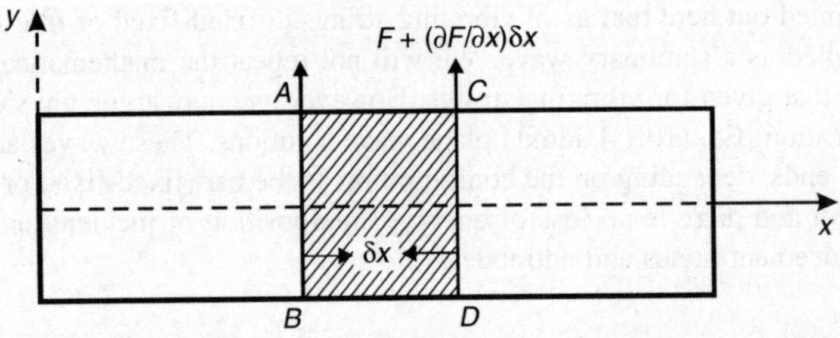

Fig. 10.4 Transverse vibration of a bar: The external force is applied normal to its neutral axis.

While studying bending of beams in your Mechanics course, you must have learnt that when a beam is bent by applying a force normal to its neutral axis, the bending moment M, which arises due to elastic changes inside the beam, is given by

AK^2 is called the a real moment of inertia of the bar about the neutral axis.

$$M = YAK^2 \frac{\partial^2 y}{\partial x^2}, \tag{10.33}$$

where Y denotes Young's modulus of the material of the bar, A is area of cross-section of the bar, K is radius of gyration about the neutral axis and y signifies depression at the point x. Furthermore, F and M are connected through the relation

$$F = -\left(\frac{\partial M}{\partial x}\right). \tag{10.34}$$

Let us now consider a thin slab $ABDC$ of thickness dx (Fig. 10.4) whose faces are perpendicular to the neutral axis. Suppose that the force acting at face AB is F. Then force at CD will be $F + (\partial F/\partial x)\,\delta x$. Since force is equal to the product of mass and acceleration, the unbalanced force $(\partial F/\partial x)\,\delta x$ can be expressed in terms of the parameters of the thin slab as

$$\left(\frac{\partial F}{\partial x}\right)\delta x = (A\rho\delta x)\frac{\partial^2 y}{\partial t^2}, \tag{10.35}$$

where ρ is density of the material of the bar.
On differentiating Eq. (10.34) with respect to x and combining with Eq. (10.33), we can write

$$\frac{\partial F}{\partial x} = -\frac{\partial^2 M}{\partial x^2} = -YAK^2\frac{\partial^4 y}{\partial x^4}.$$

On using this result in Eq. (10.35), we get

$$-YAK^2\frac{\partial^4 y}{\partial x^4}\cdot\delta x = (A\rho\delta x)\cdot\frac{\partial^2 y}{\partial t^2}$$

or

$$\frac{\partial^4 y}{\partial x^4} + \left(\frac{\rho}{YK^2}\right)\frac{\partial^2 y}{\partial t^2} = 0. \tag{10.36}$$

This equation represents the wave equation for *flexural waves* propagating along a thin bar or a rod. Note that flexural waves are not analogous to shear waves.

As before, you can solve Eq. (10.36) using the method of separation of variables. The expression for displacement shall involve trigonometric as well as hyperbolic functions. The steps are simple and we suggest that you should do Practice Exercise 10.2.

***Practice Exercise* 10.2** Using the method of separation of variables, show that the solution of Eq. (10.37) is given by

$$y(x, t) = \left[A\cosh\left(\frac{\omega x}{v}\right) + B\sinh\left(\frac{\omega x}{v}\right) + C\cos\left(\frac{\omega x}{v}\right) + D\sin\left(\frac{\omega x}{v}\right)\right]\cos(\omega t + \phi).$$

Let us express the solution of Eq. (10.37) as

$$y = Te^{j(\omega t - kx)} \tag{10.37}$$

where T is a constant. Note that the ratio ω/k defines the velocity v of the wave. On differentiating Eq. (10.37) with respect to x and t, we get

$$\frac{\partial^4 y}{\partial x^4} = k^4 y$$

and

$$\frac{\partial^2 y}{\partial t^2} = -\omega^2 y.$$

On substituting these results in Eq. (10.36), we get

$$k^4 y - \left(\frac{\rho}{YK^2}\right)\omega^2 y = 0$$

and

$$\frac{\omega^4}{v^4} y - \left(\frac{\rho}{YK^2}\right)\omega^2 y = 0. \tag{10.38}$$

We can rearrange this relation to express wave velocity in terms of elastic properties of the bar.

Since $y \neq 0$, we must have $$v^4 = \frac{\omega^4 y YK^2}{\rho\omega^2 y} = \frac{\omega^2 YK^2}{\rho}$$

or

$$v = \sqrt{\omega u K}, \tag{10.39}$$

where $u = \sqrt{Y/\rho}$ is the velocity of longitudinal waves in a bar.

Note that the wave velocity v is a function of the angular frequency ω. If a number of frequencies are simultaneously present, the speed of propagation is given by the group velocity

$$v_g = \frac{\partial \omega}{\partial k}. \tag{10.40}$$

When stationary flexural waves are set up in a bar, the boundary conditions must correspond to the given configuration. A fixed-free bar is clamped at one end and its free end is struck perpendicular to its length. This will generate transverse wave along the bar. The incident wave will be reflected at the fixed end and their superposition leads to stationary wave pattern with a node at the fixed end and an antinode at the free end.

The fundamental, first and the second overtones in a fixed-free bar are shown in Fig. 10.5. The frequency corresponding to the fundamental mode is given by

$$f_1 = \frac{\pi u K}{8l^2} \times 1.194^2 = \frac{0.56K}{l^2}\sqrt{\frac{Y}{\rho}} \tag{10.41}$$

and the overtones are given by

$$f_2 = 6.267 f_1,\ f_3 = 17.55 f_1;\ f_4 = 34.39 f_1. \tag{10.42}$$

Note that the overtones are not harmonics of the fundamental mode. We will not prove this result because mathematical steps involved in the calculation of characteristic frequencies are quite involved. However, it may be pertinent to mention that higher modes fade away very quickly.

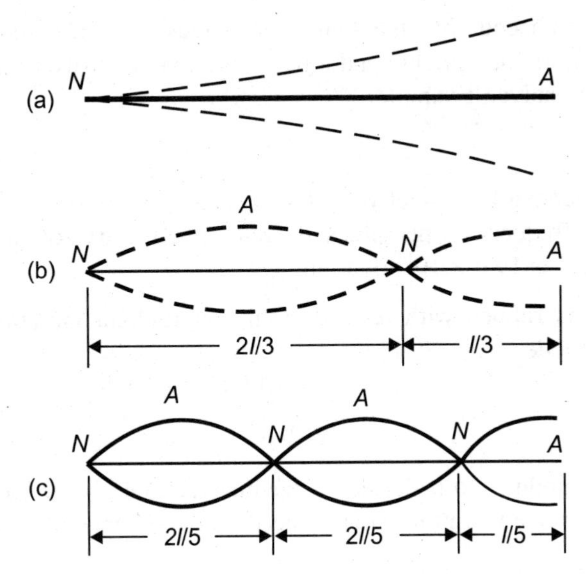

Fig. 10.5 Fundamental, first and second harmonic in a fixed-free bar executing transverse vibrations.

A free-free bar is supported properly but not clamped. It can be made to execute transverse vibrations by striking at the middle. Obviously, there will be displacement antinodes at the free ends and in-between these, there will be alternate nodes and antinodes. The frequency of fundamental mode is given by

$$f_1 = \frac{3.56K}{l^2}\sqrt{\frac{Y}{\rho}},$$

The higher harmonics are given by

$$f_2 = 2.756f_1, f_3 = 5.404f_1; f_4 = 8.933f_1 \tag{10.43}$$

The fundamental and the second harmonic are depicted in Fig. 10.6. As you will note, when the bar vibrates in the fundamental mode, there are two nodes and three antinodes. Moreover, the distance of the nodes from the nearest free end is $0.224l$. For the first overtone (second harmonic), the nodes come closer to the free ends and the node-to-node distance is not equal to $\lambda/2$ because the expression for the displacement involves trigonometric as well as hyperbolic functions.

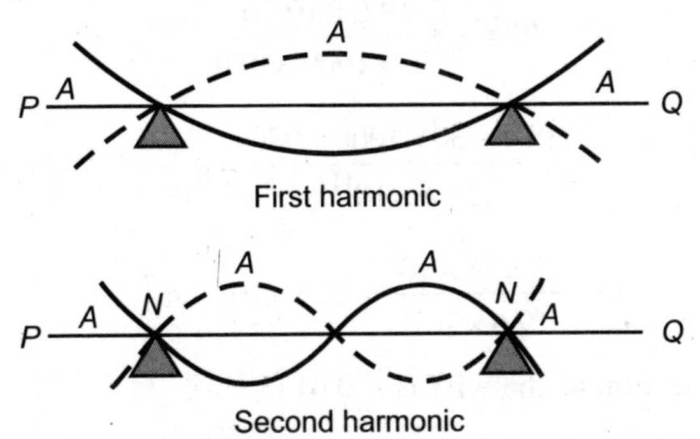

Fig. 10.6 Fundamental or first and second harmonic in a free-free bar executing transverse vibrations.

A bar clamped at both ends has the same frequencies as a free-free bar. In all cases, the overtones, when excited, die out quickly and a pure tone can be easily obtained.
Now study the following solved Example.

EXAMPLE 10.3

A load of 30 kg is suspended by a steel wire. Its frequency when rubbed with a resined cloth is found to be 30 times its frequency when plucked. Determine the area of cross-section of the wire. Young's modulus of steel is 19.6×10^{11} dyne cm^{-2}.

Solution: When wire is rubbed with resined cloth, longitudinal vibrations are produced with velocity for the lowest mode

$$f_1 = \frac{1}{2l}\sqrt{\frac{Y}{\rho}}, \tag{i}$$

where Y is the Young's modulus, ρ is density of the material and l is the length of the wire. When the wire is plucked, transverse vibrations are produced. The frequency of fundamental note is given by

$$f_2 = \frac{1}{2l}\sqrt{\frac{T}{m}}. \tag{ii}$$

Dividing Eq. (i) by Eq. (ii), we get

$$\frac{f_1}{f_2} = \sqrt{\frac{Y}{\rho}\cdot\frac{m}{T}}. \tag{iii}$$

Let a is the area of cross-section of the wire. Then

$$m = a\rho.$$

Therefore, Eq. (iii) becomes

$$\frac{f_1}{f_2} = \sqrt{\frac{Y}{T}\cdot a}.$$

From the given data, we note that

$$\frac{f_1}{f_2} = 30, T = 30 \times 1000 \times 980 \text{ dynes and } Y = 19.6 \times 10^{11} \text{ dynes cm}^{-2}$$

$$30 = \sqrt{\frac{19.6 \times 10^{11} a}{30 \times 1000 \times 980}}$$

$$\frac{900 \times 30 \times 1000 \times 980}{19.6 \times 10^{11}} = a$$

$$a = \frac{27 \times 98 \times 10^7}{19.6 \times 10^{11}} = 135 \times 10^{-4} \text{ cm}^2.$$

Hence, the area of cross-section of the wire is = 0.0135 cm^2.

10.4

TUNING FORK AND VIBRATIONS OF A BAR

In your school physics laboratory, you must have worked with a tuning fork and noted that as frequency of a tuning fork increases, it becomes smaller but thicker. Do you know the reason? To discover answer to this question, recall the motion of a free-free bar executing transverse vibrations in fundamental mode. If it is gradually bent into U shape, the nodes approach the bend, as shown in Fig. 10.7. As a result, the length of each outer portion of the vibrating system increases gradually. This lowers the frequency as well as the amplitude of vibration at the centre. (When the arms are parallel, the motion resembles the motion of a tuning fork.) Since there is nothing more that affects the original motion of the bar, its centre continues to be antinode. We attach the stem at this antinode. This forces the nodes to come still closer and reduces the amplitude of vibrations further.

You may have realised that when a tuning fork is struck mildly against a pad and pressed on a surface, its stem tends to move up. This corresponds to outward motion of the arms. This is indicated by '1' in Fig. 10.7. Similarly, the inward motion of arms lowers the stem. This is denoted by '2' in Fig. 10.7. It means that transverse vibrations of the prongs change into longitudinal vibrations of the stem. And when we prcss the stem on a surface, it is set into forced vibrations.

Now refer to Fig. 10.8, which shows the state of vibration of the tuning fork when the first overtone is excited. It corresponds to vigorous excitation of the tuning fork. As in the case of Fig. 10.7 (1, 1) and (2, 2) respectively indicate that the prongs either recede from or approach to each other. As mentioned earlier, the overtones die down quickly. For this reason, a tuning fork maintains fundamental mode for long time and is used as an indispensable tool for measurement of frequency.

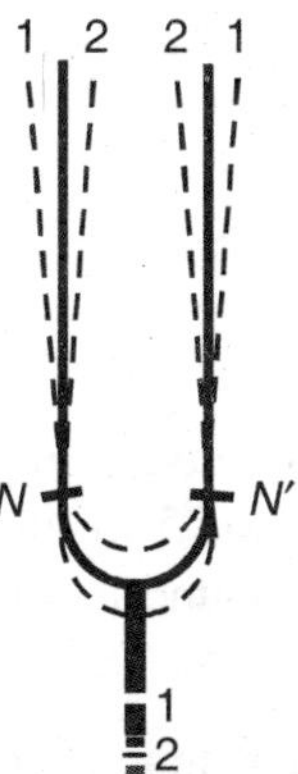

Fig. 10.7 A vibrating tuning fork (N, N'-positions of nodes).

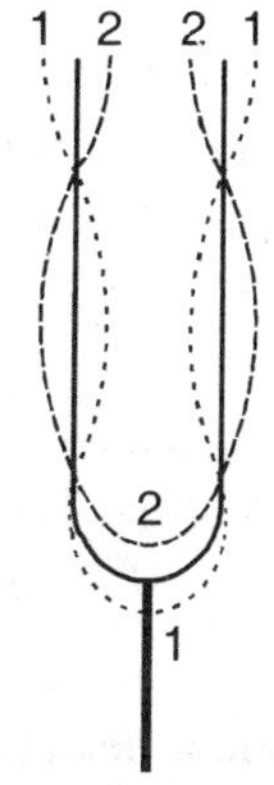

Fig. 10.8 A typical state of vibration of a tuning fork when the first overtone is excited.

REVIEW EXERCISES

10.1 A wire is held rigidly between two ends at 30 °C. In this situation it is just tight with negligible tension. Determine the coefficient of linear expansion of the material of the wire using the following data: The velocity of transverse waves in the wire is 60 ms^{-1}. Y and ρ for the material of the wire are respectively equal to 1.2×10^{11} Nm^{-2} and 6000 kgm^{-3}.

[*Ans.* $1.8 \times 10^{-5}\,°\text{C}^{-1}$]

10.2 A metallic rod of length 1 m is rigidly clamped at its mid-point. Longitudinal stationary waves are set up in the rod in such a way that there are two nodes on either side of the mid-point. Write down the equations of motion at a point 4 cm from the mid-point as also for the component waves in the rod. The amplitude of an antinode is 2×10^{-6} Nm^{-2} and 8000 kgm^{-3}.

[*Ans.* $y = 2 \times 10^{-6} \cos 2.7\pi \sin (25\pi \times 10^3 t)$ m or
$y = 2 \times 10^{-6} \cos 2.3\pi \sin (25\pi \times 10^3 t)$m;
$y_1 = 10^{-6} \sin 5\pi (5\pi \times 10^3 t - x)$ m
and $y_2 = 10^{-6} \sin 5\pi (5\pi \times 10^3 t + x)$ m]

10.3 A fixed-free bar produces the fundamental frequency f'. Calculate the length of the bar. The bar is then set in the free-free mode. What fundamental tone will it emit now? Take velocity of sound in the bar as v.

[*Ans.* $v/4f$, $2f$]

10.4 A wire is stretched between two clamps 1 m apart. The lowest frequency of transverse vibration when it is subjected to an extension of 0.05 cm is 35.3 Hz. Calculate the density of the material of the wire. Take $Y = 9 \times 10^{10}$ Nm^{-2}.

[*Ans.* 9.03×10^3 kgm^{-3}]

10.5 Write down the boundary conditions for the free-fixed-free bar of Practice Exercise 10.1, and solve the partial differential equation of transverse wave in a bar. Obtain an expression for the frequency of the nth harmonic in terms of the length l of the rod and the density ρ, Young's modulus Y of its material.

$$\left[\textit{Ans.} \quad \frac{2n-1}{l}\sqrt{\frac{Y}{\rho}}\right]$$

10.6 Six antinodes are observed in an air column when a standing wave forms in a Kundt's tube. Calculate the length of the air column if a steel bar of length 1 m is secured (a) at the middle, (b) at the end. I The velocity of sound in steel and in air are 5250 ms^{-1} and 343 ms^{-1} respectively.

[*Ans.* 0.392 m; 0.784 m]

10.7 Calculate the length of a glass bar in a Kundt's tube if five antinodes are observed in the air column when the bar is secured at the middle. The bar is 0.25 m long. It is given that $Y_{glass} = 6.9 \times 10^{10}$ Nm^{-2}, $\rho_{glass} = 2.5$ gcm^{-3} and velocity of sound in air is 343 ms^{-1}.

[*Ans.* 0.72 m]

10.8 The minimum detectable distance between antinodes in a Kundt's tube is about 4 mm. Calculate the maximum frequency that can be determined by Kundt's tube method used to determine the velocity of sound. The velocity of sound in air is 340 ms^{-1}.

[*Ans.* 43 kHz]

11

Vibrations of Air Columns

Expected Learning Outcomes

In this Chapter, you will acquire capability to:

- obtain an expression for characteristic frequencies of (i) an open and (ii) a closed pipe;
- explain the end correction for a pipe;
- obtain most general solution of wave equation for an air column; and
- derive an expression for energy of a vibrating gaseous column.

11.1 INTRODUCTION

In Chapters 9 and 10, we have discussed transverse and longitudinal vibrations of bowed, plucked and struck stretched strings and fixed-free, fixed-fixed and free-free bars. We discovered that superposition of incident and reflected waves gives rise to stationary waves, which are responsible for producing music. But all musical instruments or sound sources are not stringed. Stretched membranes, which are essentially two-dimensional analogues of stretched strings, are used in a drum, tabla, dholak, etc. to generate music. You must have enjoyed the tabla vadan of Indian maestros Ustad Zakir Hussain and Allah Rakha Khan. While drum is an integral part of every orchestra, dholak is used in folklore music, including bhajans and kirtans (From the point of view of physics, we write 2-D wave equation and solve it using the method of separation of variables, using the steps followed in solving the wave equation for a string. You will discover that nodal points appear as nodal lines.)

You also learnt that when bars are set into oscillations, in general, the overtones are not harmonics of the fundamental mode. In fact, overtones are so feeble that they decay very fast. That is why when a tuning fork is struck mildly, it acts as a standard source of frequency. In instruments like organ pipes, clarinet, flute, woodwind, trumpet and horn-loudspeakers, tube-shaped devices, called *air columns*, are used to generate music. From your school physics classes, you will recall that a column of air enclosed in a pipe constitutes an air column.

In Section 11.2, we have discussed the vibrations of air columns in hollow cylinders of uniform cross-section. Note that air columns can support only longitudinal waves and superposition of identical waves moving in opposite directions due to reflection at the boundaries leads to formation of stationary waves. If one end of the cylindrical tube is closed and the other end is open, we get what is known as *closed pipe*. In daily life, a well and a bucket under an open tap present familiar situations of closed pipes. When a tube is open at both ends, we get an *open pipe*. You will discover that the modes supported by the open and closed pipes are significantly different. However, vibrations of an air column in a pipe closed at both ends are of little practical interest and we will not consider it here. In Section 11.3, you will learn to obtain expression for energy of a vibrating air column.

11.2 VIBRATIONS OF AIR COLUMNS

When a stream of air is pushed down an air column, longitudinal waves are produced. These waves travel downward and are reflected at the other end. For perfect reflection, i.e., when there is no loss of energy and intensity of wave remains unchanged, we get two identical waves moving in opposite directions. And superposition of these waves results in stationary waves. However, the nature of the reflected wave and hence waveform of the resultant wave depends on whether the pipe is open or close ended.

If reflection occurs at a closed end, the displacement of the particles there would necessarily be zero always. That is, the closed end acts as a *node*. On the other hand, the open end acts as an *antinode*. It means that in a close-ended air column, open end acts as an antinode and the closed end acts as a node. The standing waves set up in such an air column will have only a few frequencies. These are called *resonance frequencies* of the air column. We can also say that vibrations corresponding to some definite frequencies only persist.

When reflection is not perfect, a fraction of incident energy will be lost to the surrounding medium. This invariably happens when a wave reaches the open end of a pipe. However, in our discussion here, we shall ignore all such energy losses.

To develop the theory of vibrations in pipes, we make some simplifying assumptions:

(i) The walls of the pipe are perfectly rigid.
(ii) Plane waves are produced in the pipe so that the motion is uniform throughout.
(iii) The diameter of the pipe is small compared to its length as also the wavelength of sound.
(iv) The viscous and thermal conduction effects are almost negligible.
(v) The vibrations take place quite rapidly and wave propagation can be considered as an adiabatic process.
(vi) There is no rotatory motion and vortices in the pipe.
(vii) The pressure in the pipe is steady, i.e., velocity and pressure changes are almost negligible. This ensures high quality notes.

Consider an air column in a long pipe or cylindrical tube of cross-sectional area A. Let us choose x-axis along the length of the tube. If density of the gas is ρ, you may recall from section 9.7.3 that longitudinal vibrations in the air column satisfy the differential equation

$$\frac{\partial^2 \psi}{\partial t^2} = v^2 \frac{\partial^2 \psi}{\partial x^2}, \tag{11.1}$$

where

$$v = \sqrt{\frac{E}{\rho}} \tag{11.2}$$

is the wave velocity of the waves, E is bulk modulus and $\psi(x, t)$ is displacement of a particle in the air column at position x and time t.

Using the method of separation of variables and following the steps outlined in Chapter 9, you can easily convince youself that the solution of Eq. (11.1) is given by

$$\psi(x, t) = (A_1 \cos kx + B_1 \sin kx)\,(C_1 \cos \omega t + D_1 \sin \omega t), \tag{11.3}$$

where $k = \omega/v$ is wave vector and constants A_1, B_1, C_1 and D_1 can be determined from initial and boundary conditions. Let us first consider an open pipe.

11.2.1 Vibrations in an Open Pipe

An open pipe has both ends open to the air. The familiar open pipe musical instruments include clarinet, flute, trumpet, and reed, among others. In these instruments, air is blown in through one end and sound emerges out the other end of the pipe. The motion of the medium in such instruments is practically unrestrained at the open ends. Refer to Fig. 11.1. It shows an open pipe, whose ends at $x = 0$ and $x = l$ correspond to displacement antinodes. That is, at $x = 0$ and $x = l$, $\psi(x, t)$ is maximum and $\partial\psi/\partial x = 0$. Using this condition in Eq. (11.3) at $x = 0$ for all t gives $B_1 = 0$. Then Eq. (11.3) reduces to

$$\psi(x, t) = \cos kx\,(a \cos \omega t + b \sin \omega t), \tag{11.4}$$

where a and b are new constants.

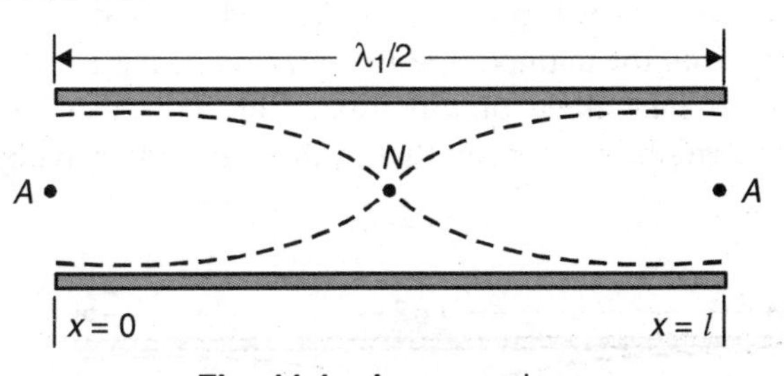

Fig. 11.1 An open pipe.

Next, we calculate the first partial derivative of displacement $\psi(x, t)$ with respect to x and apply the condition $\partial\psi/\partial x = 0$ at $x = l$ in the resultant expression. This gives

$$-\sin kl\,(a \cos \omega t + b \sin \omega t) = 0.$$

This expression will hold for all values of t only if

$$\sin kl = 0,$$

or

$$kl = \frac{\omega l}{v} = n\pi, \quad \text{for } n = 1, 2, 3 \tag{11.5}$$

So as for strings and bars, we can write the general solution as

$$\psi(x, t) = \sum_{n=1}^{\infty} \cos\frac{n\pi x}{l}\left(a_n \cos\frac{n\pi vt}{l} + b_n \sin\frac{n\pi vt}{l}\right). \tag{11.6}$$

The frequency of the nth mode of vibration of the air column in terms of bulk modulus and density ρ is given by

$$f_n = \frac{\omega_n}{2\pi} = \frac{nv}{2l} = \frac{n}{2l}\sqrt{\frac{E}{\rho}}. \tag{11.7}$$

It shows that all harmonics (odd and even) of fundamental mode can exist in this case. The frequency of fundamental mode (n = 1), is the lowest frequency. It is given by

$$f_1 = \frac{1}{2l}\sqrt{\frac{E}{\rho}}. \tag{11.8}$$

Since $\lambda_1 = v/f_1$, we find that for the fundamental mode,

$$\lambda_1 = 2l$$

or

$$l = \frac{\lambda_1}{2}. \tag{11.9}$$

Note that these expressions for displacement and frequency are similar to Eqs. (9.19) and (9.22), respectively. Equation (11.9) shows that there is only one node between the antinodes at the open ends and the distance between successive antinodes is $\lambda/2$ and a node and an antinode is quarter of a wavelength. This is illustrated in Fig. 11.2a. Physically, we say that in the fundamental mode, a pulse of compression travelling from open end 1 into the tube is reflected from open end 2 as a pulse of rarefaction, i.e., with a reversal of phase. The reflected pulse travels back in the tube and is reflected from end '1' with a phase reversal, i.e., as a pulse of compression.

Now, refer to Fig. 11.2b, which depicts the nodes and antinodes of the first overtone in an open pipe. Note that apart from the antinodes at the open ends, there are two nodes and one more antinode inside the pipe in this mode of vibration. Since distance between two consecutive antinodes is half of a wavelength, we can say that in this case, the wavelength is given by

$$\lambda_2 = l. \tag{11.10}$$

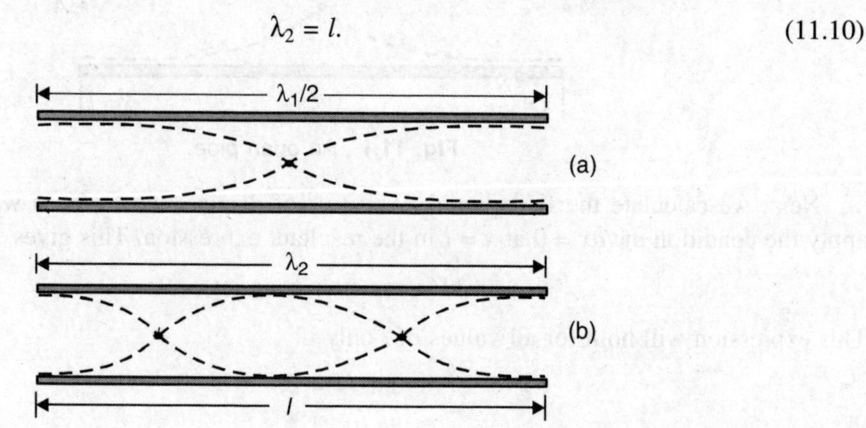

Fig. 11.2 Waveforms of (a) fundamental and (b) first harmonic in an open pipe.

In higher harmonics, there will be three, four or more integral number of nodes between the antinodes at the open ends. So, if the total number of antinodes, including those at the ends, is $n + 1$, we can say that there are n half-wavelengths between the ends of an open pipe. Then

$$l = \frac{n\lambda_n}{2}.$$

The corresponding wavelength λ_n is given by

$$\lambda_n = \frac{2l}{n} = \frac{\lambda_1}{n}$$

and frequency

$$f_n = \frac{v}{\lambda_n} = nf_1. \tag{11.11}$$

This result shows that the overtones are harmonics of fundamental mode and all frequencies are possible.

When you place a bucket under a running tap, the sound of water falling in the bucket changes as it gets filled with water. You must have looked for the possible explanation of this observed fact. It is in the physics of vibrations in a closed pipe. Let us discover it now.

11.2.2 Vibrations in a Closed Pipe

Consider a closed pipe shown in Fig. 11.3. Let us choose x-axis along the length of the pipe and assume that its ends are at $x = 0$ and $x = l$. We know that at the closed end of a rigid pipe, no motion of the medium is possible. Moreover, a closed end behaves like a denser medium and the waves reflected from it do not experience any phase change; only the direction of motion changes. It means that closed end acts like a displacement node. On the other hand, an open end acts as displacement antinode. And we expect pairs of alternating node and antinode in-between. This corresponds to a fixed-free bar.

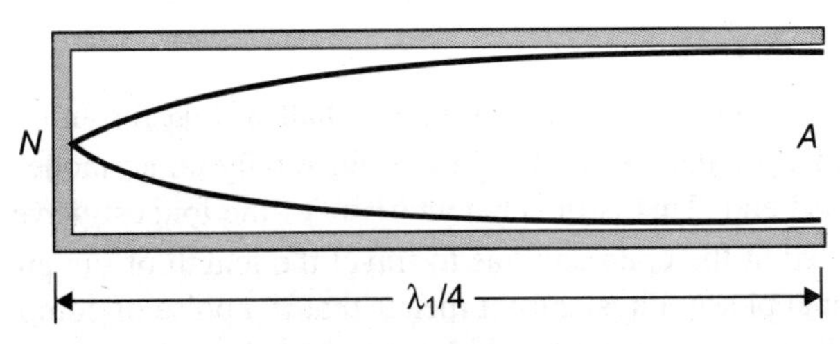

Fig. 11.3 A closed pipe.

The boundary conditions in this case are:

(i) At the closed end ($x = 0$), $\psi(0, t) = 0$ and

(ii) At the open end ($x = l$), $\psi(l, t)$ is maximum, i.e., $\partial\psi/\partial x = 0$.

Using the first condition in Eq. (11.3) for all t gives $A_1 = 0$. Then Eq. (11.3) reduces to

$$\psi(x, t) = \sin kx\,(a \cos \omega t + b \sin \omega t), \tag{11.12}$$

where a and b are new constants.

To use the second condition, we differentiate Eq. (11.12) with respect to x and put $x = l$ in the resultant expression. This gives

$$\frac{\partial \psi}{\partial x} = k \cos kx\,(a \cos \omega t + b \sin \omega t) = 0.$$

For it to hold for all values of t, we must have

$$\cos kl = 0$$

or

$$kl = (2n + 1)\frac{\pi}{2}, \quad n = 0, 1, 2, 3, \ldots \tag{11.13}$$

Hence, we can write the most general solution as

$$\psi(x, t) = \sum_{n=1}^{\infty} \sin\frac{(2n+1)\pi x}{2l}(a_n \cos \omega_n t + b_n \sin \omega_n t), \tag{11.14}$$

where frequencies are given by

$$\omega_n = (2n+1)\frac{\pi v}{2l}$$

or

$$f_n = \frac{\omega_n}{2\pi} = (2n+1)\frac{v}{4l} = \frac{(2n+1)}{4l}\sqrt{\frac{E}{\rho}}, \quad n = 0, 1, 2, 3, \ldots \tag{11.15}$$

The fundamental frequency is given by

$$f_1 = \frac{1}{4l}\sqrt{\frac{E}{\rho}} = \frac{v}{4l} \tag{11.16a}$$

and the corresponding wavelength is given by

$$\lambda_1 = 4l, \tag{11.16b}$$

since $\lambda_1 = v/f_1$.

Equation (11.16a) shows that

- The fundamental frequency of a closed pipe is half of that for an open pipe.
- In the fundamental mode, a closed pipe will have only an antinode at the open end and a node at the closed end. This is the characteristic of the longest wave possible in the pipe.
- The pulse initiated at the open end has to travel the length of the pipe four times before it attains the original phase. Physically it means that if a pulse of compression starts from the open end, it will travel in the pipe and get reflected as the pulse of compression at the closed end, which acts as the interface of a denser medium. The reflected pulse will travel back along the pipe and again get reflected at the open end as pulse of rarefaction due to a phase change of π there. The rarefaction pulse will again travel the length of the pipe and get reflected at the closed end as a pulse of rarefaction, which, in turn, is reflected at the open end as pulse of compression.

The first harmonic present will have a pair of nodes and a pair of antinodes; one node and one antinode between the nodes and antinodes at the closed and open ends. Therefore,

$$l = \frac{3\lambda_2}{4}$$

or

$$\lambda_2 = \frac{4l}{3}$$

and

$$f_2 = \frac{v}{\lambda_2} = 3f_1. \tag{11.17}$$

This result shows that the first overtone has a frequency three times the frequency of the fundamental mode. It is the third harmonic. It means that second harmonic is not excited in a closed pipe. You may now ask: Is it true only for the second harmonic or all even harmonics? To discover answer to this question, as before, we extend the argument and note that when there are n nodes and n antinodes, including the ones at the ends, the length of the pipe will be equal to $(2n - 1)$ quarter-wavelength. Hence,

$$l = \frac{(2n-1)\lambda_n}{4}$$

or

$$\lambda_n = \frac{4l}{2n-1}$$

and

$$f_n = \frac{v}{\lambda_n} = (2n - 1)f_1, \; n = 1, 2, 3... \tag{11.18}$$

This result shows that the overtones are all harmonics of the fundamental mode but unlike an open pipe, only odd harmonics will be present in a closed pipe. This is illustrated in Fig. 11.4 for the first and third harmonics of a closed pipe.

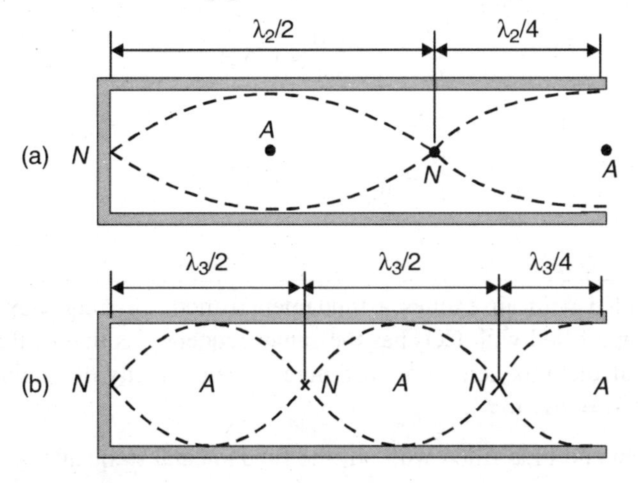

Fig. 11.4 Waveforms of (a) first and (b) third harmonics in a closed pipe.

Now, study the following solved examples carefully.

EXAMPLE 11.1

The speed of sound is 340 ms^{-1}. Calculate the first three resonance frequencies of a 2.5 m long cylindrical (i) closed pipe and (ii) open pipe.

Solution:

(i) For a closed pipe, the resonance frequencies is given by

$$f = \frac{(2n-1)v}{4l},$$

where l is the length of the cylinder and v is the speed of sound. Here l = 2.5 m and v = 340 ms^{-1}. Hence the first resonance frequency of a closed pipe is given by

$$f_1 = \frac{v}{4l} = \frac{340}{4 \times 2.5} = 34 \text{ Hz}$$

Similarly, the second resonance frequency is

$$f_2 = \frac{3v}{4l} = \frac{3 \times 340}{4 \times 2.5} = 102 \text{ Hz}$$

And third resonance frequency

$$f_3 = \frac{5v}{4l} = \frac{5 \times 340}{4 \times 2.5} = 170 \text{ Hz}.$$

(ii) For an open pipe, the resonance frequencies are given by

$$f_n = \frac{nv}{2l}$$

Therefore, the first resonance frequency is given by

$$f_1 = \frac{v}{2l} = \frac{340}{2 \times 2.5} = 68 \text{ Hz}$$

Second resonance frequency is

$$f_2 = \frac{2v}{2l} = \frac{2 \times 340}{2 \times 2.5} = 136 \text{ Hz}$$

Third resonance frequency is

$$f_3 = \frac{3v}{2l} = \frac{3 \times 340}{2 \times 2.5} = 204 \text{ Hz}$$

EXAMPLE 11.2

An open organ pipe filled with air excites a fundamental mode of frequency 400 Hz. The first harmonic of a closed pipe filled with CO_2 has the same frequency as that of the first harmonic of the open organ pipe. Calculate the length of each pipe. Given velocity of sound in air = 332 ms^{-1}, velocity of sound in CO_2 = 264 ms^{-1}.

Solution: For the open end pipe filled with air, the fundamental frequency is given by

$$f = \frac{v_{air}}{2\rho}.$$

Here, f = 400, v_{air} = 332 ms^{-1}. Hence,

$$400 = \frac{332}{2l} \quad \text{or} \quad l = 0.415 \text{ m}.$$

For the closed end pipe filled with CO_2, the fundamental frequency is given by

$$f = \frac{v_{CO2}}{4l}.$$

On substituting the given values, we get

$$400 = \frac{264}{4l} \quad \text{or} \quad l = 0.165 \text{ m}.$$

You may now like to solve Practice Exercise 11.1.

***Practice Exercise* 11.1** Calculate the ratio of fundamental mode fequencies of open and closed pipes of equal length.

[***Ans.*** 2 : 1]

In the above discussion, we assumed that air particles have maximum freedom at the open ends and these correspond to antinodes. We now relook at the situation from a different perspective.

11.2.3 End Correction

Let us consider propagation of plane waves in an open pipe. As it reaches the free end, it changes into spherical waves beyond the open end. It suggests that antinode is formed a little away from the open end. Physically it means that there is some excess pressure at the open end, but it decreases rapidly and becomes zero at some point outside the pipe. Essentially it leads to an increase in the vibrating length of the pipe over its geometrical length. So we can say that considering an antinode exactly at the open end will lead to some error. To eliminate this error, we apply a correction, called *end correction e*, at each open end. On certain assumptions, this distance has been calculated to be $0.57r$, where r is radius of the pipe. When this is done, the effective length of an open pipe becomes $l + 2e$, whereas for a closed pipe it will be $l + e$. On applying the end correction, the effective wavelength of the fundamental mode for an open pipe of length l will be given by

$$\lambda = 2(l + 2e) = 2(l + 1.14r), \tag{11.19}$$

whereas the effective wavelength of the fundamental mode for a closed pipe of same length will be given by

$$\lambda = 4(l + e) = 4(l + 0.57r). \tag{11.20}$$

This result implies that the fundamental frequency of an open pipe is slightly less than twice the fundamental frequency of a closed pipe.

11.3 ENERGY OF A VIBRATING AIR COLUMN

You now know that we can excite fundamental and desired overtones (higher harmonics) by using an open or a closed pipe. An open end corresponds to the condition of constant pressure and maximum movement of air in an oscillation. On the other hand, a closed end corresponds to zero movement and maximum pressure variation. To derive an expression for energy of a vibrating column, let us consider a pipe of length l and cross-sectional area α (Fig. 11.5). You know from your study of Elasticity that potential energy due to strain is given by

$$\text{PE} = \frac{1}{2} \times \text{Stress} \times \text{Strain} \times \text{Volume} \tag{11.21}$$

If we consider a rectangular volume element of thickness dx between the sections A and B and if the gas at pressure p has density ρ, we can obtain using Eq. (11.21) its potential energy.

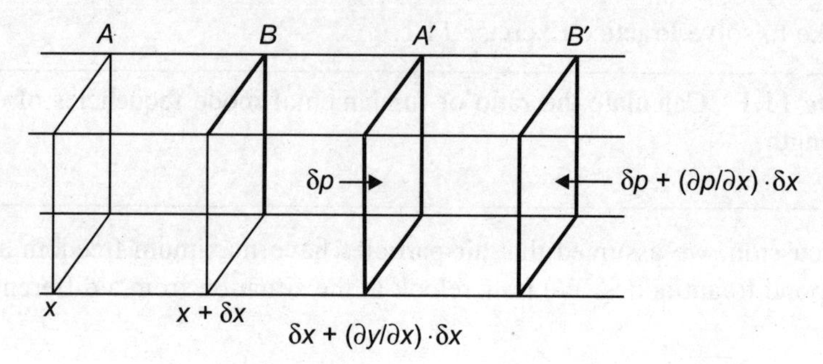

Fig. 11.5 Cross-sectional view of a vibrating column.

The volume of the element = $\alpha\, dx$

Since elasticity E is ratio of stress to strain, we can rewrite the expression of potential energy of the volume element $\alpha\, dx$ as

$$\text{PE} = \frac{1}{2} E (\text{Strain})^2 \alpha dx. \tag{11.22}$$

Recall that speed of sound in air is given by

$$v = \sqrt{\frac{E}{\rho}}.$$

So, the expression for potential energy takes the form

$$\text{PE} = \frac{1}{2} \alpha\rho v^2 dx\, (\text{Strain})^2. \tag{11.23a}$$

The volume strain is given by

$$\frac{\delta V}{V_0} = \frac{\alpha(\partial y/\partial x)\,\delta x}{\alpha \delta x} = \frac{\partial y}{\partial x}.$$

Hence, we can write the expression for potential energy as

$$\text{PE} = \frac{1}{2} \alpha\rho v^2 \left(\frac{\partial y}{\partial x}\right)^2 dx. \tag{11.23b}$$

Now, $\qquad \text{Kinetic energy} = \frac{1}{2} \text{Mass} \times (\text{Velocity})^2$

For the element considered, Mass = $\rho\alpha\, dx$ and Velocity = $\partial y/\partial t$.

$\therefore$ Its kinetic energy is given by KE $= \dfrac{1}{2}\alpha\rho\left(\dfrac{\partial y}{\partial t}\right)^2 dx.$ (11.24)

For the element considered, mass = $\rho\alpha\, dx$ and velocity = $\dfrac{\partial y}{\partial t}$.

From Eq. (9.14), we know that expression for displacement is

$$y(x,t) = \sum \left\{a_n \cos(2n+1)\frac{\pi vt}{2l} + b_n \sin(2n+1)\frac{\pi vt}{2l}\right\} \sin(2n+1)\frac{\pi x}{2l}.$$

We can rewrite this expression as

$$y = \sum \phi_n \sin(2n+1)\frac{\pi x}{2l}, \qquad (11.25)$$

where

$$\phi_n = \left\{a_n \cos(2n+1)\frac{\pi vt}{2l} + b_n \sin(2n+1)\frac{\pi vt}{2l}\right\}.$$

Now, for deriving the expression for KE we first obtain the first time derivative of displacement, which is given by

$$\frac{\partial y}{\partial t} = \sum \dot{\phi}_n \sin(2n+1)\frac{\pi x}{2l} \qquad (11.26)$$

The kinetic energy associated with sound waves propagating in a pipe of length l is obtained by way of addition of KEs due to the motion of individual gas elements of thickness dx. Hence the desired expression for kinetic energy is

$$\text{KE} = \frac{1}{2}\rho\alpha \int_0^l \left(\frac{\partial y}{\partial t}\right)^2 dx. \qquad (11.27)$$

$$\int_0^l \sin s'x \sin r'x\,dx = \frac{l}{2}\delta_{r's'}$$

On substituting for $(\partial y/\partial t)$ from Eq. (11.25), the expression for kinetic energy takes the form

$$\text{KE} = \frac{1}{2}\rho\alpha \int_0^l \sum_{r=1}^{\infty}\sum_{s=1}^{\infty} \sin(2r+1)\frac{\pi x}{2l} \sin\frac{(2s+1)\pi x}{2l}\, \dot{\phi}_r\dot{\phi}_s dx$$

$$= \frac{1}{2}\rho\alpha \sum_{r=1}^{\infty}\sum_{s=1}^{\infty} \dot{\phi}_r\dot{\phi}_s \int_0^l \sin r'x \sin s'x dx, \qquad (11.28)$$

where we have put $r' = (2r+1)\pi/2l$ and $s' = (2s+1)\pi/2l$. The integral in this expression is equal to $(l/2)\,\delta_{r's'}$ and Eq. (11. 28) reduces to

$$\text{KE} = \frac{1}{2}\rho\alpha \sum_r\sum_s \dot{\phi}_r\dot{\phi}_s \frac{l}{2}\delta_{r's'}.$$

You may now note that r' and s' effectively represent r and s respectively. So we may change the Kronecker's delta notation from $\delta_{r's'}$ to δ_{rs}.
So, the average kinetic energy is given by

$$\langle \text{KE} \rangle = \frac{1}{2}\rho\alpha \sum_r\sum_s \dot{\phi}_r\dot{\phi}_s \frac{l}{2}\delta_{rs} = \frac{1}{4}\rho\alpha \sum \dot{\phi}_r^2 = \frac{M}{4}\sum \dot{\phi}_r^2. \qquad (11.29)$$

Similarly, the potential energy is determined by way of definite integration of the RHS of Eq. (11.23b). We have

$$\text{PE} = \frac{1}{2}\int_0^l \rho\alpha v^2 \left(\frac{\partial y}{\partial x}\right)^2 dx. \tag{11.30}$$

Using Eq. (11.25), we may obtain $\partial y/\partial x$ and write,

$$\text{PE} = \frac{1}{2}\rho\alpha v^2 \int_0^l \sum_r \sum_s \phi_r \phi_s \frac{(2r+1)\pi}{2l}\frac{(2s+1)\pi}{2l}\cos\frac{(2r+1)\pi x}{2l}\cos\frac{(2s+1)\pi x}{2l}dx$$

$$= \frac{1}{2}\rho\alpha v^2 \sum_r \sum_s \phi_r \phi_s r's' \int_0^l \cos r'x \cos s'x dx$$

$$= \frac{1}{2}\rho\alpha v^2 \sum_r \sum_s \phi_r \phi_s r's' \frac{l}{2}\delta_{r's'}$$

$$= \frac{1}{4}\rho\alpha l v^2 \sum_r \phi_r^2 \frac{(2r+1)^2\pi^2}{4l^2} = \frac{Mv^2}{4}\sum\left\{\frac{(2n+1)\pi}{2l}\right\}^2 \phi_n^2. \tag{11.31}$$

Therefore, the expression for the total energy is obtained by combining Eqs. (11.29) and (11.31),

$$\text{TE} = \frac{M}{4}\left\{\sum\left[\dot{\phi}_n^2 + \left(\frac{2n+1}{2l}\right)^2 \pi^2 v^2 \phi_n^2\right]\right\}. \tag{11.32}$$

Now, substituting the values of

$$\dot{\phi}_n^2 = \left(\frac{2n+1}{2l}\right)^2 \pi^2 v^2 \left[a_n^2 \sin^2\frac{(2n+1)\pi vt}{2l} + b_n^2 \cos^2\left(\frac{2n+1}{2l}\right)\pi vt - 2a_n b_n \sin\frac{(2n+1)\pi vt}{2l}\cos\frac{(2n+1)\pi vt}{2l}\right]$$

and

$$\phi_n^2 = a_n^2 \cos^2\frac{(2n+1)\pi vt}{2l} + b_n^2 \sin^2\left(\frac{2n+1}{2l}\right)\pi vt + 2a_n b_n \cos\left(\frac{2n+1}{2l}\right)\pi vt \sin\left(\frac{2n+1}{2l}\right)\pi vt.$$

From Eq. (11.31), the expression for the total energy may be written as

$$\text{TE} = \frac{M}{4}\sum\left(\frac{2n+1}{2l}\right)^2 \pi^2 v^2 (a_n^2 + b_n^2). \tag{11.33}$$

If we put $\omega_n = [(2n+1)/2l]\,\pi v$, we find that the expression for total energy of a vibrating column is a conserved quantity, given by

$$\text{TE} = \frac{M}{4}\sum \omega_n^2 (a_n^2 + b_n^2). \tag{11.34}$$

REVIEW EXERCISES

11.1 The water level in a vertical glass tube 120 cm long can be adjusted to any position. A tuning fork vibrating at 512 Hz is held just over the top end of the tube. At which levels of water will there be resonance?

[***Ans.*** l = 103.9 cm, 71.7 cm]

11.2 The fundamental frequency of an open organ pipe is 300 Hz. The frequencies of the first overtone excited in it and a closed organ pipe are equal. Calulate the length of each pipe.

[***Ans.*** 55 cm, 41.25 cm]

11.3 A 0.3 m long wire weighs 0.01 kg. It is held fixed at both ends and vibrates in one loop. A 1 m long tube closed at one end is brought under the vibrating string. As a result, the air column in the tube is set into vibrations at its fundamental frequency by resonance. Calculate (a) the frequency of vibration of the air column and (b) the tension in the string.

[***Ans.*** 82.5 Hz, 81.7 N]

11.4 A tuning fork of unknown frequency produces three beats per second when excited along with a standard fork of frequency 216 Hz. The beat frequency decreases when a small piece of wax is put on a prong of the first fork. What is the frequency of this fork?

[***Ans.*** f = 213 Hz]

11.5 In a closed tube, the air column resonates with a tuning fork of frequency 288 Hz when its length is adjusted to 30 cm. It is given that the room temperature is 30 °C and the velocity of sound at 0 °C is 330 ms^{-1}. Calculate the temperature coefficient of velocity.

[***Ans.*** 1.6×10^{-4} °C^{-1}]

11.6 A column of air and a tuning fork produce 5 beats per second when sounded together; the fork giving the lower note. The room temperature is 30 °C. When the temperature falls to 10 °C, the number of beats reduces to 2 beats per second. Determine the frequency of the tuning fork.

[***Ans.*** 98 Hz]

11.7 A cylinder *PQ* of length 1 m is fitted with a thin flexible diaphragm *M* at the middle, and two other diaphragms at *P* and *Q*. The portions *PM* and *QM* contain two gases *p* and *q* respectively, which are set into vibrations of same frequency. What is the minimum frequency of these vibrations for which *M* is a node? The velocity of sound in *p* and *q* are 100 ms^{-1} and 300 ms^{-1} respectively.

[***Ans.*** 1650 Hz]

12

Large Amplitude Oscillations

EXPECTED LEARNING OUTCOMES

In this Chapter, you will acquire capability to:

- distinguish between a small and large amplitude oscillations on the basis of their differential equations of motion;
- differentiate between symmetric and asymmetric oscillations;
- establish the equations of motion of a body under the influence of symmetric and asymmetric restoring forces and solve them; and
- apply theories of symmetric and asymmetric oscillations to explain the formation of combination tones.

12.1 INTRODUCTION

In Chapters 2 through 6, we discussed characteristics of small oscillations under different physical conditions. You must have observed that the most significant feature of the entire discussion was that the oscillations were of very small amplitude. This helped us to fulfil the condition that the restoring force was proportional to displacement or the restoring torque was proportional to angular displacement. But in actual practice, oscillations may not necessarily be of very small amplitude. In such cases, we cannot expect simple relations like $F = -kx$ or $\tau = -k_a\theta$. In fact, we will be required to retain higher order terms in x in the expression for restoring force.

Depending on the nature of higher order terms, the restoring force will either be symmetric or asymmetric. In this Chapter, we analyse large amplitude oscillations under the influence of symmetric as well as asymmetric restoring forces. That is, we shall introduce the concept of anharmonic oscillations.

We also introduce the combination tones, which are aggregates of difference and summation tones. These find useful applications in musical performances. It should be quite interesting to study the aspects of objectivity and subjectivity of combination tones and the theory of combination tones.

12.2 FREE OSCILLATIONS OF LARGE AMPLITUDE

We know that in SHM, the restoring force is proportional to displacement and the equation of motion is of the form

$$m\ddot{x} = -kx.$$

But in case of large amplitude oscillations, the relation is not so simple. The general relation is of the form

$$m\ddot{x} = -f(x) = -(a_0 + a_1x + a_2x^2 + a_3x^3 + \cdots). \tag{12.1}$$

For small x, i.e., $x \to 0$, the second and higher order terms in x can be neglected. If we take $a_0 = 0$, Eq. (12.1) takes a compact form:

$$m\ddot{x} = -a_1x.$$

Otherwise, we have

$$m\ddot{x} + (a_1x + a_2x^2 + a_3x^3 + \cdots) = 0. \tag{12.2}$$

In case of SHM, the restoring force has a single term in which the power of displacement x is odd. So when the sign of x changes, the direction of force also changes. Then we say that the motion is symmetric about the centre of oscillation. In fact, this statement is true for all terms having odd powers of x and we may conclude that corresponding to odd powers of x, an oscillation is symmetrical about the mean equilibrium position (i.e., centre of oscillation). But if the powers of x are even, the force does not change direction with change in its sign and the oscillation is said to be asymmetric about the mean equilibrium position. This necessitates the study of anharmonic oscillations under asymmetric and symmetric restoring force. We begin by studying large amplitude oscillations under asymmetric restoring force.

12.3 LARGE AMPLITUDE OSCILLATIONS UNDER ASYMMETRIC RESTORING FORCE

In Eq. (12.2), we retain the first order term (harmonic term) and the terms having even powers of x (asymmetric terms). Then it simplifies to

$$m\ddot{x} + a_1x + a_2x^2 + a_4x^4 + \text{(Other terms containing higher even powers of } x) = 0. \tag{12.3}$$

First, let us consider terms up to x^2 only. Then we can write

$$m\ddot{x} = -a_1x - a_2x^2$$

or

$$\ddot{x} = -\frac{a_1}{m}x - \frac{a_2}{m}x^2. \tag{12.4}$$

If we put $a_1/m = \omega^2$ and $a_2/m = n$, Eq. (12.4) takes a compact form:

$$\ddot{x} + \omega^2 x + nx^2 = 0. \tag{12.5}$$

Equation (12.5) can be solved by the method of successive approximations. In this method, we first ignore the correction term nx^2 and seek solutions of the resultant equation. In case of Eq. (12.5), this gives

$$x = a \sin(\omega t + \delta).$$

Then we insert this solution (obtained in the first order approximation) in Eq. (12.5). This gives

$$nx^2 = na^2 \sin^2(\omega t + \delta)$$

$$= \frac{1}{2} na^2 \{1 - \cos 2(\omega t + \delta)\},$$

so that Eq. (12.5) now takes the form

$$\ddot{x} + \omega^2 x + \frac{1}{2} na^2 \{1 - \cos 2(\omega t + \delta)\} = 0$$

or

$$\ddot{x} + \omega^2 \left(x + \frac{1}{2}\frac{na^2}{\omega^2} \right) - \frac{1}{2} na^2 \cos\{2(\omega t + \delta)\} = 0.$$

This can be rearranged as

$$\frac{d^2}{dt^2}\left(x + \frac{1}{2}\frac{na^2}{\omega^2} \right) + \omega^2 \left(x + \frac{1}{2}\frac{na^2}{\omega^2} \right) = \frac{na^2}{2} \cos\{2(\omega t + \delta)\}.$$

Proceeding further, we introduce a change of variable by defining $u = x + (na^2/2\omega^2)$ to obtain

$$\frac{d^2 u}{dt^2} + \omega^2 u = f \cos(pt + \varepsilon), \tag{12.6}$$

where $f = (na/2) \times a$, $p = 2\omega$ and $\varepsilon = 2\delta$. We rewrite Eq. (12.6) as

$$(D^2 + \omega^2)\, u = f \cos(pt + \varepsilon),$$

where $D \equiv (d/dt)$. The solution of this equation is given by

$$u = f \frac{1}{D^2 + \omega^2} \cos(pt + \varepsilon) = \frac{f \cos(pt + \varepsilon)}{\sqrt{(\omega^2 - p^2)^2}}. \tag{12.7}$$

In terms of x, we can write

$$x + \frac{na^2}{2\omega^2} = \frac{na^2}{2} \frac{\cos\{2(\omega t + \delta)\}}{\sqrt{\{\omega^2 - (2\omega)^2\}^2}}$$

or

$$x = -\frac{na^2}{2\omega^2} + \frac{1}{2} na^2 \cdot \frac{\cos\{2(\omega t + \delta)\}}{3\omega^2}.$$

Hence, complete solution of Eq. (12.3) is given by

$$x = -\frac{na^2}{2\omega^2} + a \sin(\omega t + \delta) + \frac{na^2}{6\omega^2} \cos\{2(\omega t + \delta)\}. \tag{12.8}$$

Let us pause for a while and ask: What have we achieved so far? We note that the right hand side of Eq. (12.8) contains three terms: (i) $-(na^2/2\omega^2)$, which is constant, (ii) $a \sin(\omega t + \delta)$, which represents simple harmonic motion, and (iii) $(na^2/6\omega^2) \cos \{2(\omega t + \delta)\}$, which corresponds to an oscillation of angular frequency 2ω. Note that the first term introduces asymmetry in simple harmonic motion (represented by the second term) and the third term superposes a new vibration of angular frequency 2ω.

We now consider large amplitude oscillations under symmetric restoring force.

12.4 LARGE AMPLITUDE OSCILLATIONS UNDER SYMMETRIC RESTORING FORCE

From Section 12.2, we recall that a large amplitude oscillation is symmetrical about the mean equilibrium position if the restoring force contains only odd powers of x. Therefore, for large amplitude oscillation under symmetric restoring force, Eq. (12.3) modifies to

$$m\ddot{x} + a_1 x + a_3 x^3 + a_5 x^5 + \cdots = 0. \tag{12.9}$$

For simplicity, we retain terms only up to third power of x. Then Eq. (12.9) simplifies to

$$m\ddot{x} = -a_1 x - a_3 x^3.$$

We can rewrite it as

$$\ddot{x} + \omega^2 x + n^2 x^3 = 0, \tag{12.10}$$

where $a_1/m = \omega^2$ and $a_3/m = n^2$.

As before, we assume that $x = a \sin(\omega t + \delta)$ is a solution of Eq. (12.10) and follow the same steps as in Section 12.3. You can easily show that the complete solution of Eq. (12.10) is given by

$$x = a \sin(\omega t + \delta) - \frac{n^2 a^3 \sin\{3(\omega t + \delta)\}}{4(k_0^2 - 9\omega^2)}, \tag{12.11}$$

where

$$k_0^2 = \omega^2 - \frac{3}{4} n^2 a^3. \tag{12.12}$$

This shows that the resultant oscillation is symmetric though it has third harmonic corresponding to frequency 3ω.

You may now like to answer Practice Exercise 12.1.

***Practice Exercise* 12.1** Derive Eq. (12.11).

12.5 COMBINATION TONES

You have studied about beats in Chapter 8. If two tones of nearly equal frequencies are sounded, the resultant tone has a frequency equal to the difference of individual tones. This is known as *beat tone*. Do you know who observed this first? As per recorded evidence, the famous musician Giuseppe Tartini is credited with the observation. It proved a defining landmark in the history of instrumental music. Further interest in this field was aroused after Hermann von Helmholtz

(1821–1894) reported that he had heard a higher order tone of frequency equal to the sum of the frequencies of primary tones. He termed it as the *summation tone*. The beat tone or the difference tone and the summation tone taken together are referred to as the *combination tones*. Let us now learn more about it.

12.5.1 Production of Combination Tones

The necessary conditions for production of combination tones are:

(i) The primary tones must be sufficiently loud and sustained.
(ii) The frequencies of the primary tones should be so chosen that the frequencies of the combination tones lie in the middle of the audible range.

Combination tones are classified as *subjective* and *objective*. The former are realised only by the ear and have no existence whatsoever outside the ear. However, combination tones are endowed with objectivity also. This was established through an experiment by Rücker and Edser [A.W. Rücker and E. Edser, Phil. Mag. Ser. 5. **89**, 342–357 (1895)].

The difference tone may be heard by

(i) Sounding two tuning forks of frequencies in the range 400–500 Hz and differing in frequency by about 50 Hz.
(ii) Exciting Rudolf Koenig's double glass rods to high pitch longitudinal oscillations.
(iii) Blowing double whistle with the help of a blast of air.

12.5.2 Theories of Combination Tones

There are several theories of combination tones. But the popular ones are:

(i) Helmholtz intensity theory, and
(ii) Waetzmann's general asymmetry theory

We now discuss these one by one.

Helmholtz intensity theory

The main consideration of Helmholtz was the fact that for unequivocal production of combination tones, the primaries must be intense. His other consideration was that the ear drum or any other artificial membrane does not respond symmetrically to the double forcing applied by the generating tones. So it is a case of asymmetric vibration where the restoring force is not proportional to the displacement. Owing to asymmetry, Helmholtz added the lowest order even powered term, i.e., a term proportional to the square of the displacement. Thus, the equation of motion will have the form

$$\ddot{x} + \omega^2 x + nx^2 = a \sin \omega' t + a' \sin(\omega'' t + \delta). \tag{12.13}$$

The two terms on the right hand side represent the two forcing oscillations corresponding to the generating tones.

As before, we first leave out the term nx^2 and rewrite Eq. (12.13) as

$$\ddot{x} + \omega^2 x = a \sin \omega' t + a' \sin(\omega'' t + \delta). \tag{12.14}$$

Note that Eq. (12.14) has no term corresponding to dissipative forces. Therefore, we expect the forcing oscillations to induce in-phase displacements x_1 and x_2. Then we can write

$$x = x_1 + x_2 = a_1 \sin \omega' t + a_2 \sin (\omega'' t + \delta). \tag{12.15}$$

We use this result to calculate the effect of asymmetry in restoring force:

$$nx^2 = n\{a_1 \sin \omega' t + a_2 \sin(\omega'' t + \delta)\}^2$$

$$= n\{a_1 \sin^2 \omega' t + a_2^2 \sin^2 (\omega'' t + \delta) + 2a_1 a_2 \sin \omega' t \sin (\omega'' t + \delta)\}$$

$$= n\left[\frac{a_1^2}{2}(1 - \cos 2\omega' t) + \frac{a_2^2}{2}\{1 - \cos 2(\omega'' t + \delta)\}\right.$$

$$\left. + a_1 a_2 \{\cos (\overline{\omega' - \omega''} t - \delta) - \cos(\overline{\omega' + \omega''} t + \delta)\}\right].$$

Using this result in Eq. (12.13) we get

$$\ddot{x} + \omega^2 x = a_1 \sin \omega' t + a_2 \sin(\omega'' t + \delta) + \frac{na_1^2}{2}\cos 2\omega' t + \frac{na_2^2}{2}\cos 2(\omega'' t + \delta)$$

$$- na_1 a_2 \cos \{(\omega' - \omega'')t - \delta\} + na_1 a_2 \cos \{(\omega' + \omega'')t + \delta\} - \frac{n}{2}(a_1^2 + a_2^2). \quad (12.16)$$

From the right hand side of Eq. (12.16) we note that the resultant motion is asymmetric. In fact, the system is influenced by several harmonic forces having angular frequencies ω', ω'', $2\omega'$, $2\omega''$, $\omega' - \omega''$, $\omega' + \omega''$, etc. So, the final solution is given by

$$x = a_1 \sin \omega' t + a_2 \sin(\omega'' t + \delta) + a_3 \cos 2\omega' t + a_4 \cos 2(\omega'' t + \delta)$$

$$+ a_5 \cos \{(\omega' - \omega'')t - \delta\} + a_6 \cos \{(\omega' + \omega'')t + \delta\} + a_7. \quad (12.17)$$

You may now like to work out Practice Exercise 12.2 to fix your ideas.

***Practice Exercise* 12.2** Putting the value of x, as per Eq. (12.17) in Eq. (12.16) and equating coefficients of similar terms, obtain the values of a_1, a_2, a_3, a_4, a_5, a_6, and a_7. Show that a_5 is a fraction of aa'.

[***Ans.*** $a_1 = \dfrac{a}{\omega^2 - (\omega')^2}$, $a_2 = \dfrac{a'}{\omega^2 - (\omega'')^2}$, $a_3 = \dfrac{na^2}{2[\omega^2 - 4(\omega')^2][\omega^2 - (\omega')^2]^2}$,

$$a_4 = \frac{na'^2}{2(\omega^2 - \omega''^2)^2(\omega^2 - 4\omega''^2)}, \quad a_5 = \frac{-naa'}{(\omega^2 - \omega'^2)(\omega^2 - \omega''^2)\{\omega^2 - (\omega' - \omega'')^2\}},$$

$$a_6 = \frac{naa'}{(\omega^2 - \omega'^2)(\omega^2 - \omega''^2)\{\omega^2 - (\omega' + \omega'')^2\}}, \quad a_7 = \frac{n}{2\omega^2}\left\{\frac{a^2}{(\omega^2 - \omega'^2)^2} + \frac{a'^2}{(\omega^2 - \omega''^2)^2}\right\}$$]

Equation (12.17) indicates that the resultant oscillation has frequencies $2\omega'$, $2\omega''$, $\omega' + \omega''$, $\omega' - \omega''$, in addition to primary frequencies ω' and ω''. The amplitude of the difference tone is a fraction of aa', and hence, for it to be appreciable, the primary amplitudes should be large. But it is not corroborated by experimental observations. Even weak primaries are also found to produce combination tones. This weak point in the Helmholtz Intensity Theory was to some extent resolved by Waetzmann. We discuss it now.

Waetzmann's general asymmetry theory

Waetzmann considered general asymmetry in the combination tone producing system. He realised it in the following manner. He loaded a membrane with a weight on one side only. It was then subjected to two harmonic oscillations from two tuning forks. The resultant oscillation was recorded on a graph which resembled the formation of beats. However, a Fourier analysis of the curve showed the presence of original tones (ω', ω''), a difference tone ($\omega' - \omega''$), a weak secondary difference tone ($2\omega - \omega'$) and a summation tone ($\omega' + \omega''$).

The ear drum with ossicles on one side behaves like an asymmetrically loaded membrane and produces subjective combination tone. Similar rectifying properties (as explained through the use of Fourier analysis) may be associated with other systems also, and thus objective combination tone can be expected in air external to ear drum.

REVIEW EXERCISES

12.1 Solve Eq. (12.5) by taking $x = a \cos(\omega t + \delta)$ and interpret the solution.

$$\left[\textbf{\textit{Ans.}}\quad x = -\frac{na^2}{2\omega^2} + a\cos(\omega t + \delta) - \frac{na^2}{6\omega^2}\cos\{2(\omega t + \delta)\}\right]$$

12.2 Solve Eq. (12.10) by assuming a solution in the form $x = a \cos(\omega t + \delta)$ and show that the resultant oscillation is symmetric.

$$\left[\textbf{\textit{Ans.}}\quad x = a\cos(\omega t + \delta) - \frac{n^2 a^3}{4(k_0^2 - 9\omega^2)}\cos\{3(\omega t + \delta)\}\right]$$

12.3 Solve Eq. (12.13) if its right hand side is a combination of sinusoidal terms $b \cos \omega t$ and $b' \cos(\omega t + \phi)$. Prove that the motion is asymmetric and the system comes under the influence of several harmonic forces. Calculate the expressions for corresponding frequencies.

[***Hint:*** Proceed as in Practice Exercise 12.2]

12.4 An anharmonic oscillation is described by the equation

$$m\ddot{x} = -m\omega_0^2 x + \alpha x^2 + A\cos\omega t,$$

where α is a constant. Solve it subjected to the initial conditions: At $t = 0$, $x = 0$ and $\dot{x} = 0$. Retain terms only up to first order in α.

$$\left[\textbf{\textit{Ans.}}\quad x \simeq \frac{A[\cos(\omega t) - \cos(\omega_0 t)]}{m(\omega_0^2 - \omega^2)} + \frac{10\alpha A^2 \cos(\omega_0 t)}{3m^3\omega_0^2(\omega^2 - 4\omega_0^2)(\omega_0^2 - 4\omega^2)}\right.$$

$$\left. + \frac{\alpha A^2}{m^3(\omega_0^2 - \omega^2)^2}\left\{\frac{1}{\omega_0^2} + \frac{\cos(2\omega t)}{2(\omega_0^2 - 4\omega^2)} - \frac{\cos(2\omega_0 t)}{6\omega_0^2} + \frac{\cos[(\omega_0 - \omega)t]}{\omega^2 - 2\omega\omega_0} + \frac{\cos[(\omega_0 + \omega)t]}{\omega^2 + 2\omega\omega_0}\right\}\right]$$

Index

Acceleration, 16
Adiabatic, 215
Air column, 114
Alternating current circuit, 15

Ballistic galvanometer, 100
Beats, 78, 224, 240, 248
  beat frequency, 78, 248
  beat period, 78
  transient beats, 118, 131
Boundary condition, 230, 234

Capacitance, 15, 63
Capacitor, 15
Carbon dioxide molecule, 141, 174
Cathode ray oscilloscope, 83
Combination tones, 311, 312
  beat tone, 311
  Helmholtz intensity theory, 312–313
  subjective and objective tones, 312
  summation tone, 312
  Waetzmann's general asymmetry theory, 314
Complex variables, 3
  addition, 10
  argument, 9
  complex numbers, 7
  conjugate, 11
  division, 10
  geometrical representation, 9
  imaginary axis, 8
  imaginary number, 7
  modulus, 9
  multiplication, 10
  real number, 7
  real axis, 8
  subtraction, 10
Compound pendulum, 33, 37, 110
Compressibility, 213
Coupled oscillations, 141
  coupled masses, 142
  coupling force, 144
  coupling frequency, 145
  electrical coupling, 142
  electromagnetic coupling, 142
  energy, 166
  magnetic coupling, 142
  mechanical coupling, 142
  modulation, 149
  normal co-ordinates, 146, 147
  normal mode frequency, 146, 157, 159
  normal modes, 142, 146, 147, 149, 153, 157

Damped oscillation, 87, 91
  critically damped, 92, 93, 95
  one-dimensional, 89
  overdamped, 92, 93, 94
  weakly damped, 92, 93, 96, 97, 101, 102
Damping coefficient, 90
Damping force, 88, 90
  coulomb friction, 88
  viscous drag, 88
Dead beat, 94
Decibel, 203, 204

Degree of freedom, 35
Demoivre's theorem, 12
Differential calculus, 3
Differential coefficient, 16
Doppler effect, 181, 217–221

Eigenvalues and eigenfunctions, 261
Elasticity, 206, 209, 228
Electrical energy, 63
Electromagnetic damping, 89
End correction, 303
Euler's theorem, 14

Flexural waves, 289, 290
Forced oscillations, 111
  average power, 127, 128, 137
  driven, 111, 125
  driver, 111, 125, 133
  driving force, 111
  driving frequency, 127
  forced coupled oscillator, 169
  power factor, 128
  vibrations, 111
Fourier analysis, 3, 25
  Fourier series, 25
  Fourier's theorem, 25
  orthogonal, 26
  periodic waveform, 27
  saw-tooth wave, 31
Free oscillation, 112

Harmonic approximation, 17
Harmonics, 244, 291, 298, 301
  overtones, 245, 246

Impedance, 224
  acoustic impedance, 224, 228
  characteristic impedance, 227
  impedance matching, 236
  impedance offered by gases, 227, 228
  impedance offered by stretched string, 225
  impedance triangle, 116, 117, 128, 137
  wave impedance, 224, 225
Inductance, 63
Inertia, 206, 209
Integral calculus, 3
  average value, 3

Interatomic forces, 141, 171
Interference, 239
Inverse square law, 203
Isothermal, 215

Kronecker's delta, 26, 261
Kundt's tude, 294

Laplace's formula, 215
LC-circuit, 33, 61–63, 162
  inductive coupling, 162
LCR circuit, 99–100, 108, 133
  peak current, 135
Lenz's law, 62
Lissajous' figures, 82, 83
Logarithmic decrement, 103

Magnetic energy, 63
Mechanical reactance, 128, 135
Melde's experiment, 223, 263, 264
Modulus of
  elasticity, 56, 214
  rigidity, 42
Moment of inertia, 40
Musical sound, 246

Newton's formula, 215
Noise, 246

Ordinary differential equation, 3
  homogenous, 23
  inhomogeneous, 23, 115
  linear, 4
  order, 22
  second order, 23, 115
  transient solution, 115
Oscillation magnetometer, 33, 38
  magnetic meridian, 38
  magnetic moment, 38
Oscillations, 3
  asymmetric oscillation, 308, 309, 311
  large amplitude oscillation, 308–313
  symmetric oscillation, 308, 310
Oscillator, 23
  damped, 23
  forced, 23
  harmonic, 23

Parallelogram law, 72
Partial differential equation, 4, 23
  linear, 23
  second order, 23
  separation of variables, 24, 256–258, 282
Particle velocity, 16
Physical pendulum, 37
Pitch, 247
Power coefficients, 235, 236
Pulse, 182, 238
Pythagoras' theorem, 116

Quality factor, 105, 107, 129, 131, 136

Reduced mass, 165
Reflection of wave, 223
  reflection amplitude coefficient, 229–233
Refraction of wave, 223
  transmission amplitude coefficient, 229–233
Relaxation time, 97, 131
Resonance, 112, 122, 169
  amplitude resonance, 113, 122, 124
  natural frequency, 122, 123, 145, 153, 165
  resonant frequency, 122
  velocity resonance, 113, 122, 124
Rotating vector, 15, 67

Satellite, 65
Sharpness of resonance, 129
  bandwidth, 130, 131
  broad, 130
  fidelity, 131
  flat, 130
  half power frequency, 129–131
  selectivity, 131
  sharp, 130
Shear, 42
Simple harmonic motion, 32
  acceleration, 47, 48
  amplitude, 4, 42
  angular frequency, 46
  circular frequency, 46
  cycle, 20
  displacement, 4, 48
  energy, 58–61
  epoch, 47
  force constant, 34, 146
  frequency, 42
  kinetic energy, 58–61
  period, 20
  periodic, 20
  phase, 43
  phase difference, 5
  potential energy, 58–61
  relation with uniform circular motion, 64
  restoring force, 35
  restoring torque, 36
  time period, 34, 42
  velocity, 47, 48
Simple harmonic oscillation, 5
Simple pendulum, 33, 35, 109
Slinky approximation, 157
Sonar, 224
Sonometer, 114
Sound ranging, 216
Spring mass system, 33, 36, 52, 68, 69, 88, 110, 122, 123, 144, 148, 149, 151, 154, 171, 173–175, 182
Stiffness, 63, 126
Structural damping, 88
Superposition, 71
  principle of superposition, 71–73, 75, 77, 79
  superposition of waves, 237–240

Taylor's theorem and series, 17, 172, 207, 210, 211
Threshold of hearing, 203
Timbre, 247
Torsional pendulum, 33, 41
Trigonometry, 3
  general solution of equations, 5
  periodicity, 5
  sinusoidal, 4, 67, 186, 190, 192, 197, 246
  transformation of sums and products, 6
Tuning fork, 114, 293

Vocal cord, 179
Voltage, 15

Wave equation, 173, 209
Wave motion, 173, 179, 184, 188
Waves, 3, 182
  amplitude of a wave, 186
  compression, 185, 212, 215
  crest, 185, 186
  electromagnetic waves, 180
  energy of a wave, 199

frequency of a wave, 180, 187, 188, 189
intensity of a wave, 180, 202
longitudinal wave, 185, 234
loudness, 202, 247
matter waves, 180
microwave, 180
monochromatic wave, 190
one-dimensional wave, 185
period of a wave, 185, 186
phase of a wave, 180, 192–195
phase velocity, 196, 197, 250, 252
pressure wave, 185
progressive, 6
radio wave, 180
rarefaction, 185, 212, 215
seismic waves, 180
shock waves, 180
sound waves, 180, 215, 216, 227
spatial, 186, 187
stationary or standing wave, 6, 224, 240–246
  antinode, 242–246, 259, 293, 296–302
  node, 242–246, 259, 293, 296–302
temporal, 186
transverse wave, 185, 225, 229
trough, 185, 186
ultra-sound wave, 180
velocity amplitude, 226
water waves, 180, 181
wave formation, 181
wave group, 224, 240, 487
  group velocity, 249, 252
wave number, 187, 189
wave packet, 224, 240
wave preparation, 181
wave vector, 180
wave velocity, 180, 187, 188, 191
waveform, 186
wavelength, 186–188
waves in a string, 183, 184, 206, 255
  bowed string, 255, 274–277
  energy of a vibration string, 265–269
  law of length, 259
  law of mass, 260
  law of tension, 260
  plucked string, 255, 269–272, 277
  struck string, 255, 272–274
  vibration of strings, 254–277
waves in gases, 212
  vibration of air columns, 295–306
  energy of a vibrating air column, 303–306
waves in pipes, 245
  closed-end, 245, 246, 299–301
  open-end, 245, 246, 296–299
waves in solids, 210,
  fixed-fixed bar, 280, 285–287
  fixed-free bar, 280, 284, 285
  free-fixed-free bar, 287
  free-free bar, 280, 283, 284
  vibration of bars, 280–294
waves in two and three dimensions, 216, 217

Young's law, 272, 274
Young's modulus, 37